Concepts, Applications, and Simulations in Combinatorics

Alessio Drivet
Geogebra Institute of Turin, Italy

IGI Global
Scientific Publishing
Publishing Tomorrow's Research Today

Vice President of Editorial	Melissa Wagner
Director of Acquisitions	Mikaela Felty
Director of Book Development	Jocelynn Hessler
Production Manager	Mike Brehm
Cover Design	Jose Rosado

Published in the United States of America by
IGI Global Scientific Publishing
701 East Chocolate Avenue
Hershey, PA, 17033, USA
Tel: 717-533-8845
Fax: 717-533-7115
Website: https://www.igi-global.com E-mail: cust@igi-global.com

Library of Congress Cataloging-in-Publication Data

Names: Drivet, Alessio, 1948- author
Title: Concepts, applications, and simulations in combinatorics / Alessio Drivet.
Description: Hershey, PA : IGI Global Scientific Publishing, [2026] | Includes bibliographical references and index. | Summary: "This book aims to help teachers to understand and use the combinatorics theme in their classroom"-- Provided by publisher.
Identifiers: LCCN 2025017773 (print) | LCCN 2025017774 (ebook) | ISBN 9798337330891 hardcover | ISBN 9798337330907 paperback | ISBN 9798337330914 ebook
Subjects: LCSH: Combinatorial analysis | Functional analysis
Classification: LCC QA164 .D75 2025 (print) | LCC QA164 (ebook)
LC record available at https://lccn.loc.gov/2025017773
LC ebook record available at https://lccn.loc.gov/2025017774
British Cataloguing in Publication Data
A Cataloguing in Publication record for this book is available from the British Library.

Table of Contents

Preface

Combinatorics is the art of counting without counting

Claude Berge

This text is dedicated to those who want a sufficiently broad overview of the topic without pretending to go into excessively complex details. It is particularly dedicated to teachers who consider it important to develop the combinatorial theme.

To understand the subject, it is necessary to broaden the discussion to topics that generally do not find space in textbooks as they are considered too complex or indispensable in non-university school education. Combinatorics can be tackled at any school level, but keep in mind that the depth and complexity of its study grows as one advances the school grades. Initially, intuitive and practical concepts are addressed at the elementary and middle school level, and then more abstract theories and advanced techniques are dealt with in high school and university. At the primary school level, teachers introduce combinatorics in a very intuitive and practical way. Teachers focus on basic concepts, such as counting and sorting, using playful activities to capture pupils' interest, for example, by proposing games in which children must classify objects by colour or shape or by inventing simple stories that involve the idea of choice. Through these learning experiences, students begin to familiarize themselves with the basic principles of organization and quantification, thus laying the foundation for more advanced studies in the future. With the transition to secondary school, especially high school, combinatorics began to be explored more formally. Students learn to use specific terminology and solve problems that require techniques such as permutations and combinations. At this time, mathematical formulas are introduced that allow one to efficiently calculate the number of ways in which objects can be arranged or chosen. In this phase, it can be very formative to add a computational approach through software that allows one to visualize and simulate situations that are difficult to treat. Finally, at the university level, combinatorics becomes a complex discipline with wide margins of specialization. Just think of the advanced courses in discrete mathematics, where the students address topics such as graphs, networks, and counting techniques. At this stage, the approach is more theoretical and abstract, and the applications of combinatorics extend into various fields, from the design of algorithms in computer science to complex problems in statistics and probability. In summary, the evolution of the study of combinatorics across the various school levels is characterized by an increasing level of complexity and abstraction. Each phase prepares students to understand mathematical concepts better and fosters the development of strategic skills necessary to meet the challenges of the modern world. Combinatorics thus becomes not only a branch of mathematics but a mine of practical applications and a method of critical thinking that continues to prove indispensable in a society increasingly oriented towards computer science and data analysis.

Before continuing reading, a suggestion may be useful. The text deals with many aspects of combinatorics; however, it is possible to trace four different possible paths:

1. general
2. didactic
3. in-depth analysis
4. playful

Much of mathematics has focused on continuous models but just think of infinitesimal calculus. These models, based on concepts such as continuity, limit and infinity, are widely used to describe physical and natural phenomena. The advent of computer science and digitization has broadened the frame of reference, discrete models have gained importance, and, within mathematics, combinatorics has taken on an increasingly significant role.

What does this paradigm shift consist of? The fundamental difference lies in the object of study. Discrete mathematics deals with finite or countable objects, such as sequences and graphs. Continuous mathematics studies objects such as functions and curves that can vary without interruption. Another difference is the characteristics of the methods and techniques used. In the first case, the main techniques include counting and enumeration, while in continuous mathematics, the fundamental techniques include differential and integral calculus.

Typical applications are often diverse: discrete mathematics is widely used in computer science, cryptography, network theory, and combinatorial optimization, while continuous mathematics is essential in physics, engineering, economics, and the sciences.

For many, this is a real philosophical challenge as it raises questions about the nature of reality: Is it fundamentally discrete or continuous?

The debate is ancient. In ancient Greece, the Pythagoreans and atomists such as Democritus (IV-V century BC) supported a discreet vision. Plato (328-348 BC) and Aristotle (384-322 BC), on the other hand, developed a more continuous vision of reality.

Two giants of mathematics, Gottfried Wilhelm Leibniz (1646-1716) and Isaac Newton (1642-1726), emphasized that physical quantities such as length and temperature vary continuously without jumps (*Natura non facit saltus*). This view implies a determinism in which finite and unique values define each physical state. In physics, the principle of continuity states that variations between different states occur gradually.

In more recent times, two positions have emerged with greater force: scientific realists maintain that science provides an accurate or approximately true description of reality, and the nature of reality could be discrete and continuous depending on the scales and phenomena considered. Other philosophers of science propose a view that refers to structuralism, according to which structures and relationships exist, rather than discrete or continuous objects themselves.

The first position is important because it explains the predictive success of science. However, it is also a weakness: it can be difficult to define precisely what "approximately true" means. Structuralism avoids some problems related to the nature of objects, focusing instead on relationships. It also fits well with some interpretations of quantum mechanics and field theory, where "particles" can be seen as nodes in a network of relationships. On the other hand, everything may seem counterintuitive and distant from everyday experience.

Unfortunately (or fortunately), there is no definitive answer, but the two approaches influence how we conceive and shape the world. Some physicists have hypothesized that space-time is also discrete; to the Planck length, it looks more like a discrete net than a continuous variety. The Planck length is so small that, with the limits of measurement in our possession, we can reason in terms of a continuum, and that is why differential equations give a good description of the Universe.

At tiny scales such as those of atoms, reality seems to present itself discretely, which is what quantum mechanics suggests. While reality seems continuous and fluid in our macroscopic world, at the subatomic level, characteristics may indicate a granular structure of matter.

An interesting point of view is that of the Swedish cosmologist Max Tegmark (1967-), known for his "bold" ideas about the nature of the universe. He is famous for developing mathematical models to support the multiverse hypothesis, according to which our universe is only one of many possible universes. Concerning the subject at hand, he proposes that the concept of continuity only approximates the true essence of reality, which could be more like a set of discrete units. Furthermore, Tegmark argues that the universe can be described through a physical-mathematical structure so complex that our brain, with its limited capabilities, cannot fully understand it. It is a clear invitation to reflect on how humans perceive reality and how mathematics can offer keys to unravel its most profound mysteries.

In summary, the question of whether reality is discrete or continuous does not have a definitive and universal answer, as it depends on the context and scale of observation. On a macroscopic scale, many phenomena, such as water flow or planets' movement, seem continuous to us. On a microscopic scale, quantum mechanics describes discrete phenomena, such as the energy levels of atoms. Those who do not feel like taking sides could say that reality is a complex combination of discrete and continuous elements, and the choice between a discreet or continuous description depends on the specific context and the questions we ask ourselves. As science advances, some doubts may be dispelled.

One of the leading experts in the field, Gian Carlo Rota (1932-1999) wrote: "Almost all classical mathematics, from elementary algebra to the theory of differential equations, is applicable to the real world only in the hypothesis that it is made up of objects and events of a continuous nature. However, in many common situations in physics, chemistry and other sciences, one can realistically speak only of a collection of objects of a discrete character, which act in combination, one step at a time. The mathematics applied to such situations is called combinatorial analysis. Many of the most interesting problems of combinatorial analysis have presented themselves in the form of ingenious riddles, challenging mathematicians and non-mathematicians together. At first glance, some may even seem frivolities, yet almost all have immediate and important applications to concrete scientific problems".

Of one thing we are certain: the connection between discrete and combinatorics is profound, based on principles intrinsically linked to the nature of the objects studied. Combinatorics is, in fact, a field of mathematics that studies and analyses the different techniques of combining and organizing objects or elements into sets. These objects can be sequences, collections, tables, or any other form of discrete entities. It is an essential discipline that finds applications in many fields, including mathematics, science, and computer science. Combinatorics also studies the characteristics of graphs, which are combinatorial structures formed by nodes and edges. Graphs are used in many fields, such as social networks, game theory, and computer science.

The term combinatorics is often used as a synonym for combinatorial calculus and used interchangeably in everyday language, but there is a significant difference. Combinatorics is a more general discipline that deals with any combinatorial grouping, relationships, and properties of objects. Conversely, Combinatorics focuses on specific types of groupings and the development of algorithms and techniques for counting groupings. Combinatorial calculus is the practical application used to solve specific problems. Combinatorics deals with the "what," while combinatorial calculus refers to the "how."

Combinatorial objects, such as subsets, partitions, trees, etc., are common in mathematics. According to some authors, there are three recurring algorithmic tasks: counting, random generation, and exhaustive generation.

Combinatorics and combinatorial calculus help us understand the different possibilities and configurations in mathematics and reality, providing fundamental tools to solve a wide range of problems and applications.

A few quite different examples could illuminate the ideas and procedures involved. The problems will be analysed in more detail in the text.

1. We have an urn containing n balls (of two different colours) and draw k of them. How many ways can we do this? To answer this, we must ask ourselves a few questions. First, do we care about the order in which the balls are drawn or not? Secondly, when we have made the draw, do we put the ball back in the bowl or not? Let us formulate a trivial example, we have two balls, one black and one white, and we extract them. If order is important and repetition is allowed, there are four possibilities: (black, black), (black, white), (white, black) and (white, white). If the order is important but repetition is not allowed: there are two possibilities, (black, white) and (white, black). If the order is not important and repetition is allowed, there are three possibilities: (black, black), (black, white) and (white, white). Finally, if the order is not important and repetition is not allowed, there is only one possibility (black, white).
2. A superstitious cyclist thinks that the damage to the wheels of his bicycle, transformed into an eight, is due to the presence of this amount on the club cards. The question is, how many members would there be in the club if all the 3-digit numbers that do not contain digit 8 were used to number the new cards? It is easy to establish that the numbers without the eight are 93 and that, therefore, the club has 729 members.
3. Suppose one has three cities, A, B, and C, with two attractions each. We can represent this scenario with a graph in which each city is a node, and each attraction is an arc that connects the city to the point of interest. How many different routes can we take to visit all six attractions once? One path could be ABCACB. By applying combinatorics, we can determine the 120 possible routes, considering that you must visit each city's two attractions.
4. One has a ladder with n steps. With just one step, one can climb one or two steps. How many different ways can one climb the ladder? If one is at step n, one can get there from step n−1 with a step of 1 step or from step n−2 with two steps. It means that the number of ways to reach step n is given by the number of ways to reach the previous steps. This problem is a classic example of combinatorics and refers to the Fibonacci sequence (1170-1250): $a_n = a_{n-1} + a_{n-2}$.

5. Leonhard Euler (1707-1783) asked: Six different regiments have six officers, each holding a different rank (of six different ranks in total). Can these 36 officers be arranged in a square formation so that each row and column contains one officer of each rank and one from each regiment? Euler hypothesized that the answer was "no", and this assumption was finally proven correct in 1900. This problem is a classic example of a Latin square (an n×n matrix in which each row and column contains precisely once each of the numbers or symbols 1 through n) and can be formulated as an orthogonal Latin square, a fundamental concept in combinatorics. Orthogonality is a property that concerns the relationship between two or more Latin squares. Two Latin squares are orthogonal if, when superimposed, each possible pair of symbols appears exactly once.

As noted above, combinatorics is a vast field of mathematics that studies discrete and finite structures. Given its breadth and complexity, it is not easy to enclose it in rigid schemes; there are several possible classifications. According to the American Mathematical Society (AMS), combinatorics can be divided into numerous branches, including:

1. Enumerative combinatorics: deals with the count of possible configurations of objects. This branch includes techniques such as the inclusion-exclusion principle and recurrence formulas.
2. Algebraic Combinatorics: This course examines the relationships between combinatorics and algebra, using algebraic tools to solve combinatorial problems. It includes themes such as Bell's polynomials and Stirling's numbers.
3. Probabilistic Combinatorics: analyses the combinatorial properties of random objects, using probabilistic techniques to obtain results on combinatorial structures.
4. Topological Combinatorics: This area explores the interactions between combinatorics and topology, such as the study of triangulations of surfaces and spaces.
5. Drawing theory focuses on constructing and analysing configurations with specific properties.
6. Graph theory analyses the relationships between objects represented as vertices and edges. It has applications in computer science, engineering, and social sciences.
7. Extreme combinatorics: Investigates the properties of combinatorial structures of maximum or minimum dimensions.

Another possible subdivision, more oriented towards methods and approaches, could be the following:

1. Classical combinatorics focuses on fundamental principles, such as permutation, combination, arrangement, and the inclusion-exclusion principle. It is the starting point for the study of all the other branches of combinatorics.
2. Enumerative combinatorics involves counting the number of combinatorial structures of a certain type. To solve complex problems, counting techniques, such as generating functions and recursions, are used.
3. Analytic combinatorics study the properties and relationships between combinatorial structures. It uses mathematical tools such as linear algebra, analysis, and graph theory to achieve theoretical results.

These classifications are not mutually exclusive and often overlap. For example, enumerative combinatorics appear in both classifications, while graph theory of the AMS classification could fall under the analytic combinatorics of the second classification.

This subdivision can highlight the progression of difficulty and abstraction between the different areas of combinatorics. Classical combinatorics provides the conceptual basis, enumerative combinatorics develops tools for efficient counting, and analytic combinatorics explores combinatorial structures more in-depth and abstractly.

A wide-ranging topic is combinatorial algorithms since they have numerous practical applications in various fields. Here are some of them:

1. Resource allocation: Used to analyse the optimal distribution of financial resources to assess return and risk.
2. Optimization: Use it to solve problems such as finding the shortest path or scheduling resources.
3. Network management: In telecommunications and networks, specific algorithms solve routing, frequency assignment, and resource planning problems.

4. Chemistry: For example, combinatorics helps generate all possible isomers (structural, geometric, and optical) of a molecule.
5. Bioinformatics: Combinatorial algorithms help solve problems such as aligning genetic sequences, predicting protein structures, and determining evolutionary relationships between different species.
6. Encryption: Generating secure cryptographic keys is one of the most popular tasks.
7. Artificial Intelligence and Machine Learning: Combinatorial techniques are used to find the best configuration of the model parameters.
8. Games and puzzles: Many games and brain teaser challenges involve combinatorial concepts, so different algorithms have been developed in this field.
9. Circuit design: A typical example is designing electronic circuits to optimize the arrangement of components and connections.

The latter case can be a good example of the role played by combinatorics. Electronic circuit design leverages combinatorics in many ways, particularly optimizing component arrangement and connections. This process can cover several aspects, such as the physical arrangement of components on the board, the management of connections between the various nodes or the minimization of specific resources. Of course, an authentic design cannot be covered here, but the code could be helpful as a teaching exercise to understand the basic concepts of positioning constraints.

The program scans all permutations of electronic components and checks whether each meets certain technical constraints, such as the power limit and thermal and frequency compatibility between the components. If a solution is valid, it is added to the list of results. The program then provides a summary of the valid provisions found while also showing the top five provisions in detail.

Program 1 Circuit

```
import itertools

from dataclasses import dataclass

from typing import List, Tuple

@dataclass

class Component:

    name: str

    power: float  # Watt

    max_temperature: float  # Celsius
```

```
    frequency_range: Tuple[float, float]  # Hz

    nominal_voltage: float  # Volt

    nominal_current: float  # Ampere

# Definition of components with realistic parameters

components = {

    'Resistor': Component('Resistor', 0.25, 70, (0, 1e9), 50, 0.1),

    'Capacitor': Component('Capacitor', 0.1, 85, (1, 1e6), 25, 0.05),

    'Inductor': Component('Inductor', 0.5, 80, (0, 1e5), 30, 0.2),

    'Diode': Component('Diode', 1.0, 125, (0, 1e6), 100, 1.0),

    'Transistor': Component('Transistor', 0.5, 150, (0, 1e8), 40, 0.5),

    'LED': Component('LED', 0.1, 85, (0, 1e3), 3.3, 0.02)

def check_thermal_constraints(arrangement: List[Component]) -> bool:

    """Check the thermal constraints between adjacent components"""

    for i in range(len(arrangement) - 1):

        if arrangement[i].max_temperature > 100 and arrangement[i+1].max_temperature < 90:

            return False

    return True
```

```
def check_frequency_compatibility(arrangement: List[Component]) -> bool:

    """Check the compatibility of operating frequencies"""

    for i in range(len(arrangement) - 1):

        # Check frequency band overlap

        freq_min = max(arrangement[i].frequency_range[0], arrangement[i+1].frequency_
range[0])

        freq_max = min(arrangement[i].frequency_range[1], arrangement[i+1].frequency_
range[1])

        if freq_min > freq_max:

            return False

    return True

def check_power_distribution(arrangement: List[Component]) -> bool:

    """Check the power distribution"""

    total_power = sum(c.power for c in arrangement)

    if total_power > 2.5:  # Maximum power limit

        return False

    return True

def check_arrangement(arrangement: List[Component]) -> bool:
```

```
        return all([
            check_thermal_constraints(arrangement),
            check_frequency_compatibility(arrangement),
            check_power_distribution(arrangement)
    def main():
        component_names = list(components.keys())
        valid_arrangements = []
        for perm in itertools.permutations(component_names):
            arrangement = [components[name] for name in perm]
            if check_arrangement(arrangement):
                valid_arrangements.append(perm)
        print(f"\nDesign constraints:")
        print("1. Thermal compatibility: high-temperature components cannot be adjacent to
sensitive components.")
        print("2. Frequency compatibility: frequency bands must overlap between adjacent com-
ponents.")
        print("3. Power distribution: the total power must be < 2.5W.")
        print(f"\nValid arrangements found: {len(valid_arrangements)}")
```

```
    for i, arrangement in enumerate(valid_arrangements[:5], 1):

        print(f"\nArrangement {i}:")

        total_power = sum(components[comp].power for comp in arrangement)

        print(f"Components: {' -> '.join(arrangement)}")

        print(f"Total power: {total_power:.2f}W")

if __name__ == "__main__":

    main()
```

The most curious aspect is the possibility of combining combinatorics with literature. Combinatorial literature finds its roots in the French movement of the Oulipo (Ouvroir de Littérature Potentielle), founded by Raymond Queneau (1903-1976) and François Le Lionnais (1901-1984). In this field, one of the masters was certainly Italo Calvino (1923-1985). The fifty-five cities described in the book "Invisible Cities" are divided into eleven thematic categories, such as "Cities and Memory", "Cities and Desire", "Cities and Signs", etc. Each category explores a specific aspect of urban life and human perception. The book consists of nine chapters, each opening and closing with a dialogue between Marco Polo and Kublai Khan. Polo describes five cities within each chapter, except in the first and last chapters, which contain ten. This structure creates a network of interconnected narratives. Each city is described in a modular way, with elements that can be combined and recombined to create new interpretations. For example, cities can be seen as combinations of memory, signs, exchanges, etc. In addition to Italo Calvino, several authors have explored combinatorial literature. Jonathan Swift (1667-1745) in "Gulliver's Travels" (1726), describes a machine on the island of Balnibarni, in which a crank rotates wooden cubes with all the words of their language, conjugated in different ways and tenses and declined in various cases so that the arrangement of the words changes according to the combinations thus created.

Raymond Roussel (1877-1933) was an innovator of storytelling, using amazing techniques to create unique and visionary works. In "Locus Solus", he explores the boundary between reality and imagination through a series of eccentric episodes involving fantastical inventions and bizarre characters, creating complex plots using puns, anagrams and circular structures.

Jorge Luis Borges (1899-1986) further expanded the possibilities of combinatorial fiction, tackling complex themes in stories such as "The Library of Babel". In this story, imagine an infinite library that contains all possible combinations of books, exploring the concept of infinity. The Library of Babel offers a fascinating starting point for exploring combinatorial quantification. In this narrative, Borges describes an infinite library consisting of hexagonal rooms containing books formed by random characters. Each book has 410 pages, 40 lines per page and 80 characters per line, using an alphabet of 25 symbols (22 letters plus space, semicolon). This situation reflects the idea that a finite number of elements can generate almost infinite sentences and texts. Since each book has 1,312,000 characters (410×40×80), the total number of distinct books that can be created is 251,312,000. This number is astronomical, amounting to about 101,834,097, which means there are more possible books than atoms in the known universe, estimated at around 1080. Often his stories play with the concept of labyrinths and parallel universes, where choices and combinations influence the fate of the characters. Borges also uses literary forms such as the circular story where the beginning and the end intertwine.

Georges Perec (1936-1982) carried on the tradition of combinatorial storytelling with "Life Instructions for Use", a novel that resembles a puzzle. The novel is structured as a ten-storey Parisian building with ten rooms per floor; each chapter corresponds to a room and describes the objects contained in it and the stories of its inhabitants. Perec uses writing techniques inspired by play. He is also famous for employing the technique of "constrained writing," in which he follows specific rules, such as the lack of a letter, to explore the beauty of unexpected combinations that emerge from constraints.

Rudolph J. Wilcock (1919-1978) introduces an element of wonder in "The Synagogue of the Iconoclasts." The story is that of an eighteenth-century French watchmaker who invented the Universal Mechanical Philosopher, a device as large as an entire room capable of generating an almost infinite number of sentences. This apparatus becomes a metaphor for the human condition, trapped between automatism and the limits of human reason.

Faced with this wide choice of topics and suggestions, it is not easy to make a choice. The text tries to develop the main aspects of this field of research.

Some topics mention the use of GeoGebra, but most of the passages are supplemented by boxes that show many programs written in Python with the help of AI. To make many programs work, it is essential to have downloaded the necessary libraries. The computer approach to combinatorics offers numerous advantages, both theoretically and practically. Here are some of the principal utilities:

1. Combinatorics almost always involves complex and repetitive calculations, which can be automated, saving time and reducing errors.
2. IT tools can manage and analyse large amounts of data, which would be impractical manually. This is especially useful in bioinformatics, cryptography, and optimization applications.
3. Computer science allows combinatorial structures to be visualized, often interactively. Simulations help us better understand their properties and behaviours.
4. Having dealt with problems of quite different sizes, scalability is an added value.
5. Computers can test combinatorial conjectures; here, the possibility of testing hypotheses on many cases emerges.

The computer approach to combinatorics has become increasingly important in recent decades, especially with the expansion of artificial intelligence and computational mathematics. The computer approach allows the application of algorithms that reduce the computational cost and find solutions in a reasonable time, even for complex problems. In many cases, finding an exact solution to a combinatorial problem can be too expensive in terms of time and resources. Computer science also allows simulation or approximation of combinatorial solutions through numerical and stochastic methods. Recently, machine learning methods (especially neural networks) have been used to tackle complex combinatorial problems such as those that arise in engineering or operations research.

To understand the usefulness of this computer approach, which will be used throughout the text, we can examine four different problems.

Suppose you have to pay a certain amount using euro cents (without any limit on availability). How many ways can you get this sum?

Program 2 Coin Change

```
def coin_change(coins, n, index=0):

    # If the amount to change is 0, there is one way: not using any coins

    if n == 0:
```

```
        return 1

    # If the amount to change is negative, it's not possible to make change

    if n < 0:

        return 0

    # If we have exhausted all coins, we cannot make change

    if index >= len(coins):

        return 0

    # Sum of ways to make change including and not including the current coin

    return (coin_change(coins, n - coins[index], index) +

            coin_change(coins, n, index + 1))

# Example coins (in cents)

coins = [10, 20, 50, 100, 200]

# Ask the user to enter the desired value

value = int(input("Enter the value to obtain: "))

solutions = coin_change(coins, value)

print("Number of ways to obtain", value, ":", solutions)
```

In the second example, one shuffles a deck of 52 cards; we wonder how many operations are necessary to restore the original configuration. Based on value, the cards are divided into piles, and then 52 operations are performed. When the

13 piles are stacked in a single pile, 12 operations are performed. The deck is divided into four decks based on colour, so there will be 52 more operations. Finally, three operations are performed when the four decks are stacked to restore the initial situation. So, the total number of trades performed will be 52 + 12 + 52 + 3 = 119. If we look at this sorting algorithm, we notice that it involves combinations of operations (divisions, overlaps, and stacks), so it falls into the combinatorial category.

Program 3 Shuffle

```
import random

# Create a sorted deck

suits = ['Hearts', 'Diamonds', 'Clubs', 'Spades']

values = ['2', '3', '4', '5', '6', '7', '8', '9', '10', 'Jack', 'Queen', 'King', 'Ace']

initial_deck = [(value, suit) for suit in suits for value in values]

# Shuffle the initial deck

random.shuffle(initial_deck)

# Print the unsorted cards

print("Unsorted cards:")

for card in initial_deck:

    print(card[0], "of", card[1])

# Initialize the operation counter

operations = 0

# Divide the cards into piles based on value
```

```
piles_by_value = [[] for _ in range(len(values))]

for card in initial_deck:

    value_index = values.index(card[0])

    piles_by_value[value_index].append(card)

    operations += 1

print(f"Operations after division by value: {operations}")

# Merge the piles into a single deck ordered from King to Ace

ordered_deck_by_value = []

for i, pile in enumerate(reversed(piles_by_value)):

    if pile:

        if i < len(values) - 1:  # Count only 12 operations to merge the 13 piles

            operations += 1

        ordered_deck_by_value.extend(pile)

print(f"Operations after merging piles by value: {operations}")

# Divide the deck by suit

piles_by_suit = [[] for _ in range(len(suits))]

for card in ordered_deck_by_value:
```

```
    suit_index = suits.index(card[1])

    piles_by_suit[suit_index].append(card)

    operations += 1

print(f"Operations after division by suit: {operations}")

# Merge the 4 piles to restore the initial state

final_deck = []

for i, pile in enumerate(piles_by_suit):

    if i > 0:  # We should add only 3 operations to merge the 4 piles

        operations += 1

    final_deck.extend(pile)

print(f"Operations after final merge: {operations}")

# Print the piles divided by suit

for i, pile in enumerate(piles_by_suit):

    print(f"\nPile {suits[i]}:")

    for card in pile:

        print(card[0], "of", card[1])

# Print the total number of operations
```

```
print(f"\nTotal number of operations: {operations}")
```

The third example takes its cue from the relationship between combinatorics and bioinformatics. The two disciplines are closely interconnected, as many problems in bioinformatics can be formulated as combinatorial problems. DNA sequencing is one of the fundamental areas in which combinatorics plays a crucial role. Combinatorics is also essential in phylogenetic analysis, which studies evolutionary relationships between species: the construction of phylogenetic trees, which represent evolutionary relationships between species, can be treated by combinatorial algorithms such as Maximum Parsimony. This method is based on the principle that the most likely phylogenetic tree is the one that requires the fewest evolutionary changes.

This method tries all possible permutations, so it is effective for small sets of sequences but becomes computationally expensive for $n > 10$. We can use more advanced methods for larger datasets. In summary, the program finds the best phylogenetic order with an exhaustive approach based on maximum parsimony and the Hamming distance.

Program 4 Maximum Parsimony

```
import itertools

def hamming_distance(seq1, seq2):

    return sum(c1 != c2 for c1, c2 in zip(seq1, seq2))

def calculate_tree_score(tree, sequences):

    score = 0

    # Iterate over each position in the sequence

    for i in range(len(sequences[0])):

        # Extract the current column from all sequences

        column = [sequences[index][i] for index in tree]

        # Calculate the score for the column based on the Hamming distance

        score += min(
```

```
            hamming_distance(column, [state] * len(column))
            for state in set(column)
    return score
def maximum_parsimony(sequences):
    n = len(sequences)
    best_score = float('inf')
    best_tree = None
    # Try all possible permutations
    for tree in itertools.permutations(range(n)):
        score = calculate_tree_score(tree, sequences)
        if score < best_score:
            best_score = score
            best_tree = tree
    return best_tree, best_score
# Example sequences
sequences = [
    "ATCGGTAC",  # Sequence 0 (root)
```

```
        "CTAGTTAC",  # Sequence 1 (4 changes)

        "ATTGGTAC",  # Sequence 2 (2 changes)

        "ATCGGTAG"   # Sequence 3 (1 change)

    # Calculate the best tree and its score

    best_tree, best_score = maximum_parsimony(sequences)

    # Print the results

    print(f"Best tree: {best_tree}")

    print(f"Score: {best_score}")

    print("\nSequences in the order of the best tree:")

    for index in best_tree:

        print(f"{index}: {sequences[index]}")
```

The last program takes up the question of the 36 officers proposed by Euler. In this case, the value of n indicates the size of the square. If one poses n = 6, the answer is that one can NOT construct an orthogonal Latin square. The same is true for n = 2.

Program 5 Orthogonal Latin Square

```
import numpy as np

def generate_latin_square(n):

    """Generates a Latin square of order n."""

    latin_square = np.zeros((n, n), dtype=int)
```

```
    for i in range(n):
        for j in range(n):
            latin_square[i, j] = (i + j) % n
    return latin_square

def are_orthogonal(A, B):
    """Checks if two Latin squares A and B are orthogonal."""
    n = A.shape[0]
    pairs = set()
    for i in range(n):
        for j in range(n):
            pair = (A[i, j], B[i, j])
            if pair in pairs:
                return False
            pairs.add(pair)
    return True

def find_orthogonal_latin_squares(n):
    """Finds two orthogonal Latin squares of order n."""
```

```
    A = generate_latin_square(n)
    # Modified generation of B to include n=10
    B = np.zeros((n, n), dtype=int)
    for i in range(n):
        for j in range(n):
            B[i, j] = (i + j * (n // 2)) % n  # Modified here to better handle n=10
    if are_orthogonal(A, B):
        return A, B
    return None, None
def print_latin_square(square):
    """Prints a Latin square."""
    for row in square:
        print(" ".join(map(str, row)))
    print()
def print_orthogonal_square(A, B):
    """Prints the resulting orthogonal Latin square by overlaying A and B."""
    n = A.shape[0]
```

```
    orthogonal_square = np.zeros((n, n), dtype=object)

    for i in range(n):

        for j in range(n):

            orthogonal_square[i, j] = (A[i, j], B[i, j])

    for row in orthogonal_square:

        print(" ".join(map(str, row)))

    print()

if __name__ == "__main__":

    n = int(input("Enter the order of the Latin square: "))

    A, B = find_orthogonal_latin_squares(n)

    if A is not None and B is not None:

        print("Latin square A:")

        print_latin_square(A)

        print("Latin square B:")

        print_latin_square(B)

        print("Resulting orthogonal Latin square:")

        print_orthogonal_square(A, B)
```

```
    else:

        print(f"Unable to find two orthogonal Latin squares of order {n}.")
```

The method is not always optimal for n = 10 or other large values, and therefore, more sophisticated approaches have been studied. One could use methods such as direct construction using the Bose Cyclic Method, which works well for odd n, or combinatorial search algorithms that will be examined later. Evolutionary and probabilistic algorithms also provide a possibility.

Chapter 1
Introduction to Combinatorics

ABSTRACT

Combinatorics, a branch of mathematics that studies the possibilities of selecting, arranging, and grouping elements of a set, has ancient roots in various cultures. The chapter traces the history of combinatorics, from its origins in the ancient East to modern developments. Key figures such as Pascal, Leibniz and Fibonacci are highlighted as important contributions. The chapter illustrates the connection of combinatorics with graph theory and other complex concepts thanks to figures like Paul Erdős and Ronald Graham. The most recent developments in combinatorics are examined, focusing on the impact of artificial intelligence and machine learning. It highlighted how these technologies open new avenues for optimizing graph problems. Combinatorics is fundamental for game theory, circuit design, social networks, and drug discovery. The chapter introduces theorems such as those of Szemerédi and Green-Tao that have brought new points of view. Finally, it presents several simulations to clarify better concepts, such as the drunken path and the Sierpinsky triangle.

A BRIEF HISTORY OF COMBINATORICS

Not knowing what happened in the past would be like being a child forever

Cicerone

Combinatorics, a branch of mathematics dedicated to the study of the possibilities of selecting, arranging, and grouping the elements of a set, has a long history with roots in various cultures and historical periods (Wilson & Watkins, 2013). The I Ching, or Book of Changes, is an ancient Chinese text (thought to date back to at least 1000 BC) with deep connections to combinatorics principles (Redmond & Hon, 2014). It is based on 64 hexagrams, each consisting of six lines. Each line can be yang (whole) or yin (broken), representing a primitive binary system. The basic eight trigrams (combinations of 3 lines) generate the 64 hexagrams (6 lines). It represents one of the first systematic applications of the combinatorial principle: $2^6 = 64$. The order of the lines is significant, introducing the concept of permutations. There are 8! (40,320) possible ordering of trigrams, even if only a few are used. The I Ching represents one of the earliest examples of systematic combinatorial thinking, serving as a bridge between mathematics and divination. This text profoundly influenced Chinese thought, particularly in Confucianism, Taoism, and Western thought, with philosophers such as Leibniz and Jung being inspired by it.

The most famous example of a combinatorial arrangement is the magic square known as Lo Shu, with a written trace starting from 650 BC. According to legend, the Lo River was in flood, so a turtle with a design engraved on its back emerged to save the population from the flood. The pattern was a 3x3 grid, each containing one of the numbers between 1 and 9. Each row, each column and both diagonals of the square gave a sum of 15. This poses the problem of how to arrange the figures to achieve this result.

We find traces of combinatorial reasoning in Chinese, Babylonian, Egyptian and Indian civilizations. The Indian mathematician Pingala, in his text "Chanda śāstra", written in the fourth century BC, presented a method of combinatorial enumeration (Berstel & Perrin, 2007). He examined ways of combining syllables of different lengths (Short and

DOI: 10.4018/979-8-3373-3089-1.ch001

Long) to produce poetic meters. For example, with three syllables, one can get eight different rhythms: BBB, LLL, BLL, LBL, LLB, BLB, BBL.

While combinatorics wasn't a primary focus for the Greeks and Romans, we do find intriguing hints of its underlying principles in their work. Take Zenodorus, for instance, and his study of the isoperimetric problem. In the language of combinatorial optimisation, this challenge asks: Which configuration maximises a specific property (in Zenodorus's case, the area) while keeping a constraint (the perimeter) constant? This line of reasoning, though not directly employing formal combinatorial methods, shares a conceptual kinship. It involves evaluating multiple possibilities to pinpoint the one that optimises a given characteristic. This approach strongly foreshadows modern decision theory and operations research, disciplines that now extensively leverage combinatorics to solve complex optimisation challenges (Acerbi, 2003).

Another scholar to consider was Hipparchus; although he did not directly contribute to combinatorics, his systematic work on classification in astronomy can be seen as an embryonic form of combinatorial thinking (Acerbi, 2003).

Pappus of Alexandria, who lived between the third and fourth centuries AD, was one of the most important mathematicians of the Hellenistic era. In his "Mathematical Collection" work, Pappo explores various geometric and mathematical topics. Although he does not deal directly with combinatorics, his approach to some geometric issues includes aspects that we can consider anticipatory of enumerative combinatorics. Enumerative combinatorics is concerned with calculating the number of ways in which certain objects can be organized or counted.

For example, one of the challenges studied by Pappus concerns the calculation of circle configurations within a given geometric figure. This type of question requires identifying possible arrangements of elements (in this case, circles) that satisfy a set of conditions or constraints. One example of this geometric exploration involves tangents, where Pappus asks how many circles can be positioned to be tangent to two parallel lines and another circle. The task involves determining all possible ways to arrange the circles while respecting the given geometric conditions—an approach that, in a sense, anticipates combinatorial analysis in finding solutions within a constrained set of possibilities.

Although formalized combinatorics as we understand it today did not exist, these scholars laid the foundations for future exploration of these mathematical concepts.

One of the most illustrious mathematicians of medieval India was Bhāskara II (1114–1185). His main work, the Līlāvatī, encompasses a wide range of mathematical inquiries, covering various topics, including algebra, geometry, and, notably, combinatorics (Biggs, 1979). Bhāskara II studied arrangements and selections of elements, laying the groundwork for a more systematic understanding of these techniques. One of his contributions to combinatorial mathematics is the analysis of permutations and combinations. He introduced the concept of bhāgas, or groupings of elements appearing in a particular order. Bhāskara explains, for example, the method for calculating the number of permutations of n elements, which is reflected in the formula we know today as $n!$

The use of factorial formulas (n! being the product of all positive integers less than or equal to n) appears in his discussion of how many ways Vishnu could hold his four characteristic objects—a shell, a disc, a mace, and a lotus flower—in his four hands. He also explores advanced combinatorial techniques, such as determining the number of ways letters in a word can be arranged and selecting specific values from a set. In this, he demonstrates a mastery of combinatorial reasoning and a broader vision of mathematics as a tool for solving practical questions.

The ancient mathematicians and scientists of these civilizations faced concrete challenges, such as organizing crops, carrying out commercial transactions, and solving problems related to the calendar. The solutions to these problems required combinatorial skills, even if, seen today, they seem quite rudimentary. In reality, exciting signals will be taken up in other eras. However, we should not forget that the first mathematical bases that will be found in the most modern developments are also beginning to appear.

Even simpler puzzles from later periods can be viewed through a combinatorial lens. Consider Alcuin of York's (735–804) classic "chickens and rabbits" problem from the ninth century. It asks one to determine the count of rabbits and chickens given the total number of heads and legs. While commonly solved using a system of linear equations (e.g., with 50 heads and 140 legs, the owner has 30 chickens and 20 rabbits), it also has a clear combinatorial interpretation.

At its core, it is a counting exercise—determining how we can assign a set of features (heads and legs) to different categories (chickens and rabbits). One might consider all possible distributions of animals, such as 50 chickens and no rabbits, or vice versa. As can be quickly verified, there are 51 such combinations, but only one satisfies all given constraints. Thus, the task requires identifying the single arrangement that fulfils these conditions, making it akin to a selection process within a set of possible configurations.

In Medieval Europe, permutations and combinations can be found in astrological studies, and we found traces in the studies of Rabbi Ben Ezra (1089-1164). For example, considering the 12 zodiac signs, astrologers often had to determine all the possible arrangements of 7 planets in the 12 signs, enabling them to understand the different astral positions and configurations. In several medieval astrological treatises, such as those of John of Sacrobosco (1195-1256), we find a systematization of these planetary schemes and configurations. Although the author did not use the formal language of combinatorics, the work "De Sphaera Mundi" shows how they considered all possible arrangements and configurations to draw interpretative conclusions. In addition, medieval astrological practice also influenced the development of tables and diagrams, tools that allowed the combinations and permutations of planetary positions to be quickly calculated to produce horoscopes.

Leonardo Fibonacci (1170–1250), through his renowned "Liber Abaci", introduced to Europe the number pattern that now bears his name. This progression became a cornerstone in counting techniques and the study of combinatorial structures. Fibonacci numbers are closely linked to binary configurations. For example, the number of binary strings of length n without two consecutive 1s corresponds to the $(n+1)$th term in the Fibonacci series. This iconic pattern illustrates how specific numbers naturally emerge in the fields of combinatorics and optimisation. Thanks to its recursive structure, it finds wide application across mathematical contexts—flexibly modelling optimal arrangements, configurations, and paths.

In the 17th century, French mathematician Blaise Pascal (1623-1662) introduced the number triangle, the "Pascal triangle," which contains important information about combinations and binomial coefficients. Pascal and Fermat (1607-1665) developed a combination approach that involved counting techniques to determine the number of ways in which elements could be arranged in a set. The concept of counting remains fundamental in modern combinatorics (Lenart, 2015).

Jacob Bernoulli (1654-1705), in his "Ars Conjectandi" (1713), provides a comprehensive treatment of combinatorics and its applications to probability. In Part I, he extensively deals with combinatorics, introducing formulas and basic principles for permutations and combinations. Bernoulli uses these techniques to develop a systematic method for counting the arrangements of elements. In Part II, he applies combinatorial concepts to probability theory, addressing games of chance and probabilistic problems. For example, using combinatorics, Bernoulli calculates the odds associated with drawing prizes in a lottery. In conclusion, Ars Conjectandi is a fundamental work not only for its treatment of combinatorics but, above all, for its impact on the development of probability theory. Bernoulli's establishment of the connection between combinatorics and probability had a profound impact on mathematics and science, laying the groundwork for probability as an autonomous discipline.

Leibniz, best known as one of the inventors of infinitesimal calculus, developed the concept of permutation with repetition and circular permutation, further expanding the combinatorial tools. It is worth noting that at the age of twenty, he wrote the "Dissertatio de arte combinatoria", in which he discusses topics based on earlier works, such as those of Ramon Llull (1232-1316), who was looking for a "machine" to generate all truths by combining fundamental concepts. Leibniz conceived the *characteristica universalis* as a symbolic language that could represent every human thought in a precise way. Combinatorics played a central role in this project, as it formed the basis for building a language that could combine symbols to form more complex concepts. To create a universal language capable of capturing every nuance of knowledge, Leibniz envisioned applying combinatorial techniques to logic and philosophy.

The Swiss mathematician Leonhard Euler (1707-1783) also contributed to graph theory, an area closely related to combinatorics. His solution to the "Königsberg Bridge Problem" is considered one of the first significant achievements.

Pierre-Simon Laplace (1749-1827) played a particular role. He applied combinatorics to probability theory, thereby influencing the reasoning behind probabilistic events and their distribution through his studies.

One of the best-known personalities, also for his personal story, is Évariste Galois (1811-1832). He founded group theory, a crucial part of algebraic combinatorics. His ideas underline the modern understanding of symmetry and permutations.

Moses Ilyich Schönfinkel (1888-1942) played a significant role as a logician and mathematician. In 1920, in a conference, he presented the foundations of Combinatorial Logic, a work later published in 1924. Unfortunately, afflicted by a severe mental illness, he ended up poor and without a trace of his contributions. The rediscovery of this mathematician is mainly due to Haskell B. Curry (1900-1982), to whom we owe important contributions relating to the λ-Calculus presented for the first time by Alonzo Church (1903-1955) in the 30s of the last century. Although these are two non-overlapping topics, both systems have a relationship with proof theory. Lambda calculus is used in functional programming languages, while combinatorial logic focuses on simplifying terms without variables.

In more recent times, we can cite George Pólya (1887-1985), who has made important and innovative contributions to enumerative combinatorics and probability theory. The principle of invariance is one of the central ideas of his work. This principle states that symmetries can be ignored when counting configurations for more manageable results. Suppose one wants to count the number of ways to colour an object, considering rotations. In that case, the principle of invariance allows us to consider only unique colours. This is crucial in various fields, from art to chemistry, where symmetry plays a key role. In collaboration with Alfréd Rényi (1921-1970), Pólya developed a method for analysing random combinatorial structures. This method has been fundamental in the study of network theory and the theory of random graphs.

Paul Erdős (1913-1996) was a legendary figure, an eccentric mathematician with vast interests. He is known for his contributions to number theory, but it can be said that combinatorics was also the focus of his interests. During his career, Erdős offered cash prizes to stimulate the work of mathematicians, awards that, for a time, were managed by Ronald Graham (1935-2020), one of the leading mathematicians of the twentieth century. Known for his work in number theory and graph theory, he was a prominent figure in combinatorial analysis (Graham, 1995). Graham's number, named in his honour, is so large that it is practically impossible to understand its size. The Graham number is primarily a theoretical concept used to illustrate the remarkable growth of certain recursive functions and the complexity of solving problems in graph theory. It has no direct practical application in everyday life, but it is a fascinating example of how extraordinary mathematics can be when exploring the realm of the infinitely great.

Richard P. Stanley (1944-) is a contemporary mathematician who has made significant contributions to the advancement of algebraic combinatorics, generating functions, and graph theory. His contributions to the theory include work on polynomials, which have been used to solve several combinatorial problems, including counting partitions of integers and the theory of symmetric functions.

Additionally, we can mention László Lovász (1948–), who has made fundamental contributions to graph theory, combinatorics and combinatorial optimisation. He is known for his work on structural properties of networks and polyhedra. Among his most relevant contributions, we highlight Lovász's Theorem, which establishes a fundamental relationship between the chromatic number of a network and its stability set. This result led to significant developments in combinatorial mathematics, influencing subsequent studies on colouring and optimisation techniques (West, 2021). This brief overview would be incomplete without considering the relationship between combinatorics and computer science.

The computer has had a profound impact on combinatorics, as it enables the performance of lengthy analyses. For example, in 1977, Kenneth Appel (1932-2013) and Wolfgang Haken (1928-2022) announced the proof of the four-colour conjecture with the aid of a computer (Appel & Haken, 1977). The proof is based on reducing the infinite number of possible maps to about 2000 configurations, for which the validity of the theorem is verified on a case-by-case basis by computer. Combinatorics is essential in the design of efficient algorithms. Many of them use combinatorial techniques to generate, enumerate, or count objects. Search and sorting algorithms often use combinations and permutations. Graphs are one of the most widely used data structures in computer science. Combinatorics provides tools to analyse and optimize graph properties, such as connections and distance measurements.

RECENT DEVELOPMENTS IN COMBINATORICS

In recent years, combinatorics has gained increasing attention in various fields, including machine learning, complex structure generation, and complex network modelling (Beveridge et al., 2016). These developments have been facilitated by the intersection of combinatorics with other mathematical and computer science disciplines. Machine learning utilizes combinatorial techniques, particularly in the design of algorithms for optimisation. For example, enumeration techniques can be used to efficiently select features in predictive models. However, it is, above all, graph theory that is often applied to represent relationships between complex data, facilitating the analysis of social networks and complex systems.

A great impetus to the treatment of complex problems has been given by the development of Artificial Intelligence (AI). There are many forms of artificial intelligence (AI) learning, and they are constantly evolving. In general, we can divide AI learning into three broad categories:

1. *Supervised Learning*: In this type of learning, AI is trained on a set of input and output data that has already been labelled. The goal is for the algorithm to learn to map new inputs to the correct outputs. We can obtain a system capable of generating new sequences with desired properties by training a model on a large dataset of combinatorial sequences (permutations, combinations, etc.). A model might learn how to generate secure passwords and diagnostic trouble codes.
2. *Unsupervised Learning*: Unlike supervised learning, AI does not have labelled data in this case. The goal is to find hidden patterns and structures. Applying clustering techniques to large datasets of combinatorial objects allows for the identification of new classes of entities with interesting properties. For example, new families of graphs with particular symmetry or connectivity properties could be discovered.
3. *Reinforcement Learning*: In this type of learning, AI learns by interacting with an environment. The algorithm receives feedback (a reward or a penalty) based on its actions and tries to maximize the reward in the long term. Artificial intelligence systems can be developed to reason combinatorially and solve complex problems.

With the emergence of Machine Learning, particularly reinforcement learning and supervised learning, new avenues are opening up to address complex graph optimisation problems. Supervised learning can be used to develop predictive models that aid in solving graph optimisation tasks. Models can learn to predict node properties that affect optimisation, and they can be trained to classify the output as "optimal" or "non-optimal" based on previous examples, thereby guiding research toward promising solutions. Reinforcement learning also offers an innovative approach to combinatorial optimization on graphs (Lu et al., 2023). Each action taken by the agent (for example, choosing a node to visit) produces a reward based on the quality of the solution found. A typical application of combinatorial optimization is in the planning of transport networks.

If we want to formulate concrete examples, we can mention the development of models capable of optimizing delivery routes in real time. These models learn from interactions with the environment, continuously improving their decisions based on the rewards of more efficient routes.

As a second example, we can include the design of electronic circuits; in this case, combinatorial optimization is essential to minimize the area and energy consumption. ML techniques allow us to explore the space of possible configurations, applying genetic algorithms and neural networks to identify the best configurations.

A third example comes from the analysis of social networks, where combinatorial optimisation is used to identify significant relationships between users. In this sense, supervised learning algorithms can be employed to classify nodes based on their centrality or importance in the network. From these examples, we can deduce the fundamental role of algorithms. This is a theme that will be extensively dealt with in the following chapters.

This frontier looks promising, but it is certainly not the only possibility. Machine learning provides a powerful tool for addressing combinatorial problems, but it is not the only solution. The choice of the most suitable technique depends on the nature of the problem, the resources available and the specific objectives. We can list some particularly significant research objectives.

New results concern the combination of combinatorial theory and game theory, which analyses the strategies and interactions between these two fields. For example, game theory utilises combination theory to determine the number of possible strategies for players and to identify optimal strategies. The interaction between combinatorics and game theory has applications in various fields, including economics and finance, where combinatorics helps determine the optimal choices in complex scenarios involving multiple interacting agents. In complex decision-making contexts, combinatorial analysis of strategies (Bailey et Cameron, 2009) can provide suggestions on optimizing choices in conflict or cooperation situations. In robotics and artificial intelligence, combinatorics analyses interactions between autonomous agents that must coordinate to achieve common goals.

A fascinating field of study concerns topological combinatorics, which focuses on the topological structure of combinatorial configurations. Among the key concepts, we can mention subgroups and invariants; configurations can be analysed through topological subgroups and invariants that remain constant under continuous deformations. Simplicial spaces are constructed from simple polyhedra (simplici) and are used to represent complex configurations more easily. Some practical examples of the application of topological combinatorics include covering problems where it is critical to determine the minimum number of subsets needed to cover a set or analyse geometric configurations in Euclidean or projective spaces that can reveal interesting properties regarding the intersection and arrangement of objects.

Recently, combinatorics has been used in biology to study the structure of DNA and to develop new drugs, as in the case of the COVID-19 vaccine. Combinatorics can be used to identify new molecules that bind to a target protein, allowing for the development of a drug that targets this protein. Combinatorics can be used to generate a wide range of molecules, which can then be tested for their ability to bind to the target protein.

However, the role of proteins and their recombination does not end there.

If we want to simulate the functioning of evolution, we must keep in mind that it proceeds by exploring each time set of possible neighbours, each moving one step at a time in the adaptive landscape.

How can we show this process?

A random numerical sequence is generated from values ranging between 1 and 20, reflecting the 20 amino acids that form the 500-unit-long chain of a protein. The goal is to gradually evolve this sequence into an ordered series through a selective process.

Program 6 sequence

```
import random
def generate_sequence(n, min_val, max_val):
"""Generates a random sequence of n numbers between min_val and max_val."""
return [random.randint(min_val, max_val) for _ in range(n)]
def filter_sequence(sequence):
"""Filters the sequence, keeping only the numbers followed by their successor."""
filtered_sequence = []
for i in range(len(sequence) - 1):
if sequence[i] + 1 == sequence[i + 1]:
filtered_sequence.append(sequence[i])
# Also adds the successor
filtered_sequence.append(sequence[i + 1])
return list(set(filtered_sequence)) # Removes duplicates
def final_sequence(sequence):
"""Returns the filtered and sorted sequence."""
filtered_sequence = filter_sequence(sequence)
return sorted(filtered_sequence)
# Parameters
n = 500 # Length of the sequence
min_val = 1
max_val = 20
# Execution
random_sequence = generate_sequence(n, min_val, max_val)
print("Random sequence:", random_sequence)
final_result = final_sequence(random_sequence)
print("Final sorted sequence:", final_result)
```

The program can be related to combinatorics because it selects and filters the sequence of numbers. In particular, the filtering process can be seen as a combinatorial selection operation, where only consecutive pairs of consecutive numbers are chosen. In other words, the program explores combinations of numbers, but only those that satisfy a specific condition (being consecutive).

When duplicates are removed from the filtered sequence, the result is essentially reduced to a set (set of consecutive numbers without repetitions), which is connected to the concept of "sets" in combinatorics. In summary, the program does not perform complex combinatorial calculations; however, selecting elements from a set based on a rule (consecutive pairs) can be seen as a form of applied combinatorics.

The field of evolution can be imagined as a drunken walk in morphospace. Interestingly, entire regions of morphospace have never been reached due to structural or developmental genetic constraints. In other words, evolution leaves gaps that could be filled in an infinite time, but still longer.

A couple of simulations give this idea.

Program 7 3D Drunk in motion

```
import matplotlib.pyplot as plt
from mpl_toolkits.mplot3d import Axes3D
import numpy as np
import random
# Function for the 3D drunkard's walk with boundaries
def drunkard_walk_3d_with_limits(steps, limit):
x, y, z = [0], [0], [0]
fig = plt.figure()
ax = fig.add_subplot(111, projection='3d')
ax.set_xlim([-limit, limit])
ax.set_ylim([-limit, limit])
ax.set_zlim([-limit, limit])
ax.set_xlabel('X')
ax.set_ylabel('Y')
ax.set_zlabel('Z')
# Add step counter
step_text = ax.text2D(0.05, 0.95, "", transform=ax.transAxes, fontsize=12, color='red')
for i in range(steps):
direction = random.choice(['x', 'y', 'z'])
step = random.choice([-1, 1])
if direction == 'x':
new_x = x[-1] + step
if -limit <= new_x <= limit:
x.append(new_x)
y.append(y[-1])
z.append(z[-1])
elif direction == 'y':
new_y = y[-1] + step
if -limit <= new_y <= limit:
x.append(x[-1])
y.append(new_y)
z.append(z[-1])
else:
```

```
new_z = z[-1] + step
if -limit <= new_z <= limit:
x.append(x[-1])
y.append(y[-1])
z.append(new_z)
ax.plot(x, y, z, marker='o', color='black', markersize=2)
step_text.set_text(f"Steps: {i+1}")
plt.draw()
plt.pause(0.01) # Reduce the pause time to speed up the movement
plt.show()
# Ask the user to input the parameters
steps = int(input("Enter the number of steps: "))
limit = int(input("Enter the dimensions of the space: "))
# Generate the walk with limits
drunkard_walk_3d_with_limits(steps, limit)
```

This program has several connections with combinatorics, particularly with the concepts of random walk and state space. The drunken walk is an example of *Random Walk*, a fundamental concept in combinatorics and probability. At each step, the drunk can choose between 3 directions (x, y, z axes) and two possible steps (+1 or −1), so the total number of possible combinations of steps for *n* steps is 6n (3 directions × 2 steps for each direction). Every possible configuration of the position of the walk can be seen as a point in a three-dimensional space, and the walk is a sequence of steps between these points. The limit of the size of the space prevents the path from "exiting" this space, adding a combinatorial restriction to the problem.

The discovery of theorems gives a big boost that demonstrates the vivacity with which combinatorics is developing: one of these (although not very recent) is Szemerédi's theorem, which concerns the problem of the sum of subsequences in which the sum is maximum. In other words, no matter how chaotic the set of numbers is, one can always find an arithmetic progression within it (Szemerédi, 1975). This theorem has important implications for number theory and graph theory. In the natural sciences, the theorem is used to study the properties of biological systems.

Program 8 Szemerédi's theorem

```
import random
def generate_random_array(length, min_value, max_value):
return [random.randint(min_value, max_value) for _ in range(length)]
def contains_arithmetic_progression(arr, k):
n = len(arr)
if k <= 1:
return True, arr # Any array contains an arithmetic progression of length <= 1
if n < k:
return False, None # The array is too short to contain a progression of length k
arr_sorted = sorted(arr)
elements = set(arr_sorted)
for i in range(n):
for j in range(i + 1, n):
d = arr_sorted[j] - arr_sorted[i] # Calculate the difference
progression = [arr_sorted[i], arr_sorted[j]] # Start the progression with the two elements
next_term = arr_sorted[j] + d
while next_term in elements:
```

```
progression.append(next_term)
next_term += d
if len(progression) == k:
return True, progression
return False, None
# Example usage
length = 20 # Length of the array
min_value = 1 # Minimum value for random numbers
max_value = 50 # Maximum value for random numbers
k = 4 # Length of the arithmetic progression to search for
# Generate the random array
array = generate_random_array(length, min_value, max_value)
# Check if it contains an arithmetic progression
found, progression = contains_arithmetic_progression(array, k)
# Print the results
print(f"Initial array: {array}")
if found:
print(f"The array contains an arithmetic progression of length {k}: {progression}")
else:
print(f"The array does NOT contain an arithmetic progression of length {k}.")
```

A possible outcome can be:
Initial Array: [44, 43, 36, 7, 50, 3, 43, 16, 24, 22, 47, 20, 27, 9, 5, 16, 28, 37, 39, 1]
The array contains an arithmetic progression of length 4: [1, 3, 5, 7].
Of course, there may be situations like the following:
Initial Array: [10, 18, 25, 29, 39, 31, 43, 22, 29, 49, 35, 4, 49, 27, 42, 27, 35, 20, 10, 48]
The array contains an arithmetic progression of length 4: [10, 10, 10, 10].

The presence of four consecutive 10s means that the array contains repeated numbers. In this case, the progression is trivial because there are no variations.

Another example is the Green-Tao theorem, which concerns the presence of arbitrarily long arithmetic progressions in sequences of prime numbers and has had a massive impact on number theory and additive combinatorics. The formal definition is as follows: for every natural number *k*, there are arithmetic progressions of prime numbers with *k* terms. This theorem extends Szemerédi's theorem (Green et Tao, 2008).

Program 9 Green-Tao theorem

```
import itertools
def sieve_of_eratosthenes(limit):
sieve = [True] * (limit + 1)
sieve[0] = sieve[1] = False
for num in range(2, int(limit**0.5) + 1):
if sieve[num]:
sieve[num*num: limit+1: num] = [False] * len(range(num*num, limit+1, num))
return [num for num, is_prime in enumerate(sieve) if is_prime]
def find_arithmetic_progression(prime_numbers, k):
prime_set = set(prime_numbers)
n = len(prime_numbers)
for i in range(n):
```

```
for j in range(i + 1, n):
diff = prime_numbers[j] - prime_numbers[i]
progression = [prime_numbers[i] + diff * m for m in range(k)]
if all(num in prime_set for num in progression):
return progression
return None
# Parameters
limit = 1000 # Upper limit for prime search
k = 5 # Length of the arithmetic progression
# Generate prime numbers and find the progression
prime_numbers = sieve_of_eratosthenes(limit)
progression = find_arithmetic_progression(prime_numbers, k)
# Results
print(f"Primes up to {limit}: {prime_numbers[:10]}...")
if progression:
print(f"Arithmetic progression of length {k} found: {progression}")
else:
print(f"No arithmetic progression of length {k} found.")
```

The outcome is:

Prime numbers up to 1000: [2, 3, 5, 7, 11, 13, 17, 19, 23, 29]...

Arithmetic progression of length 5 found: [5, 11, 17, 23, 29].

This program is related to the Green-Tao theorem, which states that arbitrarily long arithmetic progressions of prime numbers exist. The theorem proves the existence of arbitrarily long prime arithmetic progressions but does not explicitly construct them. This program searches for such progressions in a finite range (up to a limit). The limitation is that it may not always find more extended progressions due to the constraints on limit.

In 2022, a group of researchers proved a new version of Ramsey's theorem for hypercomplex graphs; Ramsey's theorem states that for every colouring of the edges of a complete graph with a large enough number of vertices, it is possible to find a complete monochromatic subgraph. This result has important implications for graph theory and probability theory.

A fascinating field of research is combinatorial fractals, i.e., combinatorial structures with an infinite dimension (Mandelbrot, 1983; Rutar, 2023; Bunimovich & Skums, 2024). In mathematics, a fractal is a geometric object that exhibits a scale-invariance property.

The result has implications for understanding the structure of complex systems, such as neural networks, and can be used to explain their ability to learn and adapt. A group of researchers from the Max Planck Institute and Phillips University in Marburg has recently discovered the first regular molecular fractal in nature. A microbial enzyme spontaneously assembles into a regular fractal pattern known as the Sierpinsky triangle. It is a series of triangles that repeats itself endlessly. A base shape is replaced recursively with a combination of more complex shapes. Each recursion level applies to a production rule that creates a more complex framework.

Program 10 Sierpinski triangle

```
import matplotlib.pyplot as plt
import matplotlib.animation as animation
import numpy as np
def sierpinski_triangle(vertices, depth):
if depth == 0:
return [vertices]
else:
```

```
triangles = []
# Calculate the midpoints of the triangle's sides
mid1 = (vertices[0] + vertices[1]) / 2
mid2 = (vertices[1] + vertices[2]) / 2
mid3 = (vertices[2] + vertices[0]) / 2
# Recursively generate triangles for the three new sections
triangles.extend(sierpinski_triangle([vertices[0], mid1, mid3], depth - 1))
triangles.extend(sierpinski_triangle([mid1, vertices[1], mid2], depth - 1))
triangles.extend(sierpinski_triangle([mid3, mid2, vertices[2]], depth - 1))
return triangles
def update(frame, ax, vertices):
ax.clear()
ax.set_aspect('equal')
ax.axis('off')
# Generate triangles for the current depth
triangles = sierpinski_triangle(vertices, frame)
# Add each triangle to the axes
for triangle in triangles:
ax.add_patch(plt.Polygon(triangle, edgecolor='black'))
return ax.patches
def animate_sierpinski(depth):
fig, ax = plt.subplots()
vertices = np.array([[0, 0], [1, 0], [0.5, np.sqrt(3) / 2]])
# Create the animation
ani = animation.FuncAnimation(
fig, update, frames=range(depth + 1),
fargs=(ax, vertices), repeat=False, interval=1000
)
plt.show()
# Generate and show the Sierpinski Triangle animation with depth 6
animate_sierpinski(depth=6)
```

Figure 1. Sierpinski triangle

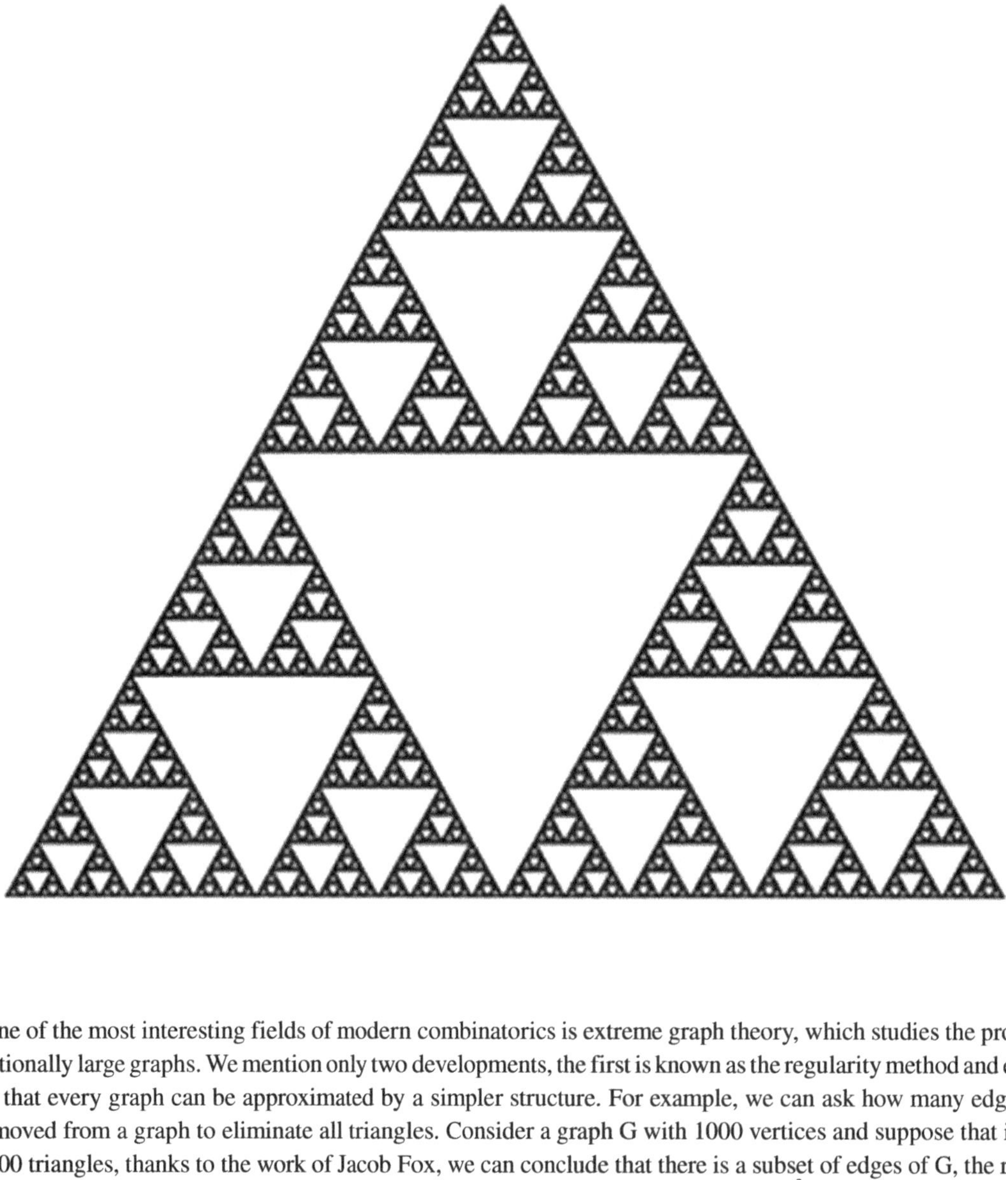

One of the most interesting fields of modern combinatorics is extreme graph theory, which studies the properties of exceptionally large graphs. We mention only two developments, the first is known as the regularity method and essentially states that every graph can be approximated by a simpler structure. For example, we can ask how many edges need to be removed from a graph to eliminate all triangles. Consider a graph G with 1000 vertices and suppose that it contains 100,000 triangles, thanks to the work of Jacob Fox, we can conclude that there is a subset of edges of G, the removal of which eliminates all triangles, and the size of this subset is at most: $T(n,r) = (1 - \frac{1}{r})\frac{n^2}{2}$ edges where *c* it is a constant. This number is significantly less than the total number of edges in the graph, which could be up to (1000 • 999) / 2 ≈ 500,000. In practice, this result tells us that we can make the graph triangle-free by removing a number of edges that is much less than the total number of edges in the graph.

Another example comes from the so-called Turán problem (Keevash, 2011). Paul (Pál) Turán (1910-1976) was a Hungarian mathematician known for his contributions to number theory and for long collaborations with Paul Erdős. More precisely, the task is to determine the maximum number of edges in a graph *G* that does not contain a complete subgraph K_{r+1} (a complete graph with *r+1 vertices*), where *r* is a positive integer. The solution is given by Turan's Theorem, which states that the maximum number of arcs in such a graph is given by: $T(n,r) = (1 - \frac{1}{r})\frac{n^2}{2}$.

Turán's problem marked the beginning of a new area in graph theory called extremal graph theory (Jukna, 2011). This field is concerned with determining the upper and lower limits of graph properties about specific conditions, such as the presence or absence of prohibited subgraphs. The derived ideas have found applications in various fields, such as algorithm design and data structure development, as well as in experimental design, to ensure that specific conditions are met.

The Turan problem can be implemented as a calculator program to verify and graphically explore the above.

Program 11 Turan's problem

```
import itertools
import matplotlib.pyplot as plt
import networkx as nx
import math
def turan_number(n, r):
return math.floor((r - 1) / r * (n ** 2) / 2)
def create_bipartite_graph(n, r):
G = nx.Graph()
# Divide the vertices into two groups
group_size = n // 2
for i in range(group_size):
for j in range(group_size, n):
G.add_edge(i, j)
return G
def contains_k_r_plus_1_subgraph(graph, r):
for node in graph.nodes:
neighbors = list(graph.neighbors(node))
if len(neighbors) < r + 1:
continue # Not enough neighbors to form K_(r+1)
# Check combinations among the neighbors
for subset in itertools.combinations(neighbors, r):
if all(graph.has_edge(subset[i], subset[j]) for i in range(r) for j in range(i + 1, r)):
return True
return False
def draw_graph(graph, title):
pos = nx.spring_layout(graph) # Layout for a good visual arrangement
plt.figure(figsize=(8, 6))
# Customize node and edge colors
node_color = 'lightblue'
edge_color = 'gray'
nx.draw(graph, pos, with_labels=True, node_color=node_color,
edge_color=edge_color, node_size=700,
font_size=16, font_weight='bold')
plt.title(title)
plt.show()
def main():
try:
n = int(input("Enter the number of vertices (n): "))
r = int(input("Enter the value of r: "))
if n <= 0 or r < 1:
```

```
print("Error: n must be greater than 0 and r must be at least 1.")
return
# Calculate the maximum number of edges
max_edges = turan_number(n, r)
print(f"Maximum number of edges in a graph with {n} vertices without K_{r+1}: {max_edg-
es}")
# Create a bipartite graph
G = create_bipartite_graph(n, r)
# Check if the graph contains K_(r+1)
contains_k_r_plus_1 = contains_k_r_plus_1_subgraph(G, r)
if contains_k_r_plus_1:
print(f"The graph contains a complete subgraph K_{r+1}.")
else:
print(f"The graph does NOT contain a complete subgraph K_{r+1}.")
print(f"Number of edges in the graph: {G.number_of_edges()}")
# Draw the graph
draw_graph(G, f"Graph with {n} vertices and {G.number_of_edges()} edges")
except ValueError:
print("Error: please enter valid numeric values.")
if __name__ == "__main__":
main()
```

The function calculates the maximum number of edges a graph with n vertices can have without containing a complete subgraph. The formula used is $floor(\frac{r-1}{r} \cdot \frac{n^2}{2})$, which gives the maximum number of edges. The output will depend on the values provided by the user. The program checks if the bipartite graph created contains a complete subgraph, then prints the number of edges and draws the bipartite graph with the labelled nodes and visible connections.

Figure 2. Turán graph

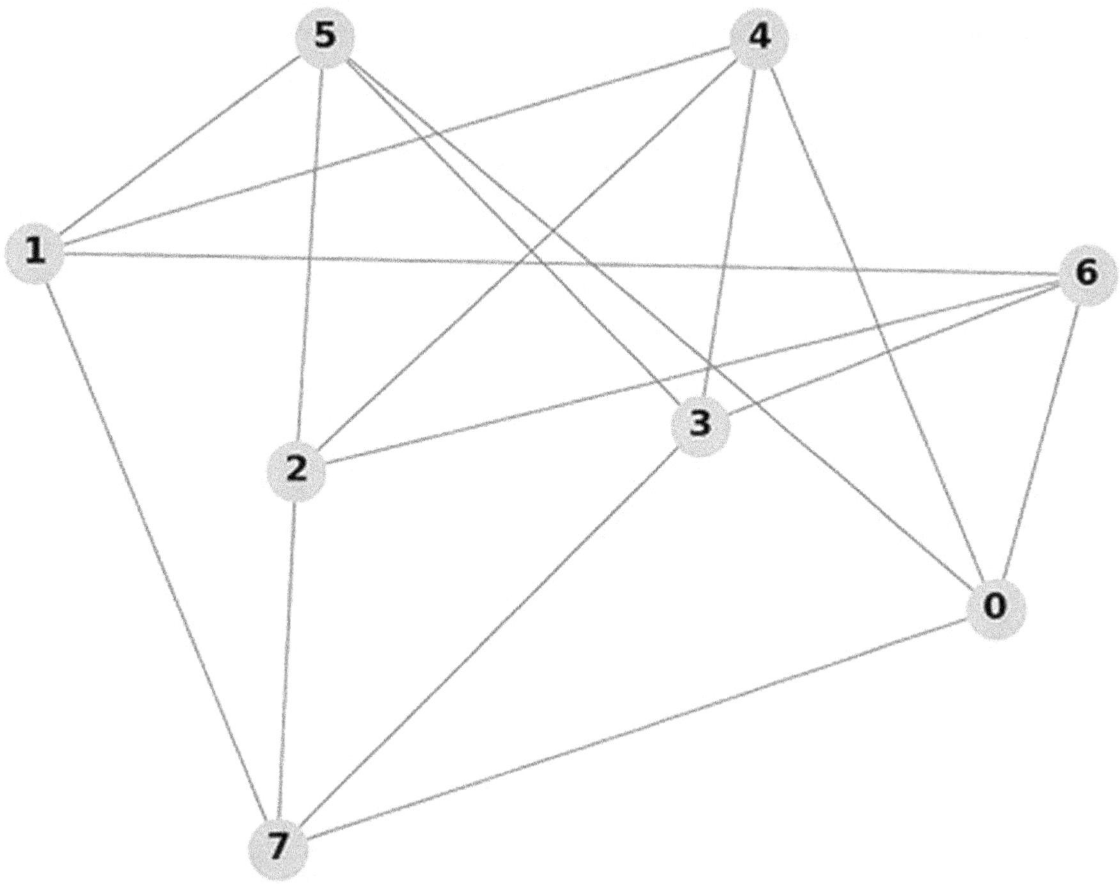

The recent contribution (2021) to the problem is relevant thanks to the work of David Conlon and Joonkyung Lee, who used probabilistic methods to obtain new upper bounds (Conlon et Lee, 2021).

Many other significant developments in combinatorics are quite recent. In 2022, Andrew Childs and his collaborators unveiled new quantum algorithms designed to tackle flow problems in networks and matchings in graphs. His work (Childs, 2009) is particularly noteworthy as it demonstrates how quantum computing can offer substantial advantages over classical algorithms for solving these well-established combinatorial challenges.

Network flows involve distributing a specific quantity of a commodity (such as water, energy, or information) across a network of nodes and edges. Determining a network's maximum or minimum flow is a fundamental challenge with wide-ranging practical applications.

A matching in a graph (often referred to as "coupling" in this context) is a collection of edges where no two share a common vertex. Finding the largest possible matching is another classic task, essential in various fields, including task assignment and circuit design. The potential for applying these quantum advancements is vast, with one promising avenue being the design of quantum algorithms based on graph scattering.

Lutz Warnke and Matthew Kwan have made significant advances in understanding the structure of sparse random graphs with applications to complex network theory. Their work on the configuration of sparse random graphs is crucial for understanding the behaviour of complex networks, such as social networks or biological networks. Sparse random graphs are graphs with few connections compared to the maximum possible number of connections. Complex network theory examines the emergent properties of large networks, including robustness, resilience, and the ability to disseminate information.

Robustness and reliability are crucial for combinatorial procedures employed in artificial intelligence, particularly in real-world applications where data are often imperfect and conditions change dynamically.

REFERENCES

Acerbi, F. (2003). On the shoulders of Hipparchus: A reappraisal of ancient Greek combinatorics. *Archive for History of Exact Sciences, 57*(6), 465–502. DOI: 10.1007/s00407-003-0067-0

Appel, K., & Haken, W. (1977). The solution of the four-color-map problem. *Scientific American, 237*(4), 108–121. DOI: 10.1038/scientificamerican1077-108

. Bailey, R. A., & Cameron, P. J. (2009). Combinatorics of optimal designs. *Surveys in combinatorics, 365*(19-73), 3.

Berstel, J., & Perrin, D. (2007). The origins of combinatorics on words. *European Journal of Combinatorics, 28*(3), 996–1022. DOI: 10.1016/j.ejc.2005.07.019

Beveridge, A., Griggs, J. R., Hogben, L., Musiker, G., & Tetali, P. (Eds.). (2016). *Recent trends in combinatorics* (Vol. 159). Springer. DOI: 10.1007/978-3-319-24298-9

Biggs, N. L. (1979). The roots of combinatorics. *Historia Mathematica, 6*(2), 109–136. DOI: 10.1016/0315-0860(79)90074-0

Bunimovich, L., & Skums, P. (2024). Fractal networks: Topology, dimension, and complexity. *Chaos (Woodbury, N.Y.), 34*(4), 042101. DOI: 10.1063/5.0200632 PMID: 38598678

Childs, A. M. (2009). Universal computation by quantum walk. *Physical Review Letters, 102*(18), 180501. DOI: 10.1103/PhysRevLett.102.180501 PMID: 19518851

Conlon, D., & Lee, J. (2021). On the extremal number of subdivisions. *International Mathematics Research Notices, 2021*(12), 9122–9145. DOI: 10.1093/imrn/rnz088

Graham, R. L. (1995). *Handbook of combinatorics.* Elsevier.

Green, B., & Tao, T. (2008). The primes contain arbitrarily long arithmetic progressions. *Annals of Mathematics, 167*(2), 481–547. DOI: 10.4007/annals.2008.167.481

Jukna, S. (2011). *Extremal combinatorics: with applications in computer science* (Vol. 571). Springer. DOI: 10.1007/978-3-642-17364-6

Keevash, P. (2011). Hypergraph turan problems. *Surveys in combinatorics, 392*, 83-140.

Lenart, C. (2015). The many faces of modern combinatorics. *arXiv preprint arXiv:1503.04240.*

Lu, H., Li, Z., Wang, R., Ren, Q., Li, X., Yuan, M., & Yan, J. (2023). Roco: A general framework for evaluating robustness of combinatorial optimization solvers on graphs. In *The Eleventh International Conference on Learning Representations.*

Mandelbrot, B. B. (1983). The fractal geometry of nature/Revised and enlarged edition. *New York.*

Redmond, G., & Hon, T. K. (2014). *Teaching the I Ching (Book of Changes).* Oxford University Press. DOI: 10.1093/acprof:oso/9780199766819.001.0001

Rutar, A. (2023). Geometric and combinatorial properties of self-similar multifractal measures. *Ergodic Theory and Dynamical Systems, 43*(6), 2028–2072. DOI: 10.1017/etds.2022.28

Szemerèdi, E. (1975). On sets of integers containing no k elements in arithmetic progression. *Acta Arith, 27*(199-245), 2.

West, D. B. (2021). *Combinatorial mathematics.* Cambridge University Press.

Wilson, R., & Watkins, J. J. (Eds.). (2013). *Combinatorics: ancient & modern.* OUP Oxford. DOI: 10.1093/acprof:oso/9780199656592.001.0001

Chapter 2
Methods

ABSTRACT

The chapter explores the fundamental concepts of combinatorics, presenting essential methods and models for the analysis of combinatorial structures. The chapter discusses classical combinatorics, focusing on permutations, combinations, and partitions. It introduces the perspective of J.G. Dubois, who emphasizes combinatorial structures rather than mere enumeration, classifying problems into selections, distributions, and partitions. The fundamental principle of product and sum, the principle of equality and inclusion-exclusion, which allow us to manage counting problems and set structures, are examined in detail. Additional combinatorial tools discussed include generating functions, which transform combinatorial problems into algebraic problems, and binomial coefficients, which are fundamental in the calculus of combinations and related to Pascal's triangle. The chapter also highlights combinatorial models such as shafts and the pigeonhole principle. Finally, the computational aspects of combinatorics are explored through algorithms.

INTRODUCTION

Combinatorics is the mathematics of patterns
Persi Diaconis

If combinatorics were a toolbox, what tools would be indispensable? It is not easy to answer because this field of mathematics is wide and varied. Given the breadth of the subject, it is not easy to exhaustively define the essential methods of combinatorics, and not all scholars propose the same approach (Jukna, 2001; Bogart, 2004). Classical combinatorics (Dubois, 1984; Roberts & Tesman, 2009; Stanley, 2011), generally focuses on problems such as Permutations (ordered arrangements of objects), Combinations (choices of subsets of objects), and Partitions (division of a set into subsets).

The purpose of combinatorics is oversimplified in these terms. Combinatorics transcends the role of a simple counting tool to become a methodological cornerstone in the decision-making process. This discipline can be effectively employed for three primary purposes, each corresponding to a distinct stage of the decision-making process.

- *Identifying Possibilities*: Firstly, combinatorics is essential for identifying and defining the number of possibilities within a given choice or probability scenario. This initial phase is critical, as an accurate understanding of the entire "solution space" is a prerequisite for an informed decision. Through tools like permutations, arrangements, and combinations, it is possible to precisely quantify the total number of possible outcomes or configurations. For instance, when calculating the probability of an event, combinatorics provides the crucial denominator—the number of all possible cases—allowing for the mapping of the universe of available options before any evaluation.
- *Narrowing Down Choices*: Subsequently, combinatorics helps to reduce this number of possibilities to facilitate a choice. While combinatorics does not inherently eliminate options, its ability to enumerate and characterise different configurations is instrumental in defining manageable subsets upon which to apply selection criteria. Knowing the exact number of alternatives that meet specific constraints or requirements allows for the analysis to be focused on a more restricted and pertinent set of options. For example, in logistics, once all possible routes

DOI: 10.4018/979-8-3373-3089-1.ch002

for a delivery are calculated, combinatorics enables the identification and isolation of only those that fall within predefined time or cost parameters, making the choice easier and more targeted.

- *Verifying and Confirming Decisions*: Finally, combinatorics supports the stage of verifying and confirming the choice as the best one. After making a decision, it is crucial to assess its robustness and optimality. Combinatorial techniques can be employed to analyse alternative scenarios or to calculate the probability that the chosen decision will indeed be the most advantageous given the conditions. This allows not only for post-decision validation but also for continuous optimisation, providing a quantitative framework for comparing the chosen option with other potential configurations and for testing its resilience to variations or uncertainties.

An interesting position is that of J.G. Dubois; in his 1984 work, he explores combinatorial structures and tries to classify configurations through a particular model, shifting the focus from simple enumeration to a deeper structural analysis. Dubois suggests that we think of problems not just as "counts" but as real "structures." Let us imagine people are organising a party. The challenge is not merely counting guests; it quickly extends to determining the optimal arrangement of tables, seating everyone, and even the sequence in which courses are served.

According to Dubois, simple combinatorial configurations can be classified into three models: *selections*, which emphasize the concept of sampling; *distributions*, related to the concept of mapping, and *partitions* or divisions of a set into subsets. This subdivision does not exactly correspond to the classical classification of combinatorics, but it offers a particular point of view that emphasizes specific aspects. The selections focus on the concept of sampling, which is related to classical combinatorics, particularly combinations. The distribution model can be associated with partitions of objects or distributions of objects across categories. Partitions are about dividing a set into disjoint subsets. This concept is certainly similar to that of classical combinatorics. However, the fundamental difference lies in the fact that Dubois emphasizes structuring the set, where each subset is treated as a distinct "class".

In the latter case, an example can make this approach more explicit. Suppose we divide students in a school into different groupings (e.g., math, science, art, literature), where each student can belong to more than one subset. The interest lies not only in the number of ways to divide students but also in the shape of the network to which they belong and the interactions between the different groupings.

The Dubois approach offers significant advantages when dealing with complex combinatorial problems, where it is necessary to count and understand the structure and relationships between objects. It is advantageous in abstract, theoretical and interdisciplinary contexts. However, its conceptual complexity and greater computational difficulty make it less suitable for simpler situations.

In addition to Dubois' combinatorial method, other significant and non-traditional positions in combinatorics methods deserve attention. These positions differ in approaches and applications, expanding the field of combinatorics beyond classical methods (Bender & Williamson, 2013) . Extreme combinatorics deals with situations in which it is necessary to determine whether a particular arrangement of objects is possible, often addressing complex problems such as those related to the configurations of graphs or sets (Katona, 1975; Erdős & Simonovits, 1982; Simonovits, 1984; Bollobás, 1998).

Enumerative combinatorics focuses on counting objects, employing techniques to determine asymptotic values.

Probabilistic combinatorics uses probability techniques to study the asymptotic properties of combinatorial elements.

Traditional methods of combinatorics focus primarily on simple problems and explicit counts.

However, extreme and enumerative combinatorics extend in more complex directions, utilising asymptotic counting and determining upper and lower bounds to understand combinatorial structures better and optimise their classification.

The topic could be further expanded and, later, several ideas will be taken up, for now, we can limit ourselves to a more traditional overview of the methods of combinatorics, but useful as it is sufficiently articulated:

1. *The counting principle* – This principle is the basis for calculating the number of ways in which objects can be organized or selected. It is based on fundamental concepts such as the principle of multiplication (Lockwood et al., 2017), the principle of addition, and the principle of inclusion-exclusion, all of which are typical examples of enumerative methods.
2. *Generating functions* – Generating functions are powerful mathematical tools that enable the manipulation of combinatorial sequences by converting complex counting tasks into algebraic expressions, thus simplifying calculations.

There are different types: ordinary, exponential and recursive. Recursive generating functions, for example, are used to represent sequences that follow a recurrence relation, such as Fibonacci numbers or Catalan numbers.

3. *Combinatorial models* – There are two main models. The tree model is a diagram consisting of an initial node from which successive branches branch off until the terminal leaves are reached. The tree structure is particularly useful when dealing with problems with multiple levels of choice and complexity, as it allows one to visualize and follow the different alternatives clearly and organized. The principle of boxes (or pigeon house) states that if you have more objects that need to be placed in fewer containers, then at least one of the containers will contain more than one object; this principle helps solve distribution and assignment problems.
4. *Binomial coefficients* – Binomial coefficients indicate the number of ways in which it is possible to select k elements from a set of n, ignoring the order. Newton's binomial theorem describes the development of the power of a binomial and finds important applications in solving combinatorial problems, often in combination with other techniques (Wolfram, 1984).
5. *Algorithms* – Computer science and other disciplines utilise combinatorial algorithms to address computational challenges involving the combinations and arrangements of elements. Search algorithms also play a key role in finding answers to these tasks. The backtracking, a typical construction method, is used to identify all possible solutions. In contrast, the genetic algorithm is used to find approximate solutions to optimisation challenges.
6. *Graphs* – These are combinatorial structures that consist of nodes (vertices) and arcs (edges) that connect nodes. We can distinguish them based on some salient characteristics: oriented graphs have a direction, representing asymmetric relationships (e.g. flows in a network); weighted graphs have associated a weight, representing a distance or a cost (e.g. road networks); planar graphs can be drawn on a plane without the arcs crossing each other; Bipartite graphs have vertices divided into two sets, and edges only connect vertices of different sets. They are combinatorial systems used to model a wide range of problems, including transportation planning, electrical circuit design, and social network analysis.
7. *Recursive counting* – Recursive counting is a fundamental technique in mathematics and computer science used to solve problems that can be divided into smaller subproblems. Recursion allows a function to be defined in terms of itself, simplifying the computation of complex results. This approach enables solving counting challenges by breaking them into smaller parts, recursively determining solutions, and combining them to obtain the final result.

Only some aspects will be examined in this chapter; the others will be developed in the following chapters.

THE COUNTING PRINCIPLE

The Basic Principle of the Product

It is a principle in combinatorial theory that states that if there are m ways of doing one thing and n ways of doing another, then there are $m \cdot n$ ways of doing both things simultaneously. It is important to note that the multiplication principle applies to independent events. Two events are considered independent if the occurrence of one does not affect the occurrence of the other. If the events are dependent, the calculation may require different methods.

Example 1: Suppose one has a combination lock with four numbers from 0 to 9. How many different combinations of possible numbers are there to open the lock?

- For each numeric position in the lock, we have 10 options (0-9). Since there are 4 number positions, the total number of possible combinations is given by $10 \bullet 10 \bullet 10 \bullet 10 = 10{,}000$.

Example 2: Four mountaineers rope up to cross a glacier. Two are beginners, so you want them to be neither in first nor last place. How many ways can the rope team be formed?

- The two beginners and the two experts can arrange themselves in two different ways. Therefore, we will have 2 • 2 ways of forming the roped party.

Example 3: One of the most common barcodes is EAN-13, which is used to identify products. This code consists of 13 digits; some have a specific function, and the last one is calculated based on a control algorithm. Suppose we want to determine how many combinations are possible for the first 12 digits, ignoring, for simplicity, the check digit and the specifications that refer to offset and overflow.

- The number is obtained by multiplying 10 by 12 times; therefore, the result is 10^{12}.

Example 4: Imagine organizing a chess tournament, where each game can have one of the following outcomes: White's win, Black's win, or a draw. If there are four matches in the tournament, how many possible outcomes?

- Since there are three possible outcomes per game, there are $3^4 = 81$ combinations.

Example 5: An electronics store offers four smartphones, three tablets, and two laptop models. If one wants to create a package that includes a smartphone, a tablet and a laptop, how many packages are possible?

- 4 smartphones • 3 tablets • 2 laptops = 24 possible packages.

Example 6: In the decimal numbering system, how many 4-digit natural numbers start and end with an even digit?

- As for the three internal digits, the number of places is k = 4, and the digits to be arranged are n = 10. The even digits are 0, 2, 4, 6, and 8. For the first digit, however, the value cannot be 0 (otherwise, the number would not have 4 digits), so there are only 4 possible options: 2, 4, 6, 8. On the other hand, the last digit can be any even digit, so there are 5 options (0, 2, 4, 6, 8). Multiply the possibilities for each position: 4 • 10 • 10 • 5 = 2000.

Fundamental Principle of Summation

It states that if there are *m* ways of doing one thing and *n* ways of doing another thing, where the two options cannot happen simultaneously (mutually exclusive), then there are $m + n$ ways in which one of the two things can happen.

Example 1: Consider the case where a person has to choose how to get home. Suppose there are two possible alternatives: one main and one secondary. If there are three ways to get to the house via the main option (by bike, bus, car) and two ways to get to it through the secondary option (by scooter, on foot), what is the total number of ways to get to the house if the options to get to the house are independent of each other?

- 3 + 2 = 5.

Example 2: Consider the roll of two six-sided dice. In this case, how many ways can one get a sum of 7?

- 7 by rolling a 1 and a 6, a 2 and a 5, a 3 and a 4, a 4 and a 3, a 5 and a 2, or a 6 and a 1. These are disjointed events; therefore, there are six diverse ways to achieve the required result.

Example 3: DNA is a long chain of molecules called nucleotides: Adenine (A), Cytosine (C), Guanine (G), and Thymine (T). The order of these nucleotides determines the genetic code. We want to find the total number of possible DNA sequences for lengths ranging from 4 to 8.

- Using the product rule, the number of possible sequences for a length *k* is 4k. We can apply the sum principle to find the total number of possible sequences for all lengths 4 to 8. There are 87,296 cases.

Example 4: Let us consider the set V formed by vowels and the set C formed by consonants; how many 2-letter words do we have? In English, there are 5 vowels and 21 consonants. A 2-letter word can be made by combining VV, VC, CV, and CC.

- According to the principle of multiplication, there are 25 words made up of only vowels, 441 words made up of only consonants, and 210 words made up of vowels-consonants or consonants-vowels. Applying the sum principle, we have $25 + 441 + 210 = 676$ possible 2-letter words.

Example 5: We want to create a PIN that must comply with these rules: it must be made up of 4 digits, or it must be made up of 3 capital letters of the English alphabet, or it must be made up of 2 digits followed by a symbol between @, #, $. What is the total possible password?

- In the first case there are 10,000 possibilities, in the second 17,576, in the third 300. For the principle of sum, the total of possible passwords will be: $10{,}000 + 17{,}576 + 300 = 27{,}876$ possible different PINs.

Principle of Equality

The principle refers to the fact that if two sets have the same number of elements, they can be put in one-to-one correspondence. In other words, if there are as many ways to do something as there are to do another, then the number of ways is the same.

Example 1: Suppose you need to count the number of ways you can choose a team of 3 players from a group of 6 players. That is, we want to calculate the number of ways we can match players with teams so that each team contains exactly three players.

- As we will see later, there are 20 possible combinations. We can also look at this problem from another perspective. If we choose a team of 3 players, we automatically exclude the other 3 players who have not been selected. So, we can also count the number of ways we can form a team of 3 players by choosing the 3 players who will not be part of the team. Also, in this case, the answer is 20.

Example 2: Let us say we have a company that manufactures electronic components. We want to organize a work team to develop new products. The company has 8 engineers, and we want to form a team of 4 people to manage the project.

- The problem of selecting the team is primarily a matter of calculating combinations. However, it indirectly reflects the principle of equality through the idea that subsets have the same number of possible combinations if the number of elements is the same.

Example 3: Let us say we have five different books and want to select two to take on vacation. We want to calculate the number of ways we can choose these two books.

- The calculation of the combinations for the 2 chosen books and the three excluded books shows 10 ways to make the selection.

Example 4: Is the number of anagrams of the word BREAD beginning with A the same as the number of anagrams beginning with D?

- The word BREAD is composed of 5 distinct letters (B, R, E, A, D), so the total number of anagrams of the word BREAD is given by $5 \cdot 4 \cdot 3 \cdot 2 \cdot 1 = 120$. If the anagram starts with A, then we need to permute the remaining 4 letters (B, R, E, D). The number of ways we can arrange these 4 letters is $4 \cdot 3 \cdot 2 \cdot 1 = 24$. Similarly, if the anagram starts with D, we need to permute the remaining 4 letters (B, R, E, A). Again, the number of ways we can arrange these 4 letters is 24.

Example 5: In a grid size 4 × 4, are the possible paths from the lower left corner to the upper right corner the same as those obtained by joining the paths that pass through the centre?

- In a 4 × 4 grid, the total number of paths from the lower left corner to the upper right corner is given by the combination of movements to the right (D) and up (A). In this case, we have 4 D-moves and 4 A-moves; in total, there are 70 routes. If we consider the paths that pass through the centre of the grid, we must divide the problem into two parts: from the starting point (0.0) to the centre (2.2) and from the centre (2.2) to the endpoint (4.4), in total there are 6 • 6 = 36 paths. The routes that pass through the centre are a part of the total routes, but they do not coincide with the total number of routes.

Inclusion-and-Exclusion Principle

The principle of inclusion and exclusion is a powerful tool used in combinatorics to calculate the number of ways in which a property can be satisfied by a set of objects, considering the intersections between the different properties. It allows for the accurate counting of cardinalities (a measure of the "size" of a set) of sets that can be obtained by combining smaller sets, taking into account their overlaps.

The principle of inclusion and exclusion can be applied to a wide range of combinatorial problems, such as calculating the number of sequences that meet specific conditions, counting restricted combinations, and other situations where overlapping subsets need to be considered.

The key idea of the principle is to add the number of elements that satisfy the individual properties and subtract the number of elements that satisfy the intersections between property pairs to compensate for the repeated count. Next, add the intersections between property triplets, subtract the intersections between property quadruples, and so on, alternating between adding and subtracting until all possible combinations of the properties are included. The last term is corrected with a negative sign to ensure multiple overlaps are counted correctly.

Example 1: Let us consider a set of 30 students. Of these, 12 students speak English, 18 speak French, and 8 speak both languages. How many students speak at least one of the two languages?

- We can apply the inclusion-exclusion principle here. We denote by A the set of students who speak English and by B the set of students who speak French. We want to find $|A \cup B|$, i.e., the number of students who speak at least one of the two languages. $|A \cup B| = |A| + |B| - |A \cap B|$. Since $|A| = 12$, $|B| = 18$ and $|A \cap B| = 8$, we can calculate $|A \cup B|$: $|A \cup B| = 12 + 18 - 8 = 22$. So, 22 students speak at least one of the two languages.

Example 2: Determines the number of positive integers up to 30 divisible by 2, 3, or 5.

- Using the principle of inclusion and exclusion, we need to calculate: $|A \cup B \cup C| = |A| + |B| + |C| - |A \cap B| - |A \cap C| - |B \cap C| + |A \cap B \cap C|$, $|A|$ represents the number of integers divisible by 2 to 30: 30/2 = 15, $|B|$ represents the number of integers divisible by 3 to 30: 30/3 = 10, $|C|$ represents the number of integers divisible by 5 to 30: 30/5 = 6, $|A \cap B|$ represents the number of integers divisible by both 2 and 3, i.e. 6: 30/6 = 5, $|A \cap C|$ represents the number of integers divisible by both 2 and 5, i.e. 10: 30/10 = 3, $|B \cap C|$ represents the number of integers divisible by both 3 and 5, i.e. 15: 30/15 = 2, $|A \cap B \cap C|$ represents the number of integers divisible by both 2, 3 and 5, i.e. 30: 30/30 = 1. We can now calculate the number of positive integers up to 30 that are divisible by at least one of the numbers 2, 3, or 5 using the principle of inclusion and exclusion: $|A \cup B \cup C| = 15 + 10 + 6 - 5 - 3 - 2 + 1 = 22$

Example 3: Let us consider 50 students in a school. Of these, 30 are members of the chess club, 25 are members of the football group, 15 are members of both the chess club and the football group, and 5 are not members. We want to determine the number of students enrolled in at least one of the two organizations.

- The inclusion-exclusion principle tells us that the number of students enrolled in at least one of the two organizations is given by: $|A \cup B| = |A| + |B| - |A \cap B|$, substituting the values we get $|A \cup B| = 30 + 25 - 15 = 40$.

Example 4: There are red, blue and green balls in a shop. There are: red balls: 30, blue balls: 25, green balls: 20, red and blue balls: 10, blue and green balls: 5, green and red balls: 3, all three coloured balls: 2. How many different balls are there in the store?

- Using the inclusion-exclusion principle for three sets: $|R \cup B \cup V| = 30 + 25 + 20 - 10 - 5 - 3 + 2 = 59$.

Example 5: How many 4-digit PINs contain at least a 7 and at least a 9?

- We want to determine $|S_7 \cap S_9|$. Total PIN = 10^4, PIN without 7 = 9^4, PIN without 9 = 9^4, PIN without 7 or 9 = 8^4, from which we get $10^4 - (9^4 + 9^4) + 8^4 = 974$.

GENERATING FUNCTIONS

Generating functions are powerful mathematical tools used to represent and manipulate numerical sequences. They allow us to transform combinatorial questions into algebraic expressions, making it easier to analyse and solve them.

The two primary types are:

- Ordinary Generating Functions (OGFs): Typically used when the order of elements matters and there is no factorial scaling in the coefficients.
- Exponential Generating Functions (EGFs): Ideal for problems involving labelled structures, where each term is divided by the factorial of its index.

While both encode sequences, they exhibit some significant differences in their encoding methods and applications.

Ordinary Generating Functions

An Ordinary Generating Function is a powerful mathematical technique used in combinatorics to represent and manipulate numerical sequences and is an infinite power series associated with a numerical sequence. Each coefficient in the series corresponds to one end of the series. This representation, seemingly simple, conceals a profound connection to the world of combinatorics. For example, we can use it to count the number of permutations, combinations, partitions of a set, binary trees, and many other discrete structures (Rosen, 1999). Many combinatorial problems show recurrence relations, and generating functions transform these relationships into equations, which are often easier to solve.

An OGF for a sequence {an} is defined as the series:

$$F(x) = \sum_{n=0}^{\infty} a_n x^n$$

Here, x is a variable and an represents the end at position n.

Generating functions have several useful properties; the sum of two generating functions F(x) and G(x) is given by:

$$F(x) + G(x) = \sum_{n=0}^{\infty} (a_n + b_n) x^n$$

While the product is defined as:

$$F(x) \cdot G(x) = \sum_{n=0}^{\infty} c_n x^n$$

$$\text{with } c_n = \sum_{k=0}^{\infty} a_k b_{n-k}$$

This product allows us to combine information from two sequences. One can apply it to find the distribution of sums from dice rolls. When two dice are rolled, each face contributes to a generating expression, and the product of their ordinary generating functions (OGFs) yields the function for the total outcomes. This approach allows us to calculate the probability of each possible result.

Another example: given two sequences $\{a_n\}$ and $\{b_n\}$, the generating expression for their convolution is the product of their respective OGFs. This principle is useful in both series analysis and combinatorics.

OGFs are particularly effective in solving enumeration tasks (Aigner & Axler, 2007). The general formula for constructing an ordinary generating function depends on the specific context and constraints of the task. For example, they can be used to calculate the number of ways to choose objects from a set, as in the case of binomial coefficients.

The generating function associated with binomial coefficients is:

$$G(x) = (1 + x)^n$$

which generates, as we will see later, the binomial coefficients $\binom{n}{k}$.

Ordinary generating functions can solve various problems. If we are faced with a geometric series, we have:

$$G(x) = 1 + x^n + x^{2n} + x^{3n} + \ldots = \frac{1}{1 - x^n}$$

Suppose one has to pay a certain amount using 1 or 2 euro coins and 5 or 10 euro notes. How many ways to pay? For example, if the sum is €10, we could pay with a ten-euro note, with two five-euro notes, with ten €1 coins, with a five-euro note, two €2 coins and a €1 note, and so on. The combinations are many; using OGFs, we could write:

$$\frac{1}{1 - x} \cdot \frac{1}{1 - x^2} \cdot \frac{1}{1 - x^5} \cdot \frac{1}{1 - x^{10}}$$

Expanding the total OGF up to x^{10}, we get $1 + x + 2x^2 + 2x^3 + 3x^4 + 4x^5 + 5x^6 + 6x^7 + 7x^8 + 8x^9 + 11x^{10}$

The coefficient of x^{10} is 11, so there are eleven ways to get €10.

Exponential Generating Functions

Exponential generating functions are mathematical tools used to represent and analyse numerical sequences, especially with a certain type of exponential behaviour. The general form is:

$$G(x) = \sum_{n=0}^{\infty} a_n \frac{x^n}{n!}$$

x is an independent variable, a_n is the n-th term of the numerical sequence.

Exponential generating functions are an advanced topic requiring a solid understanding of series theory and operations to be used effectively in combinatorics and sequence analysis. Suppose we want to count the number of ways in which it is possible to partition a set of n elements into non-empty subsets. The exponential generating function for this problem is

$$G(x) = \frac{1 - x}{1 + x + x^2 + x^3 + \ldots}$$

The term $x^n/n!$ represents the number of ways to form a subset of *n* elements. For example, $x^2/2!$ represents the number of ways to form two disjoint subsets.

The former are more general and easier to manipulate algebraically but may not be optimal for specific cases. The latter are tailored to handle exponential growth, thanks to the factorial in the denominator, which tames such behaviour. Exponential generating functions are especially useful in combinatorics, where coefficients count configurations and often yield more concise, manageable expressions.

A classic example is the Fibonacci numbers, defined by a simple recurrence where each term is the sum of the two preceding ones: 1, 1, 2, 3, 5, 8, 13, 21, …

Mathematically, the recursive relation defines the sequence: $F_n = F_{n-1} + F_{n-2}$ for $n \geq 2$ with initial conditions $F_0 = 0$ and $F_1 = 1$. Generating functions allow us to encode these numbers into a compact algebraic expression. In the Fibonacci case, both ordinary and exponential generating functions offer insightful results.

The ordinary generating function is:

$$F(x) = x + x^2 + 2x^3 + 3x^4 + \dots$$

It is a simple definition, but it becomes complicated to manipulate algebraically to extract specific terms.

A more robust and compact alternative is the exponential generating function, which, for the Fibonacci sequence, takes the form: $G(x) = \frac{1}{1 - x - x^2}$

This expression is much simpler and more compact. We can expand the function into a power series to understand how this exponential-generating function represents the Fibonacci sequence. In this case, the exponential function captures the recursive relationship between the terms of the sequence. The coefficient of x_n corresponds to the term F_n of the sequence.

Here is a summary table comparing ordinary generating functions (OGFs) and exponential generating functions (EGF):

Table 1. Comparison between OGF and EGF

Characteristic	Ordinary Generating Functions (OGF)	Exponential Generating Functions (EGF)
General form	$F(x) = \sum_{n=0} a_n x^n$	$G(x) = \sum_{n=0} a_n \frac{x^n}{n!}$
Coefficients	They directly represent the terms of the sequence.	They represent the terms of the sequence divided by n!.
Applications	Widely used in combinatorics to count discrete objects, such as sequences of permutations, combinations, and partitions.	It is particularly suitable for sequences with exponential growth or for problems that require the use of continuous expressions or that involve stochastic processes.
Advantages	It is simple to manipulate algebraically and can be used in direct calculations. They allow one to get a clear and immediate view of the sequence.	Suitable for sequences with exponential growth, they often lead to more compact and easily tractable expressions when the sequence growth is exponential or when trying to solve differential equations or partitioning problems.
Disadvantages	It can be less efficient for fast-growing sequences as high-order terms become dominant.	It is more complex to manipulate algebraically since they include n! in the coefficients. They are not always the best choice when the sequence does not show exponential growth or when direct manipulation of the ordinary function is preferable.

In conclusion, OGFs are particularly simple and intuitive when analysing discrete sequences in basic combinatorial contexts, but they become less valuable when they grow particularly rapidly. EGFs, although more complex to manage algebraically, offer significant advantages for fast-growing sequences or in contexts where factorial normalization is necessary, such as in optimization problems, probabilistic distributions, or stochastic models.

Overall, the choice between the two depends on the nature of the sequence and the specific challenge being addressed.

COMBINATORIAL MODELS

Tree Template

The combinatorics tree model is a graphical representation used to visualise and analyse situations involving sequential choices. In this model, each node represents a decision point or state, while the branches extending from a node depict the available choices or options at that moment. Every path from the tree's root to a terminal leaf corresponds to a specific sequence of decisions that leads to a particular outcome.

Consider the "wardrobe" scenario as a two-step experiment. We can visualise it by denoting trousers with 't' and jackets with 'j':

Figure 1. Choice tree

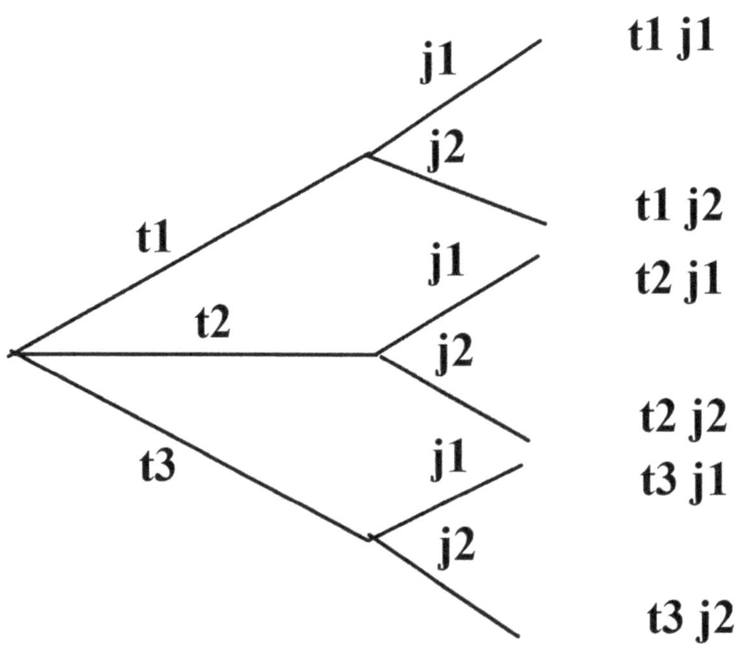

In general, the tree model proves particularly useful in combinatorics for scenarios involving sequential choices, such as those concerning arrangements, partitions, or search tasks where each step depends on previous selections.

The tree model is advantageous in combinatorics for problems with sequential choices, such as arrangements, partitions, or search problems where each step depends on previous choices.

Principle of Boxes

It is a mathematical concept also known as the "pigeonhole principle" (Ajtai, 1994). It owes its name to the analogy with the pigeon problem in pigeon houses. Let us imagine that we have a certain number of pigeons (the "objects") and a certain number of pigeon houses (the "containers"). Each pigeon must be assigned to a pigeon house. If there are more pigeons than pigeon houses, then at least one must contain more than one pigeon. The key idea is that when there are more objects to distribute than can fit in the containers, a "collision" occurs, i.e., at least two objects will end up in the same container. This principle is often used to demonstrate non-uniqueness results or to establish upper limits on the number of elements distributed in containers.

In this case, we assume we have a certain number of boxes and a set of objects to insert, such as vowels. How many ways to insert all five letters with three boxes available? Six of the 60 possible solutions are:

Figure 2. Letters in boxes

a	e	i	e	a	i	i	a	e
a	e	o	e	a	o	i	a	o

In fact, in the first box, one of the five vowels can be placed; in the second, there is room for only one of the remaining four; the third can be filled using one of the three remaining letters, so $5 \bullet 4 \bullet 3 = 60$.

Let us look at another famous example: in a city of 1 million inhabitants, is it guaranteed that at least two people have the same amount of hair? The answer is positive because the maximum number of hairs on a human head is much less than 1 million, generally no more than 150,000.

Let us conclude with an example that requires some thought: if one places 10 points within a 3×3 side square, at least two points will be less than $\sqrt{2}$ units apart. Let us imagine dividing the 3×3 square into nine smaller squares with a 1×1 side. We have nine smaller squares (pigeon houses) and 10 points (pigeons). So, by the pigeon house principle, at least two points must fall in the same 1×1 square. Since these two points are in the same 1×1 square, their maximum possible distance is $\sqrt{2}$, which is the diagonal of this square.

In conclusion, the Pigeonhole Principle (often referred to as the "boxes principle") is a powerful tool used to address various combinatorial challenges. These include distribution scenarios, such as placing indistinguishable objects into distinct containers, partition tasks where a set is divided into subsets and combination challenges with restrictions.

BINOMIAL COEFFICIENTS

Binomial or combinatorial coefficients appear in the binomial expression raised to a specific power. In particular, the binomial coefficient is the number of ways in which it is possible to select *k* distinct elements from a set of *n* without considering the order. The binomial coefficient is denoted as $C_{n,k}$, C(n, k) or (n on k).

Pascal's triangle is a graphical representation of binomial coefficients organized in a triangular shape. Pascal's triangle starts with a row containing only number 1, and each subsequent number in subsequent rows is obtained by adding the two adjacent upper numbers.

For example, the first lines of Pascal's triangle are:

Figure 3. Pascal's triangle

```
                        1
                    1       1
                1       2       1
            1       3       3       1
        1       4       6       4       1
    1       5       10      10      5       1
1       6       15      20      15      6       1
1       7       21      35      35      21      7       1
```

The first detailed description of a triangle of binomial coefficients appeared in an Indian book by Pingala between the fifth and second centuries BC.

The numbers within Pascal's triangle are precisely the binomial coefficients. For example, in Pascal's triangle, the number 6 corresponds to the binomial coefficient C(4, 2), representing the number of ways two elements can be selected from a set of four. The binomial expansion of $(a+b)^4$ follows the following scheme: $\binom{4}{0}a^4b^0 + \binom{4}{1}a^3b^1 + \binom{4}{2}a^2b^2 + \binom{4}{3}a^1b^3 + \binom{4}{4}a^0b^4$.

Examining the simple combinations below, we will see how it is possible to calculate the terms C(n, k).

Pascal's triangle has numerous interesting properties. For example, the numbers within Pascal's triangle are symmetric to the central vertical axis, so C(n, k) = C(n, n-k). This feature has been applied to solve examples of the principle of equality. The sum of the coefficients of each row expresses the powers of 2. This property can be applied to combinatorial problems in which one has to count the subsets of a set. The difference between odd and even coefficients is zero. The famous Fibonacci numbers are found as the sum of the numbers on the diagonals.

Figure 4. The Fibonacci numbers

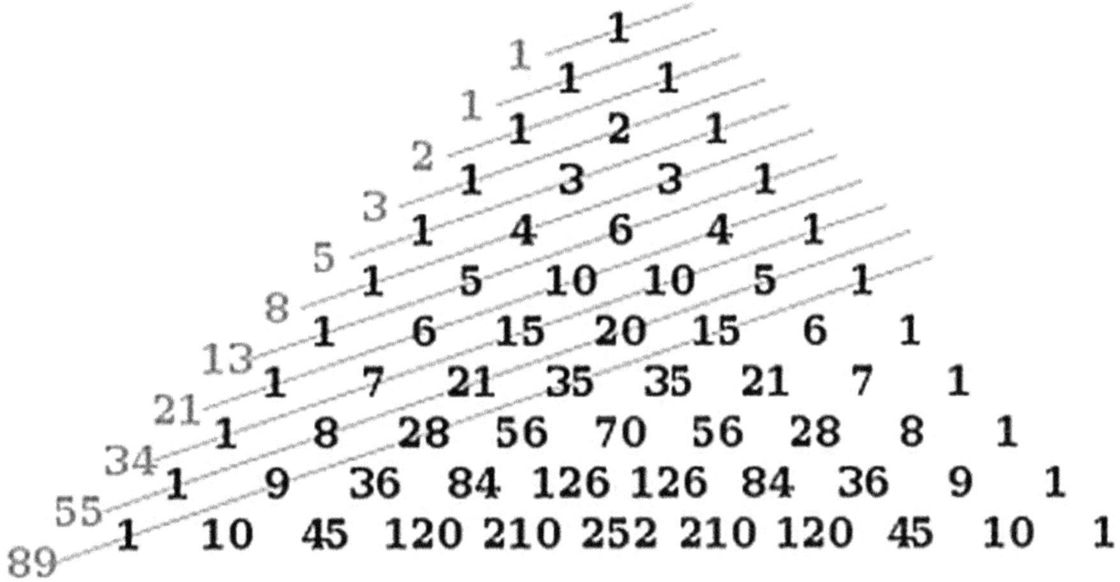

Scrolling the opposite diagonals, we find the natural, triangular, tetrahedral, and pyramidal numbers.

Pascal's triangle and binomial coefficients are useful for solving many combinatorial problems. For example, Pascal's triangle directly gives the number of possible combinations. It is used to calculate probabilities in experiments with binary outcomes (e.g., coin toss). It appears in several contexts of graph theory, for example, in counting paths in specific graphs. Catalan numerals, found in different combinatorial structures, can be expressed in binomial coefficients and thus found in Pascal's triangle.

In conclusion, binomial coefficients and Pascal's triangle are essential tools in combinatorial theory. Binomial coefficients provide the number of ways to select elements from a set. At the same time, Pascal's triangle displays these coefficients in an orderly fashion, making it easier to understand combinatorial patterns.

A program to display some properties is as follows:

Program 12 Pascal's triangle

```
import math
import itertools
def binomial_coefficient(n, k):
"""Calculates the binomial coefficient 'n choose k'."""
if k < 0 or k > n:
return 0
retûrn math.factorial(n) // (math.factorial(k) * math.factorial(n - k))
def print_pascal_triangle(rows):
"""Prints the first 'rows' rows of Pascal's triangle."""
for i in range(rows):
row = [binomial_coefficient(i, k) for k in range(i + 1)]
print(row)
```

```
print("Pascal's Triangle (first 10 rows):")
print_pascal_triangle(10)
def combinations(lst, k):
"""Generates all combinations of 'k' elements from a list."""
return itertools.combinations(lst, k)
# Example: All combinations of 2 elements from [1, 2, 3, 4]
lst = [1, 2, 3, 4]
k = 2
print("\nCombinations of 2 elements from [1, 2, 3, 4]:")
for comb in combinations(lst, k):
print(comb)
def binomial_probability(n, k, p):
"""Calculates the probability of exactly 'k' successes in 'n' trials with probability 'p'
of success."""
binom_coeff = binomial_coefficient(n, k)
return binom_coeff * p**k * (1 - p)**(n - k)
# Example: Probability of getting 2 heads in 3 coin tosses
n = 3
k = 2
p = 0.5 # Probability of heads
probability = binomial_probability(n, k, p)
print(f"\nProbability of getting exactly {k} heads in {n} tosses = {probability}")
def lattice_paths(rows, cols):
"""Calculates the number of distinct paths on a 'rows' x 'cols' grid."""
return binomial_coefficient(rows + cols - 2, rows - 1)
# Example: Calculate paths on a 3x4 grid
rows = 3
cols = 4
paths = lattice_paths(rows, cols)
print(f"\nNumber of paths on a {rows}x{cols} grid: {paths}")
def catalan_number(n):
"""Calculates the 'n'-th Catalan number."""
return binomial_coefficient(2 * n, n) // (n + 1)
# Example: Calculate the 4th Catalan number
n = 4
catalan = catalan_number(n)
print(f"\n{n}-th Catalan number: {catalan}")
```

Let us look at the output:

1. The first ten lines of Pascal's triangle are printed.
2. Then, one gets the printing of all combinations of 2 elements from the list [1, 2, 3, 4].
3. It is verified that the probability of obtaining exactly 2 successes (in this case, heads) in 3 trials (in this case, throws) with a probability p = 0.5 of success in each trial can be determined using Newton's binomial. Pascal's triangle is closely related to the binomial coefficients that are used to calculate this probability. We use the term 3^a2^b from the triangle, so the probability is $3 \bullet 0.5^2 \bullet 0.5$.
4. The binomial coefficient is then used to calculate the paths on a 3x4 grid. The number of paths is given by $\binom{3+4-2}{3-1}$
5. Finally, the 4th Catalan number is displayed.

Catalan numbers (a sequence of natural numbers that appear in numerous combinatorial counting problems) relate to Pascal's triangle in that each Catalan number can be derived from a particular binomial coefficient of the form $\binom{2n}{n}$ and then divided by n+1. The first term represents the number of ways to choose elements from a set of $2n$ elements.

It is significant to emphasize that Catalan's numbers and Pascal's triangle are connected through combinatorics. Later, we will see how to generate some Catalan numbers and how these are applied to paths in a grid. As further examples, we can cite the number of ways of organizing different structures, such as binary trees or parenthesized expressions.

Table 2. Number of parentheses in an expression

Factors (*n*)	**Number of Ways to parenting expressions**
2	1
3	2
4	5
5	14
6	42
7	132
8	429
9	1430
10	4862

REFERENCES

Aigner, M., & Axler, S. (2007). *A course in enumeration* (Vol. 238). Springer.

Ajtai, M. (1994). The complexity of the pigeonhole principle. *Combinatorica, 14*(4), 417–433. DOI: 10.1007/BF01302964

Bender, E. A., & Williamson, S. G. (2013). *Foundations of combinatorics with applications.* Courier Corporation.

Bogart, K. P. (2004). *Combinatorics through guided discovery.* Kenneth P. Bogart.

Bollobás, B. (1998). *Random graphs.* Springer New York.

Dubois, J. G. (1984). Une systématique des configurations combinatoires simples (A Systematic Study of Simple Combinatorial Configurations). *Educational Studies in Mathematics, 15*(1), 37–57. DOI: 10.1007/BF00380438

Erdős, P., & Simonovits, M. (1982). Compactness results in extremal graph theory. *Combinatorica, 2*(3), 275–288. DOI: 10.1007/BF02579234

Jukna, S. (2001). *Extremal combinatorics: with applications in computer science* (Vol. 29). Springer. DOI: 10.1007/978-3-662-04650-0

Katona, G. O. (1975, April). Extremal problems for hypergraphs. In *Combinatorics: Proceedings of the NATO Advanced Study Institute held at Nijenrode Castle,* Breukelen, The Netherlands 8–20 July 1974 (pp. 215-244). Dordrecht: Springer Netherlands.

Lockwood, E., Reed, Z., & Caughman, J. S. (2017). An analysis of statements of the multiplication principle in combinatorics, discrete, and finite mathematics textbooks. *International Journal of Research in Undergraduate Mathematics Education, 3*(3), 381–416. DOI: 10.1007/s40753-016-0045-y

Roberts, F. S., & Tesman, B. (2009). Applied Combinatorics SECOND EDITION. In *Applied Combinatorics* (2nd ed., pp. 1–848). Taylor and Francis. DOI: 10.1201/b12335

Rosen, K. H. (Ed.). (1999). *Handbook of discrete and combinatorial mathematics.* CRC press. DOI: 10.1201/9781439832905

Simonovits, M. (1984). Extremal graph problems, degenerate extremal problems, and supersaturated graphs. *Progress in graph theory (Waterloo, Ont., 1982)*, 419-437.

Stanley, R. P. (2011). Enumerative Combinatorics Volume 1 second edition. *Cambridge studies in advanced mathematics.* DOI: 10.1017/CBO9781139058520

Wolfram, S. (1984). Geometry of binomial coefficients. *The American Mathematical Monthly, 91*(9), 566–571. DOI: 10.1080/00029890.1984.11971496

Chapter 3
From Ordinary Generating Functions to Combinatorial Calculus

ABSTRACT

The chapter explores fundamental concepts of combinatorics, focusing on ordinary generating functions (FGOs) and their applications in counting problems. FGOs are powerful tools for analyzing sequences by transforming them into algebraic expressions. They help in solving problems related to permutations, combinations, and arrangements. The chapter demonstrates how FGOs can generate sequences like Fibonacci numbers and solve real-world problems like binary sequence formation. The chapter then discusses different types of permutations, including simple permutations, circular permutations, and permutations with repetition. It also covers arrangements, which are ordered selections of elements and their variations with repetition. Combinations, where order does not matter, are analyzed, particularly in games like poker and probability calculations. Special cases include combinations with repetition and dismutations, a concept where no element remains in its original position.

INTRODUCTION

Ordinary generating functions are a natural language for combinatorics

Richard Stanley

Ordinary generating functions (OGFs) are among the most powerful and flexible tools in combinatorics (Cameron, 1994; Lovász, 2007; Harris et al., 2009; Roberts & Tesman, 2009; Erickson, 2013; Mladenović et al., 2019; Bóna, 2023). They provide an elegant framework for analysing and manipulating numerical data, transforming complex counting problems (Sagan, 2020) into straightforward algebraic expressions. Practically speaking, an OGF associates a series of powers with a list of values, allowing their properties to be studied through algebraic techniques and providing a compact and unified way to handle an entire set of terms as a single object.

OGFs go far beyond solving isolated problems; they offer a broader perspective for exploring structurally similar families of combinatorial objects. Through their use, it becomes possible to uncover relationships between different configurations, develop advanced methods beyond elementary techniques, and apply insights to areas like number theory, probability, and the structural analysis of algorithms.

Let us consider the Fibonacci sequence: the number that grows by adding the previous ones: 1, 1, 2, 3, 5... Formally, it is defined by $F_0 = 0, F_1 = 1, F_n = F_{n-1} + F_{n-2}$; the ordinary generating function is $G(x) = \sum_{n=0}^{\infty} F_n x^n = \frac{x}{1-x-x^2}$ it allows us to quickly calculate the terms of the sequence without having to resort to recursive methods.

This ordinary generating function (OGF) not only produces the Fibonacci numbers but also provides a versatile framework for tackling combinatorial problems. Consider the task of counting binary strings of length n that do not

DOI: 10.4018/979-8-3373-3089-1.ch003

contain consecutive 1s. Such binary patterns are particularly relevant in computer science applications, such as data compression and encoding schemes.

Under the stated constraint, if the final digit is "0," the preceding part can be any valid binary string of length n−1. If it ends in "1," the digit before that must be "0," which limits the preceding part to a valid configuration of length n−2. This recurrence mirrors the Fibonacci relation, establishing a direct link between the count of such valid binary strings and the Fibonacci numbers.

Ordinary generating functions help us to directly calculate sequence terms and provide a powerful methodology for solving combinatorial problems, making it easier to count configurations that would otherwise be complex to manage.

Permutations (Bóna, 2022), arrangements, combinations, and dismutations examined in this chapter are not directly ordinary generating functions (OGFs) but can be represented through them.

For example, the generating function for permutations is $P(x) = \sum_{n=0}^{\infty} n!x^n = \frac{1}{1-x}$.

To obtain the formula of simple permutations, *n!*, we can use the power series development of the ordinary generating function. The power series is $P(x) = 0!x^0 + 1!x^1 + 2!x^2 + 3!x^3 + \cdots = 1 + x + 2x^2 + 6x^3 + \ldots$ the constant term 1 represents the permutation of 0 elements, the subsequent terms represent those of 1, 2, 3, ... elements. The coefficient of x_n in this series is n! because each term $n!x_n$ represents the number of permutations of *n* objects. This follows from the very definition of the generating series, constructed in such a way that each term represents the number of permutations of *n* objects. Consequently, the number of permutations of *n* distinct objects is given by the coefficient of x_n in the power series associated with its ordinary generating function.

For combinations, the generating function is $C(x) = (1 + x)^n$. We expand the power series by means of the binomial theorem by obtaining $C(x) = \sum_{n=0}^{\infty} \binom{n}{k} x^n$. The term in the summation represents the number of combinations of *n* objects taken *k* at a time multiplied by the corresponding power. In this series, the coefficient of x^k is exactly $\binom{n}{k}$.

The same can be said for all the other formulas.

It should be emphasized that to enumerate permutations of a set with repeated elements, exponential generating functions must be used. In summary, ordinary and exponential-generating functions are valuable tools for solving combinatorial problems (Lamanna et al., 2022). The choice of the most suitable depends on the specific characteristics of the task at hand.

Having clarified these aspects, we can proceed to the examination of the fundamental relationships of combinatorial calculus.

SIMPLE PERMUTATIONS

Permutations refer to the order in which the elements of a set are arranged. For example, if we have a set of letters like {A, B, C, D}, there are 24 cases:

ABCD ABDC ACBD ACDB ADBC ADCB BACD BADC BCAD BCDA BDAC BDCA

CABD CADB CBAD CBDA CDAB CDBA DABC DACB DBAC DBCA DCAB DCBA

In a permutation, each element of the set appears only once, and the order is important.

To better understand the meaning of the term, we can refer to a famous photo in which eleven workers can be seen eating lunch, sitting on a beam suspended 250 meters high.

Figure 1. Workers on pause

It was 1932 when this photo was printed on the pages of the New York Herald Tribune. It is known as *Lunch atop a skyscraper*. The image recalls the workers, known as the "Cowboys of the Sky" or "Ironworkers," sitting on a beam during a break in the construction of New York's first skyscrapers. Among these, many were Mohicans, particularly skilled at working at heights without protection.

How many ways can the eleven Cowboys of the Sky be arranged? The reasoning is quite simple: initially, we can choose one of the 11 workers; next, we can select one of the 10 remaining workers, then add one of the 9 remaining workers, and so on.; The result is 11 • 10 • 9 • ... • 1.

In this case we will speak of simple permutations, sometimes indicated by the symbol P_n, but more frequently indicated as n!. This symbol, called factorial, is precisely the product of the natural numbers from n to 1. It should be emphasized that 0!, by definition, is worth 1.

$$n! = n \cdot (n-1) \cdot (n-2) \cdot \ldots \cdot 1$$

applying the formula we get: 11! = 39,916,800

In one of the most famous depictions, Snow White stands next to a rectangular table with dwarfs sitting around her (Drivet, 2020) . Dopey sits on her right.. How many ways can this event occur? Since it is possible to assume that Snow White always stands at the head of the table, the first dwarf has seven free chairs available, the second 6, etc., while the last can only occupy the last remaining chair. In practice, the positions are 7 • 6 •... • 1. This is the factorial. The seven diners can then arrange themselves around the table in 7! = 5040 different ways. Let us now examine the position of Dopey: he can sit to the left or right of Snow White, and the other dwarfs can occupy the six positions left free. This leads to 2 • 6! = 1440 different ways favourable to the phenomenon's occurrence.

As an exercise, some simple examples can be proposed:

1. Let us consider a group of three friends: A, B, and C. We want to determine the number of ways they can sit on a bench. [6]
2. Six people have to make a presentation at a conference; how many possible sequences in which these presentations can be organized. [720]
3. How many anagrams can be obtained from the word *computer*? [40320]
4. How many contain the syllable *co*? [5040]
5. Suppose we have a production chain in a manufacturing industry. This chain includes seven machines numbered from 1 to 7. The product must pass through these machines. How many ways is it possible to organize the order in which the product passes through the machines? [5040]

There is also a generating function for *Circular permutations*, a variant of classical permutations in which, however, the elements arranged in a circle instead of in a straight line are considered. In other words, instead of considering the linear order of the elements, we only consider the relative order between them (Wilf, 2005).

In this case, two arrangements are considered equal if they can be obtained by rotation. For example, consider the circular permutation of A, B, and C: ABC. If we rotate the elements of a position clockwise, we get BCA. Rotating one position more, we get CAB. All these circular arrangements are equivalent and represent the same circular permutation.

The total number of possible circular permutations with n elements is $(n-1)!$ because the first element can be fixed arbitrarily, while the remaining arrangements are determined by the circular arrangement of the remaining $(n-1)$.

A typical application of this typology is in group theory. Circular permutations constitute a group known as a cyclic group, in which the composition operations represent the rotations of individual elements.

Circular permutations are a fascinating concept in combinatorics with numerous theoretical and practical applications. Circular permutations are an interesting concept in combinatorics with several theoretical and practical applications. They can be used to solve problems involving cyclic rotations, such as arranging elements in a cyclic sequence or describing loops of events in a process.

As an exercise, some simple examples can be proposed:

1. Suppose four people {A, B, C, D} sitting around a round table. Determine the possible circular arrangements of these people. [6]
2. A jeweller has five gemstones of assorted colours and wants to create a circular necklace with these stones. How many ways are there to arrange the stones on the necklace? [24]
3. Let us imagine that we have five cities that a courier must visit, departing from one city and returning to the city of departure, forming a circular route. We want to determine all the possible unique routes that the courier can follow, considering that the route can be travelled clockwise or counterclockwise. [12]
4. A football team of eleven players is arranged in a circle to “charge” before a match. How many different ways are there to organise players in a circle? [3,628,800]
5. Having a deck of 52 playing cards, we want to arrange the cards in a circle for a card game. How many ways are there to organize cards? [$1{,}55 \cdot 10^{66}$]

PERMUTATIONS WITH REPETITION

If the elements can be repeated, as in the case of anagrams of words containing identical letters, they are Permutations with repetition or Permutations with indistinguishable objects.

For example, the word ABCA always has four letters as ABCD, but there are only 12 anagrams:

AABC AACB ABAC ABCA ACAB ACBA

BAAC BACA BCAA AAB AAB CBAA

For the calculation, where a, b, c, ... represent the number of repetitions of each element, the formula is used:

$$P_n^{a,b,c\ldots} = \frac{n!}{a!b!c!\ldots}$$

The word CAMERA has 11 letters, two *a* and two *e*, so there are 9,979,200 distinct configurations. Suppose we have a jury that must formulate a vote to judge a sport or singing performance. The four members of the jury raise 4 paddles used to give the votes: two show the 7, one the 6, and the other the 8. How many possible ways can they be arranged?

$$P_4^{2,1,1} = \frac{4!}{2!1!1!} = 12$$

Permutations can be found in areas that are not immediately imaginable: in mathematics, the development of a binomial of the type $(a + b)^n$ is widely treated. If we wanted to know the coefficient of a^5b^7 in the development of $(a + b)^{12}$, it would be enough to consider the anagrams of aaaaabbbbbbb. These are permutations with repetition, and applying the formula, it turns out that their total number is 12!/(5!•7!) = 792.

A similar example is the following: Suppose that the city's plan is similar to that of Manhattan, with streets parallel to each other and perpendicular. How many ways can one reach point B, starting from point A four blocks away to the East (E) and two to the North (N), by choosing to go up or right at each intersection? The permutations of EEEENN give the number of possible paths, so 6!/(4!•2!)=15. This is nothing more than a problem like the anagram of the ABCA.

This example is nothing more than the concretization of the Polya lattice.

Figure 2. Polya lattice

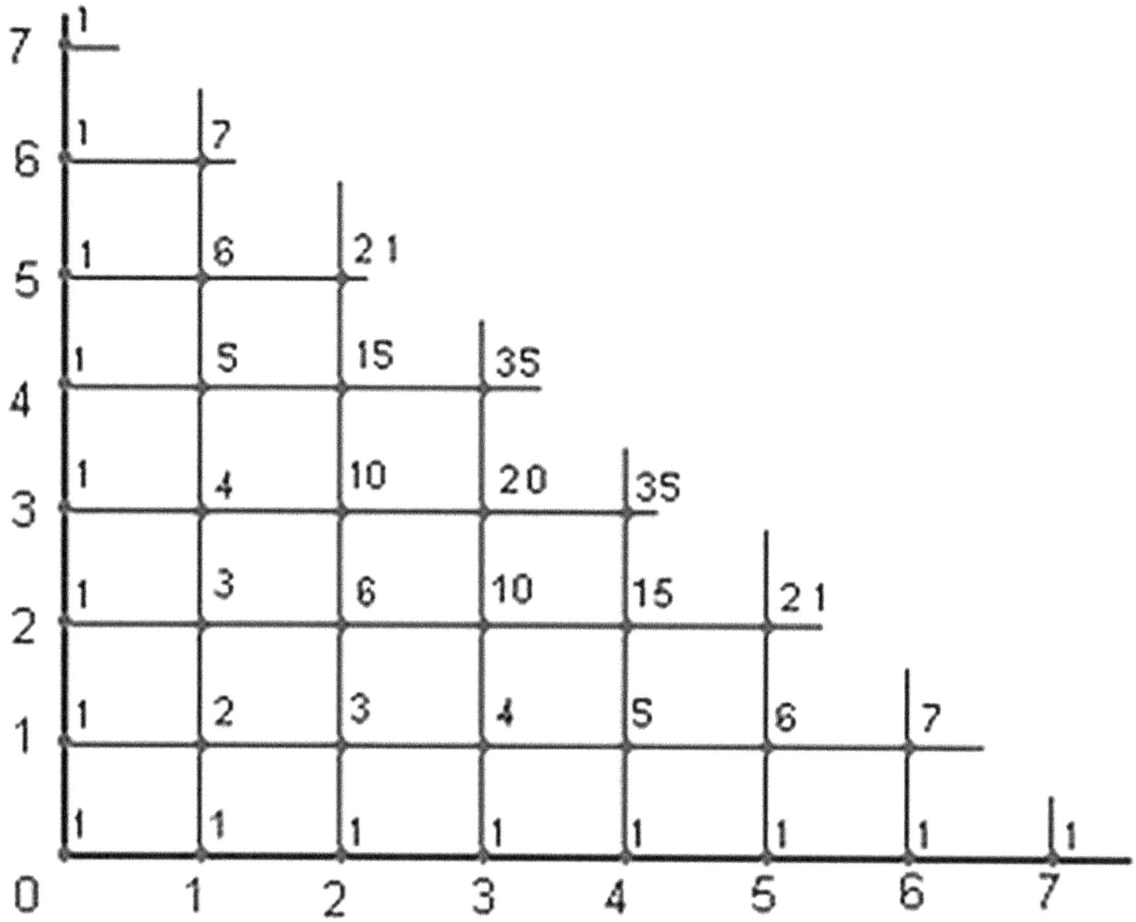

The numbers of the minimum paths to go from point (0,0) to point (x, y), placed in any of the grid nodes, are calculated on this grid. They were calculated by determining the number of steps to the right (D) and up (A) starting from (0,0) to reach the point considered and then counting the anagrams of the word thus formed.

As an exercise, some simple examples can be proposed:

1. Suppose we have a dress consisting of 4 buttons, of which two are red and two are blue. How many permutations are possible? [6]
2. A football match between teams A and B ended 3-2. How many ways could the networks have succeeded? [10]
3. Let us consider a product code consisting of 6 characters, where 2 are '3' numbers, 3 are '5' numbers, and 1 is a special 'X' character. How many unique codes are there? [60]
4. Six pizzas, 3 margarita pizzas, 2 with vegetables and 1 four-season pizza were ordered in a pizzeria. The waiter does not remember who to give them to; how many different ways can they be distributed? [60]
5. How many anagrams one obtain from the word *permutation*? [19,958,400]

SIMPLE ARRANGEMENTS

On the other hand, arrangements are similar to permutations, but order is important in this case. For example, given the set of letters {A, B, C, D} in which you want to determine the arrangements of 2 elements, you will have 12 arrangements:

AB AC AD BA BC BD CA CB CD DA DB DC

Let us resume the discussion on the workers. The situation changes if we are satisfied with taking only *k* workers. In this case, we are talking about simple provisions. For the calculation, the formula is:

$$D_{n,k} = n \cdot (n-1) \cdot (n-2) \cdot \ldots \cdot (n-k+1) = \frac{n!}{(n-k)!}$$

Choosing any three workers each time yields 990 dispositions.

Taking up a problem presented in the first chapter, we can expand it in this way: 8 mountaineers tie up in a rope team to cross a glacier. Two are beginners, so one wants them to be neither in first nor last place. How many ways can the rope team be formed?

First, let us determine how many ways I can fit beginner mountaineer A and beginner mountaineer B into the correct positions. $D_{6,2}$ gives the modes; after placing the two beginner mountaineers, it is necessary to place the remaining 6 in the other 6 available places and therefore $D_{6,6}$ or, equivalently, 6!. Applying the multiplication principle, the different roped parties are given by $D_{6,2} \cdot 6! = 6 \cdot 5 \cdot 6 \cdot 5 \cdot 4 \cdot 3 \cdot 2 = 21{,}600$.

As an exercise, some simple examples can be proposed:

1. Imagine five team members; the CEO wants to assign three leadership roles: Team Leader, Project Manager and Designer. Order is important because each role is unique. How many ways can roles be assigned? [60]
2. There are 8 participants in the race, and we want to determine the order of the first 3 classifieds (gold, silver, bronze). [336]
3. In the hypothesis that the 195 states recognized by the UN decided to equip themselves with a flag of three distinct colours with vertical bands, would the seven colours of the rainbow (red, orange, yellow, green, blue, indigo, violet) be enough to carry out such a project? [yes, the possible flags would be 210]
4. A race in which bettors must guess the horses that will finish in 1st, 2nd and 3rd place is called "Tic-Tac-Toe Race". Assuming ten horses start, how many possible arrival orders are in the first three positions? [720]
5. How many ways can seven books chosen from 20 books be arranged on a bookcase? [390,700,800]

ARRANGEMENTS WITH REPETITION

Let us examine a practical case. The padlock is a portable lock that protects the contents of suitcases, bags, and lockers from theft. It consists of a case with a locking mechanism and a metal ring. The mechanism is controlled by a key or using numbered rollers.

To open a combination lock, one needs to try all possible cases; in this case, the order counts (1234 cannot open a lock with combination 4231), and the elements are repeatable. The first roll can take one of the ten values; this can be associated with one of the ten values of the second roll, and the same goes for the other cases. To open a combination lock, one must try all possible cases. With four rolls, we will have 10 • 10 • 10 • 10 = 10,000 different positions.

These are provisions with repetition.

For the calculation, the reasoning made can be summarized by condensing it into a single formula:

$$D'_{n,k} = n^k$$

Similar examples include entering a personal identification number (PIN) to withdraw money at an ATM, accessing a bank account, unlocking a smartphone or PC, or gaining access to buildings or restricted areas. In these cases, forgetting the code renders the desired service unusable, as access is blocked after a few attempts.

As an exercise, some simple examples can be proposed:

1. If we toss a coin four times, how many possible outcomes are there, considering the order of the tosses? [16]
2. How many different three-digit numbers can be formed with the digits 1 to 9? [729]
3. Consider 4 colours (red, blue, green, yellow) to design a pattern of 5 elements, where each element can be one of 4 colours, and the colours can repeat. [1024]
4. Car license plates consist of 2 letters, 3 digits, 2 letters. Knowing that letters can be chosen from the 26 letters of the Anglo-Saxon alphabet, calculate how many cars can be registered. [456.976.000]
5. A 6-character password consists of uppercase letters (26 options), lowercase letters (26 options), and numbers (10 options). Each character can be an uppercase letter, lowercase letter, or number, and the characters can repeat each other. How many passwords are possible? [56.800.235.584]

To give an idea of how arrangements with repetition can take on enormous value, let us try to refer to the famous text by Jorge Borges (1889-1996), The Library of Babel (Borges, 1995): “Each wall of each hexagon corresponds to five shelves; each shelf contains thirty-two books of uniform format; each book is four hundred and ten pages long; each page, of forty lines; each line, of forty black letters [...] The number of orthographic symbols is twenty-five.” Given the possible arrangements with repetition, there are $25^{410 \cdot 40 \cdot 40} = 25,656,000$ different works.

SIMPLE COMBINATIONS

Combinations represent a collection of objects selected from a more extensive collection in which the order in which they are selected is not relevant. For example, considering the usual set of letters {A, B, C, D}, you want to find all the combinations of two letters from this set. The cases will be: {A, B}, {A, C}, {A, D}, {B, C}, {B, D}, {C, D}. We notice that the order of the letters within each combination is not relevant, so {A, B} is the same as {B, A}.

One of the fields in which combinations are widely studied and utilised is card games.

Poker is one of the most famous card games. There are many variations: Draw poker, in which each player is dealt five cards and can decide to change from one to four; Stud poker, in which the cards are dealt one at a time, some face down and others face up; Texas hold 'em, in which the player is dealt with two hole cards and can use a certain number of the five community cards face up. The game aims to get the best combination or the other players to quit. The face value of the singular cards is as follows, from the strongest to the weakest: Ace, King, Queen, Jack, Ten, Nine, Eight, Seven, Six, Five, Four, Three, Two. For the suits, the sequence, from strongest to weakest, is Hearts, Diamonds, Clubs, and Spades.

It is not a simple game of chance because, although there is a probabilistic basis, the player's skill counts. We can examine the main aspects of American poker, i.e., played with 52 cards. We can arrange the possible outcomes in order of importance, from the strongest hand to the weakest.

1. Royal Flush: sequence of the five highest cards belonging to the same suit
2. Straight Flush: five cards in sequence of the same suit
3. Four of a kind: four cards of the same value
4. Full House: Three of the same cards and one equal pair
5. Flush: Five cards of the same suit, but not in sequence
6. Straight: five cards in sequence but not of the same suit
7. Three of a kind: three of the same cards and two different ones
8. Two Pair: Two pairs of cards of the same rank plus a different card
9. Pair: a pair of cards of the same rank and the others different
10. High Card: A hand that does not contain any of the previous combinations

To define the probabilities of each hand, it is necessary to resort to the study of combinations, that is, to that part of combinatorial calculus in which the order of the elements does not matter. The number of combinations, i.e., the number of subsets of *k* elements that can be formed from a set containing *n* elements, often referred to as the binomial coefficient, is given by:

$$C_{n,k} = C(n,k) = \binom{n}{k} = \frac{n!}{k!(n-k)!}$$

If one uses all the cards, the possible combinations of five cards can be calculated as follows: C(52,5) = 52!/(5! • 47!) = 259,860.

Once the possible cases have been calculated, the favourable ones remain to be defined. In order not to be long-winded, only two cases can be considered:

Three of a kind: choosing a card among the 13 possible cards of the same rank multiplies by how one can choose three cards in a set of 4 identical cards C(4,3), then you multiply how you can choose two cards in a set of 12 C(12,2), finally having four suits for each type, between the two chosen, 4 must be multiplied by itself twice (4^2): P = 13 • C(4,3) • C(12,2) • 4^2 = 54,912. Alternatively, the formula 52 • C(48,1) • C(44,1)/2 could be used!.

The second formula involves a different approach: We choose any card from the 52 available for the first card of three of a kind, then we choose a second card from the same rank among the remaining 48 cards, and finally, we choose a third card from the same rank among the remaining 44 cards. Finally, we divide it into 2! Because the order in which we choose the cards does not matter.

Pair: One card among the 13 possible cards is multiplied by how one can choose two cards in a set of 4 identical cards C(4,2), then you multiply how it is possible to choose three cards in a set of 12 C(12,3), finally having four suits for each type, among the three chosen you must multiply 4 by yourself three times (4^3): P = 13 • C(4,2) • C(12,3) • 4^3 = 1,098,240. In this case, the alternative formula is 52 • C(48,1) • C(44,1) • C(40,1)/2!

Leaving aside the world of cards, we can give other examples. Let us imagine a test or questionnaire with a multiple-choice question that offers four answer options (A, B, C, D) and another with three (X, Y, Z). One needs to calculate how many distinct answer sets are possible across the two questions. For the first, we choose two answers out of four: $C_{4,2}$ = 4! / (2! • (4 - 2)!) = 6. In the second, where only one choice is allowed among three, the count is 3. Since the responses to the two questions are independent, the total number of possible answer pairings is 6 • 3 = 18.

In the game of 8-ball pool, balls fall into two distinct categories (solid and striped) and are arranged in a triangle following specific rules: ball number 1 is placed at the tip, ball number 8 at the centre, and one solid and one striped at the bottom corners. Out of 15 balls, three are in fixed positions, leaving 12 to be freely organized. The number of valid arrangements is 12! / (7! • 5!) = 792 since order within each category does not matter.

Earlier, we referred to the problem of sub-sequences. In the array [1, -2, 3, -4, 5, -6, 7, -8, 9, -10], the number of non-empty sub-sequences is given by the formula 2^N - 1, where *n* is the number of elements. Thus, 2^{10} - 1 = 1023 possible subsets.

$$\sum_{k=1}^{10}\binom{10}{k} = 10 + 45 + 120 + 210 + 252 + 210 + 120 + 45 + 10 + 1 = 1023$$

Speaking of the Binomial Theorem, we have already mentioned the problem of combinations and its expression C(n,k), an integer that represents the number of ways in which it is possible to select a subset of *k* elements from a set of *n*, without considering the order. The formula may seem complex, but it has an intuitive explanation. To obtain $\binom{n}{k}$, we are essentially dividing the number of ways we can organize *n* objects in sequence (n!) by the number of ways we can organize the first *k* objects in sequence (k!) and the number of ways we can organize the remaining n-k objects (n–k)!.

An interesting example of a combination is the one relating to the behaviour of balls descending on different paths in a machine developed by Francis Galton (1822-1911) for the study of binomial distribution.

Figure 3. Galton's machine

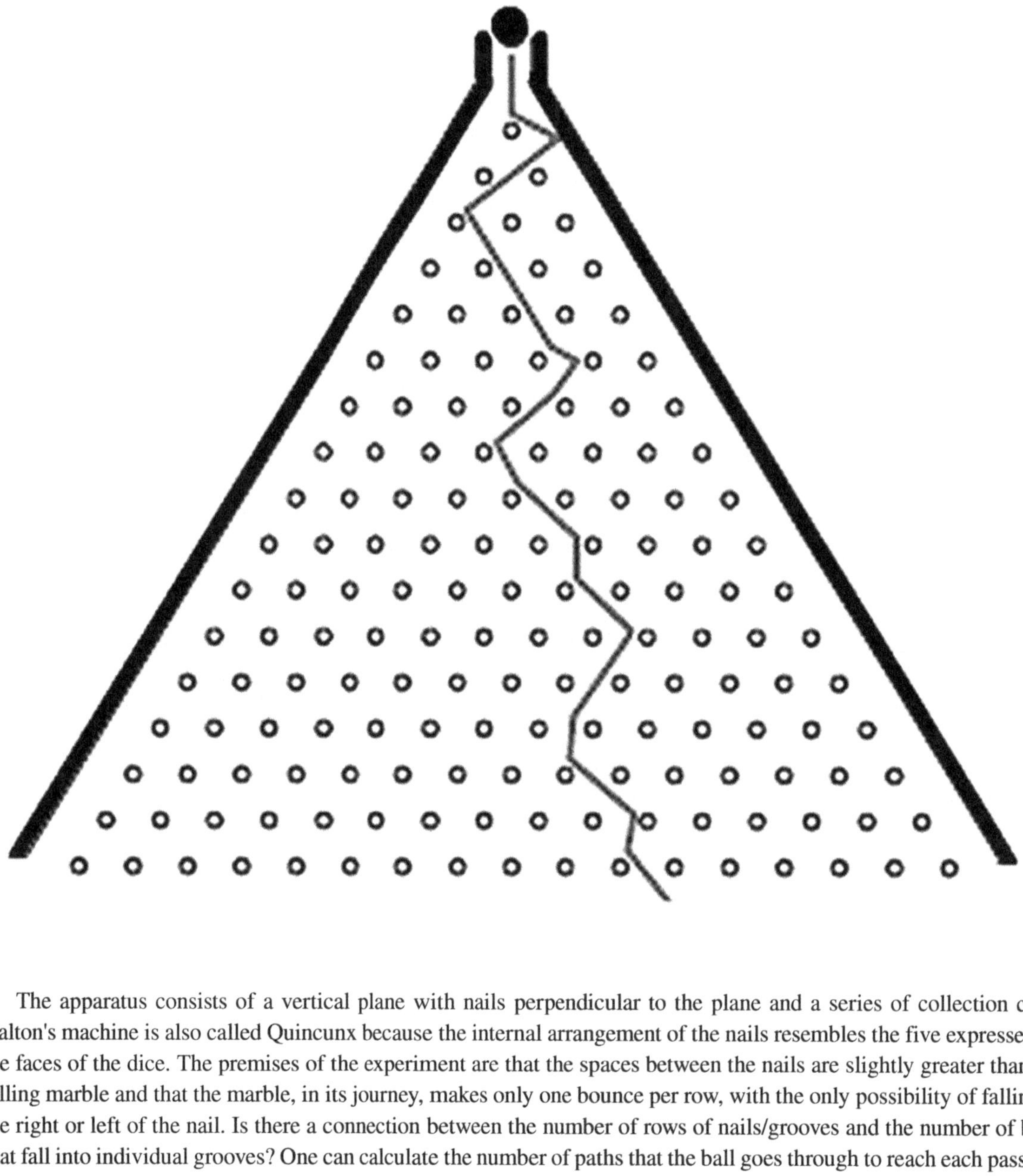

The apparatus consists of a vertical plane with nails perpendicular to the plane and a series of collection cells. Galton's machine is also called Quincunx because the internal arrangement of the nails resembles the five expressed on the faces of the dice. The premises of the experiment are that the spaces between the nails are slightly greater than the falling marble and that the marble, in its journey, makes only one bounce per row, with the only possibility of falling to the right or left of the nail. Is there a connection between the number of rows of nails/grooves and the number of balls that fall into individual grooves? One can calculate the number of paths that the ball goes through to reach each passage. It is given by C(n,k) (n = row and k = column).

As an exercise, some simple examples can be proposed:

1. During a toast between six friends, each crosses his glass only once with all the others. How many jingles do we hear? [15]
2. In a lottery, we can choose three numbers from a set of 15; how many distinct ways are there? [455]
3. In Texas Hold'em, two cards are initially dealt from a deck of 52. How many ways can this pair be achieved? [1326]

4. To play the Italian game SuperEnalotto, one must select six numbers from 90. How many tickets should one fill out to be sure to make 6? [622,614,630, but the playable combinations are, at most, 27,132]
5. Twenty friends and former schoolmates reunite and organise a dinner. Initially, they greet each other by shaking hands. How many handshakes will there be? [190]
6. The Italian Chamber of Deputies is composed of 400 members, who must elect a commission of 6 members. How many ways? [5,478,557,838,600]

COMBINATIONS WITH REPETITION

This type of selection is presented less frequently because it requires a deeper understanding of mathematical reasoning, making it more challenging for educators to incorporate it into lessons. It also tends to appear in more specialized, less common applications. In many classrooms, diagrams or visual tools help clarify concepts such as ordinary combinations and permutations. However, repeated choices—due to their abstract nature—do not always lend themselves as easily to visual representation. That said, these concerns are primarily based on misconceptions.

Let us examine some useful scenarios.

Repeated selections prove helpful whenever the same item can be picked more than once, and the order of selection is irrelevant.

First, however, let us define the formula:

$$C'_{n,k} = \binom{n+k-1}{k}$$

We want to fill a tray with *k* pastes (even of the same type), choosing from the n pastes available. How many different ways can we fill the tray?

For example, with three types of pastes available (n = 3), we want to fill the tray with four pastes (k = 4); the number of different ways is given by: $C'_{3+4-1,4} = C'_{6,4} = 15$.

A similar model applies when generating a 4-digit smartphone PIN using the 10 digits 0–9. We find the number of distinct codes where digits may repeat (and order does not matter): $C'_{10+4-1,4} = C'_{13,4} = 715$.

Now imagine someone crafting a custom fragrance by selecting three scents from a set of 6 (jasmine, lavender, vanilla, rose, musk, citrus), possibly repeating a scent to intensify its presence. Some scent sets could be: lavender–lavender–musk; rose–citrus–jasmine; or even vanilla–vanilla–vanilla. The total number of distinct sets is $C'_{6+3-1,3} = C'_{8,3} = 56$.

Finally, consider the number of quadruples of natural numbers whose sum is 10. Since the sum of the four natural numbers must equal ten, the problem concerns arranging the first ten natural numbers in a sequence. We, therefore, want to find all the possible solutions for the equation: a + b + c + d = 10 where a, b, c, d are natural numbers (including zero). We observe that these numbers can be repeated, and the order in which they appear does not matter. That is, this is a problem of combinations with repetitions. Based on these considerations, one must calculate the combinations with repetition with n = 4 and k = 10:

$$C_{4,10} = \binom{4+10-1}{4} = 715$$

This aligns with asking how many ways one can distribute 10 indistinguishable balls into 4 distinct boxes.

As further exercises, some simple examples can be proposed:

1. There are five types of sweets (candies, cookies, chocolate, muffins, cake), and you want to choose 3 sweets to create an assortment. Each type of dessert can be chosen more than once. So many distinct ways to choose three desserts from five types available. [35]
2. Imagine that one have 4 types of activities (excursion, cultural visit, relaxation, sports) and want to plan 5 activities during your trip. Each type of activity can be chosen more than once. [56]

3. An urn contains ten numbered balls. How many ways can we re-enter and draw them without considering the order? [92,378]
4. How many ways to distribute 12 white socks in 5 drawers? (can each drawer contain from 0 to 12 socks)? [1820]
5. Is there a condition for at least one sock in each drawer? [330]

DISMUTATIONS

A dismutation, also known as a permutation without fixed points, is a permutation of the elements of a set in which no element remains in its original position. In other words, it is an arrangement of the elements of a whole in which no element occupies its original position.

Considering a set of n elements, a dismutation of these is a permutation in which no element is in its original position. If we consider the set $\{1, 2, 3\}$, the possible dismutations are: (2, 3, 1), (3, 1, 2). In either case, nothing is in its original position.

The cardinality or number of dismutations of a set of n elements is often denoted by the symbol $!n$ and is given by the formula:

$$!n = \sum_{i=0}^{n}(-1)^i \frac{n!}{i!}$$

There is also the recursive formula

$$!n = (n-1)(!(n-1)+!(n-2))$$

Dismutations find applications in various mathematical contexts, such as error correction code theory, cryptography, algorithm analysis, and game theory. They are used to solve arrangement problems where no element remains in its original position, such as randomly occupied chairs or the creation of random permutations that meet specific restrictions.

A common item is the keychain—a rigid metal ring often adorned with an ornamental pendant. This simple object can inspire a combinatorial exploration, leading to intriguing mathematical questions. Four friends have grown tired of their keychains and decided to redistribute them so that each has a different one from the one they had before. How many ways can keychains be exchanged?

Figure 4. Keychain

The solution is greatly simplified if we assign a number to each keychain. So, it is a matter of permuting the numbers 1, 2, 3, and 4 so that the number i does not appear at position i. Being a Derangement, we can easily calculate the nine possibilities to do so. In this case, the verification is simple. The cases are 2143, 2341, 2413, 3142, 3421, 3412, 4123, 4312, and 4321.

The main difference between circular permutations and dismutations lies in the condition of fixed positions. In the first case, some elements may remain in their original position, while in dismutations, none remain stationary. In short, circular permutations involve arranging elements in a circle, considering only the relative order of the elements. At the same time, in dismutations, no element remains in its original position.

It is interesting to note that the ratio of the number of dismutations to the factorial !n/n! tends to 1/e (approximately 0.36788) when *n* tends to infinity.

Why does this happen?

Dividing both sides of the Derangement formula by n!, we observe that the summation on the right converges to 1/e in the limit as *n* approaches infinity.

Let us look at some simple examples:

1. How many dismutations are there using the numbers 1, 2, 3, 4? [9]
2. In a box, there are five balls numbered from 1 to 5. How many ways are there to draw balls so that no balls are placed in their corresponding position? [44]
3. A company must reassign seven employees to 7 different workstations, ensuring that no employee remains at their original workstation. Calculates the number of ways this reassignment can be done. [1854]
4. We have eight books and want to arrange them so that no book remains in its original position. In other words, if we number objects from 1 to 8, we want to know how many ways we can arrange these objects in a sequence in which no object is in the position corresponding to its original number. [14,833]
5. In a row of ten seats at the cinema, how many different ways are there to assign as many people as possible to the seats so that no one person is assigned to the seat with the same number? [1,334,961]

In summary, one can identify the type of combinatorics to be used with a simple algorithmic structure (Flajolet & Sedgewick, 2009).

Figure 5. Combinatorial choices

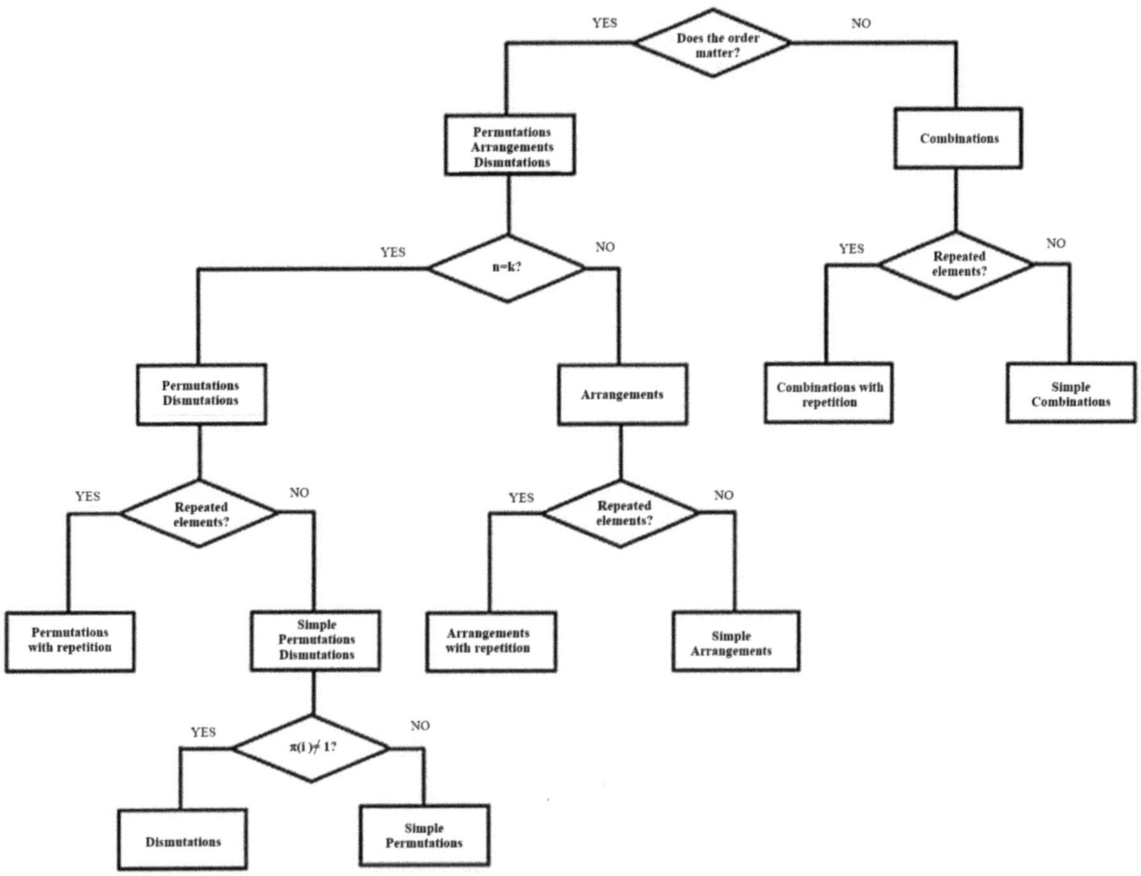

All possibilities can be managed with a program that allows you to choose the desired type and make the calculations.

Program 13 generating functions

```
# Generating functions
from sympy import symbols, Function, factorial
# Declaration of symbolic variables
n, a, b, c, x = symbols('n a b c x')
# Ordinary generating function for permutations with repetition
P_rep = Function('P_rep')(x)
# Function to calculate permutations with repetition
def permutations_with_repetition(a_value, b_value, c_value, n_value):
denominator = factorial(a_value) * factorial(b_value) * factorial(c_value)
return factorial(n_value) / denominator
# Function to calculate dismutations
def count_dismutations(n_value):
result = 0
```

```
for i in range(n_value + 1):
result += ((-1)**i) / factorial(i)
return factorial(n_value) * result
while True:
# Menu for selection
print("Choose the type of combinatorial object:")
print("1. Simple permutations")
print("2. Permutations with repetition")
print("3. Circular permutations")
print("4. Simple arrangements")
print("5. Arrangements with repetition")
print("6. Simple combinations")
print("7. Combinations with repetition")
print("8. Dismutations")
print("9. Exit")
choice = int(input("Enter the number corresponding to your choice: "))
if choice == 9:
break # Exit the loop if the choice is "Exit"
n_value = int(input("Enter the value of n: "))
result = None
if choice == 1:
# Calculation of simple permutations
result = factorial(n_value)
elif choice == 2:
# Calculation of permutations with repetition
a_value = int(input("Enter the value of a: "))
b_value = int(input("Enter the value of b: "))
c_value = int(input("Enter the value of c: "))
result = permutations_with_repetition(a_value, b_value, c_value, n_value)
elif choice == 3:
# Calculation of circular permutations
result = factorial(n_value - 1)
elif choice == 4:
# Calculation of simple arrangements
k_value = int(input("Enter the value of k: "))
result = factorial(n_value) / factorial(n_value - k_value)
elif choice == 5:
# Calculation of arrangements with repetition
a_value = int(input("Enter the value of a: "))
result = n_value**a_value
elif choice == 6:
# Calculation of simple combinations
a_value = int(input("Enter the value of a: "))
result = factorial(n_value) / (factorial(a_value) * factorial(n_value - a_value))
elif choice == 7:
# Calculation of combinations with repetition
a_value = int(input("Enter the value of a: "))
result = factorial(n_value + a_value - 1) / (factorial(a_value) * factorial(n_value - 1))
elif choice == 8:
```

REFERENCES

Bóna, M. (2022). *Combinatorics of permutations*. CRC Press. DOI: 10.1201/9780429274107

Bóna, M. (2023). *Walk Through Combinatorics, A: An Introduction To Enumeration, Graph Theory, And Selected Other Topics*. World Scientific. DOI: 10.1142/13455

Borges, J. L. (1995). *La biblioteca di Babele* [The Library of Babel]. Mondadori.

Cameron, P. J. (1994). *Combinatorics: topics, techniques, algorithms*. Cambridge University Press.

Drivet, A. (Ed.). (2020). *Examining an Operational Approach to Teaching Probability*. IGI Global.

Erickson, M. J. (2013). *Introduction to combinatorics*. John Wiley & Sons.

Flajolet, P., & Sedgewick, R. (2009). *Analytic combinatorics*. Cambridge University press. DOI: 10.1017/CBO9780511801655

Harris, J., Hirst, J. L., & Mossinghoff, M. (2009). *Combinatorics and Graph Theory*. Springer Science & Business Media.

Lamanna, L., Gea, M. M., & Batanero, C. (2022, February). Secondary school students' strategies in solving combination problems. In *Twelfth Congress of the European Society for Research in Mathematics Education.*

Lovász, L. (2007). *Combinatorial problems and exercises* (Vol. 361). American Mathematical Soc.

Mladenović, P. Mladenović, & Chernyk. (2019). *Combinatorics*. Springer International Publishing.

Roberts, F., & Tesman, B. (2009). *Applied combinatorics*. CRC Press. DOI: 10.1201/b12335

Sagan, B. E. (2020). *Combinatorics: The art of counting* (Vol. 210). American Mathematical Soc. DOI: 10.1090/gsm/210

Wilf, H. S. (2005). *generatingfunctionology*. CRC press.

Chapter 4
Algorithms and Conjectures

ABSTRACT

The chapter deals with combinatorial algorithms and their role in analyzing discrete structures such as graphs, sets, and permutations. The sieve of Eratosthenes is introduced to find prime numbers, highlighting the inclusion-exclusion principle to count the elements of a set without certain divisors. Next, we analyze the Atkin sieve, which uses modular congruences to identify prime numbers more efficiently. Another part of the chapter explores the distribution of twin primes and proposes a statistical analysis of their distances through combinatorial models. The Ulam spiral is discussed, revealing interesting patterns in the former distribution. Finally, mathematical conjectures are explored, such as the Erdős–Szekeres conjecture on the presence of convex polygons between sets of points and hypotheses on the distribution of permutations with prime jumps, and a simple example starting from the Erdős–Ginzburg–Ziv theorem is exposed. The chapter highlights the link between combinatorics, algorithms and conjectures in mathematical progress.

ALGORITHMS

Programming computer languages allows you to reflect on the ability to think

Nicholas Negroponte

Previously, we emphasised the importance of combinatorial algorithms. These algorithms are designed to solve problems involving discrete structures, such as graphs, sets, strings, or permutations, and they are essential in many fields of computer science and applied mathematics (Graham et al., 1989; Cameron, 1994).

A simple example may clarify.

The sieve of Eratosthenes, one of the oldest methods for identifying prime numbers, exemplifies an algorithm rooted in arithmetic logic yet closely intertwined with combinatorics. More advanced extensions of this sieve have been developed, revealing deep connections to principles of enumerative combinatorics. In general, combinatorial sieves operate through a structured process of inclusion and exclusion, closely reflecting the principle of inclusion-exclusion.

The fundamental idea is to begin with a universal set of elements and progressively eliminate those that fail to satisfy specific desired properties. Combinatorics, for its part, is primarily concerned with counting such elements, and sieve methods contribute meaningfully in this context by offering systematic tools for filtering and selecting subsets—such as in the generation of prime numbers (Berman & Fryer, 2014). It is worth emphasizing that the problem of counting prime numbers is of significant interest in both combinatorics and number theory.

The general sieve constitutes a powerful generalisation of the sieve of Eratosthenes, relying, as previously mentioned, on the inclusion-exclusion principle, a foundational concept in enumerative combinatorics that will be discussed in greater detail later. Instead of removing multiples directly, it focuses on counting the number of elements in a set that are not divisible by any of a given set of prime numbers *P*. In summary, given a set of integers and a set of primes *P*, the inclusion-exclusion principle is applied to determine how many elements are not divisible by any prime in *P*.

DOI: 10.4018/979-8-3373-3089-1.ch004

Program 14 General Sieve

```
from math import sqrt, prod
def general_sieve(limit):
def count_multiples(primes):
return limit // prod(primes)
# Calculate the square root of the limit
root = int(sqrt(limit))
# Initialize the list of prime numbers up to the square root of the limit
primes = [True] * (root + 1)
primes[0] = primes[1] = False # 0 and 1 are not prime numbers
# Implement the Sieve of Eratosthenes to find prime numbers up to the square root
for i in range(2, int(sqrt(root)) + 1):
if primes[i]:
for j in range(i*i, root + 1, i):
primes[j] = False
base_primes = [i for i in range(len(primes)) if primes[i]]
# Apply the principle of inclusion-exclusion
count = limit
for k in range(1, len(base_primes) + 1):
for combination in combinations(base_primes, k):
if k % 2 == 1:
count -= count_multiples(combination)
else:
count += count_multiples(combination)
return count - 1 # Subtract 1 to exclude the number 1
def combinations(iterable, r):
pool = tuple(iterable)
n = len(pool)
if r > n:
return
indices = list(range(r))
yield tuple(pool[i] for i in indices)
while True:
for i in reversed(range(r)):
if indices[i] != i + n - r:
break
else:
return
indices[i] += 1
for j in range(i + 1, r):
indices[j] = indices[j - 1] + 1
yield tuple(pool[i] for i in indices)
limit = int(input("Upper limit"))
prime_count = general_sieve(limit)
print(f"The number of prime numbers up to {limit} is: {prime_count}")
```

Atkin's sieve is an enhanced version of Eratosthenes' sieve, utilising a slightly different method to find prime numbers like an investigator uses different clues to solve a mystery (Ghidarcea & Popescu, 2024). This sieve is considered a combinatorial algorithm because it uses specific numerical properties and congruences to determine the primality of numbers in a systematic and structured way (Atkin & Bernstein, 2004). The sieve relies on a series of modular congruences to identify prime numbers. The algorithm structures the selection of candidate numbers using quadratic formulas.

This systematic approach ensures that every possible combination is considered. When a number satisfies one of the above-mentioned congruences, its primality state is reversed. Finally, the algorithm eliminates multiples of squares of found prime numbers, ensuring that only actual prime numbers remain marked as such (Hahn & Sachs, 2008).

Let us take a more detailed look. Instead of eliminating multiples of prime numbers as Eratosthenes' sieve does, Atkin's sieve uses modulo 60 arithmetic properties and quadratic forms to identify potential prime numbers.

The following equations for x and y positive integers (with x > y) are considered:

$$4x^2 + y^2 = n$$
$$3x^2 + y^2 = n$$
$$3x^2 - y^2 = n$$

For every n < limit, it is checked whether the number of solutions to the above equations is odd. If it is, the primality state of n. The squares of the prime numbers are found, and their multiples are eliminated.

The proposed code indicates these combinatorial properties. The program is suitable for educational purposes and for generating prime numbers within reasonable size ranges.

Program 15 Atkin's Sieve

```
def atkin_sieve(limit):
# Create a list of booleans initially set to False
is_prime = [False] * (limit + 1)
sqrt_limit = int(limit**0.5)
# Apply the conditions of the Atkin sieve
for x in range(1, sqrt_limit + 1):
for y in range(1, sqrt_limit + 1):
# Congruence for n % 12 == 1 or n % 12 == 5
n = 4 * x**2 + y**2
if n <= limit and (n % 12 == 1 or n % 12 == 5):
is_prime[n] = not is_prime[n]
# Congruence for n % 12 == 7
n = 3 * x**2 + y**2
if n <= limit and n % 12 == 7:
is_prime[n] = not is_prime[n]
# Congruence for n % 12 == 11
n = 3 * x**2 - y**2
if x > y and n <= limit and n % 12 == 11:
is_prime[n] = not is_prime[n]
# Eliminate the multiples of the squares of primes
for n in range(5, sqrt_limit + 1):
if is_prime[n]:
for k in range(n**2, limit + 1, n**2):
is_prime[k] = False
# The numbers 2 and 3 are handled separately
```

```
primes = [2, 3] if limit >= 2 else []
primes.extend([n for n in range(5, limit + 1) if is_prime[n]])
return primes
# Example usage
n = int(input("Limit: "))
primes = atkin_sieve(n)
print(f"Prime numbers up to {n}: {primes}")
```

One last consideration: Atkin's algorithm filters candidates for prime numbers using modular conditions. Generating functions can significantly enhance the understanding of algorithms by representing prime sequences, performing asymptotic analysis, solving recurrences, and proving combinatorial identities. Generating functions can significantly improve understanding of the algorithm by representing prime sequences, performing asymptotic analysis, solving recurrences, and proving combinatorial identities.

Before continuing the discussion of algorithms, it is helpful to anticipate some aspects that connect them to combinatorics. We will explore these topics in more detail later. The classification of algorithms, traditionally divided into exact, heuristic, and analytic, fully applies to combinatorial ones, as they typically address complex, often NP-hard problems where combinatorics plays a central role.

The first typology is used when it is possible (and practical) to explore or manage the entire combinatorial space of solutions, for example, in backtracking, dynamic programming, and graph-based algorithms.

The second is essential for significant combinatorial problems where full exploration is computationally intractable, for example, in TSP, graph colouring, and scheduling. Analytical methods employ formal mathematical tools to model, approximate, or analyse combinatorial problems, including complexity analysis, asymptotic estimation, and linear and mixed-integer programming (Colbourne & Dinitz, 2007). A table can well summarise this.

Table 1. Algorithms and combinatorics

Type of algorithm	Connection to combinatorics	Typical use
Exact	Explore the complete combinatorial space	Small problems, optimal solutions
Heuristic	Exploit combinatorial patterns to find good solutions	Large problems, approximations
Analytical	Use mathematical tools to model or simplify the combinatorial structure	Theoretical analysis, relaxation, estimates

The following challenges can be classified as analytic algorithms with exact constructive support.

Studying how prime numbers are distributed within a specific range requires combinatorial techniques to analyse patterns and frequencies.

Let us take as an example twin prime number pairs, i.e., those prime numbers that differ by two units. The distribution follows a pattern, with the groups of twins becoming increasingly distant as one proceeds through the sequence. The conjecture could hypothesize that the distribution of distances between twins D(n) follows an exact statistical distribution. In this sense, one could hypothesise that the distribution of distances between the first twin pairs follows a specific combinatorial scheme. The idea that the distribution of distances between pairs of twins can follow a specific combinatorial scheme is a fascinating hypothesis. Although twin primes appear to be irregularly distributed, there may be underlying structures that follow combinatorial patterns. These patterns may not be immediately apparent but may emerge through in-depth analysis. However, it is essential to note that this remains an area of mathematical speculation.

Building a program that provides fascinating insights into the potential connection between the observed distribution of twin first distances and combinatorial models is possible. The code extends the search to one million, calculates and prints the median, mode, and moving average, and creates and displays a histogram of distances and cumulative distribution. It also prints the correlation between consecutive distances, creates and displays a logarithmic histogram, and adds a Q-Q plot to compare the distribution with a normal distribution.

Program 16 Distance Between Twins

```
import math
import numpy as np
import matplotlib.pyplot as plt
from statistics import mode
def is_prime(num):
if num <= 1:
return False
for i in range(2, int(math.sqrt(num)) + 1):
if num % i == 0:
return False
return True
def calculate_twin_prime_distances(primes):
distances = []
for i in range(len(primes) - 2):
if primes[i + 1] - primes[i] == 2:
distances.append(primes[i + 2] - primes[i + 1])
return distances
def analyse_distances(distances):
num_cases = len(distances)
print("Number of distances:", num_cases)
if num_cases > 0:
# Partial statistics
for counter in [100, 200, 300, 400, 500, 600, 700, 800, 900, 1000]:
if counter <= num_cases:
mean = round(np.mean(distances[:counter]), 2)
std_dev = round(np.std(distances[:counter]), 2)
print(f"Mean after {counter} cases: {mean}")
print(f"Standard deviation after {counter} cases: {std_dev}")
# Final statistics
final_mean = round(np.mean(distances), 2)
final_std_dev = round(np.std(distances), 2)
print(f"Final mean: {final_mean}")
print(f"Final standard deviation: {final_std_dev}")
print("Minimum value:", min(distances))
print("Maximum value:", max(distances))
print(f"Median: {np.median(distances)}")
print(f"Mode: {mode(distances)}")
# Correlation between consecutive distances
if len(distances) > 1:
correlation = np.corrcoef(distances[:-1], distances[1:])[0, 1]
print(f"Correlation between consecutive distances: {correlation:.4f}")
# Histogram of distances
plt.figure(figsize=(10, 6))
plt.hist(distances, bins=50)
plt.title("Histogram of distances between twin primes")
plt.xlabel("Distance")
```

```
plt.ylabel("Frequency")
plt.show()
# Cumulative distribution
sorted_distances = sorted(distances)
cdf = np.arange(len(distances)) / len(distances)
plt.figure(figsize=(10, 6))
plt.plot(sorted_distances, cdf)
plt.title("Cumulative distribution of distances")
plt.xlabel("Distance")
plt.ylabel("Cumulative probability")
plt.show()
# Histogram on logarithmic scale
plt.figure(figsize=(10, 6))
plt.hist(np.log(distances), bins=50)
plt.title("Histogram of distances (logarithmic scale)")
plt.xlabel("Log(Distance)")
plt.ylabel("Frequency")
plt.show()
# Moving average of distances
window = 100
if len(distances) >= window:
moving_average = np.convolve(distances, np.ones(window) / window, mode='valid')
plt.figure(figsize=(10, 6))
plt.plot(range(len(moving_average)), moving_average)
plt.title("Moving average of distances")
plt.xlabel("Index")
plt.ylabel(f"Moving average (window size {window})")
plt.show()
# Generate a list of prime numbers up to 1,000,000
primes = [number for number in range(2, 1000001) if is_prime(number)]
# Calculate the distances between pairs of twin primes
twin_prime_distances = calculate_twin_prime_distances(primes)
# Analyse the distribution of distances
analyse_distances(twin_prime_distances)
```

Number of distances: 8169
Mean after 100 cases: 8.36
Standard deviation after 100 cases: 5.15
Mean after 200 cases: 9.05
Standard deviation after 200 cases: 5.43
Mean after 300 cases: 10.0
Standard deviation after 300 cases: 6.56
Mean after 400 cases: 10.29
Standard deviation after 400 cases: 6.88
Mean after 500 cases: 10.48
Standard deviation after 500 cases: 7.18
Mean after 600 cases: 10.71
Standard deviation after 600 cases: 7.34

Mean after 700 cases: 10.93
Standard deviation after 700 cases: 7.55
Mean after 800 cases: 11.13
Standard deviation after 800 cases: 7.57
Mean after 900 cases: 11.4
Standard deviation after 900 cases: 7.84
Mean after 1000 cases: 11.37
Standard deviation after 1000 cases: 7.7
Final mean: 14.13
Final standard deviation: 10.63
Minimum value: 2
Maximum value: 100
Median: 10.0
Mode: 10
Correlation between consecutive distances: -0.0119

Let us limit ourselves for now to the analysis of the first graph. The graph displays an asymmetric distribution to the right, characterised by a long tail. It is typical of many natural phenomena and suggests that shorter distances are much more common than long ones. A noticeable spike in the first few bars of the histogram indicates that shorter distances are the most frequent. The frequency decreases rapidly as the distance increases, suggesting that it becomes increasingly rare to find pairs of twins as the distance increases. Although most of the distances are concentrated on the left side of the graph, the distribution has a long tail, indicating that, although rare, exceptionally large distances can occur between pairs of first twins.

Figure 1. Histogram of the distances between twin primes

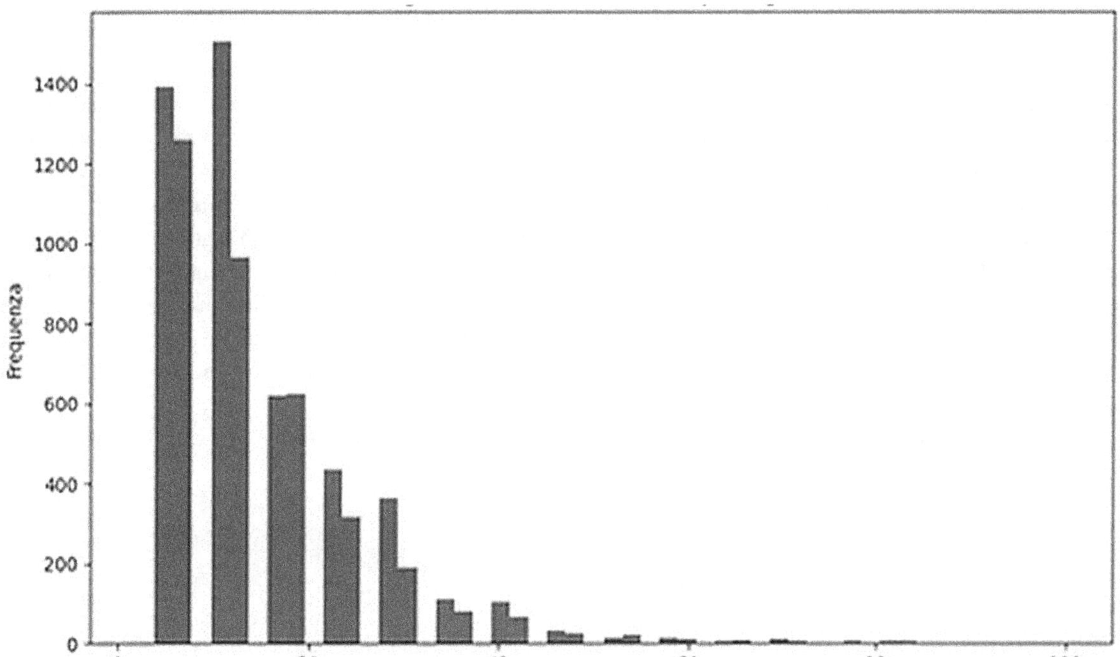

The trend of averages seems to indicate a link with a function. The results may suggest the usefulness of logarithmic-scale analysis, which is common in many combinatorial models.

Figure 2. Histogram of distances (log scale)

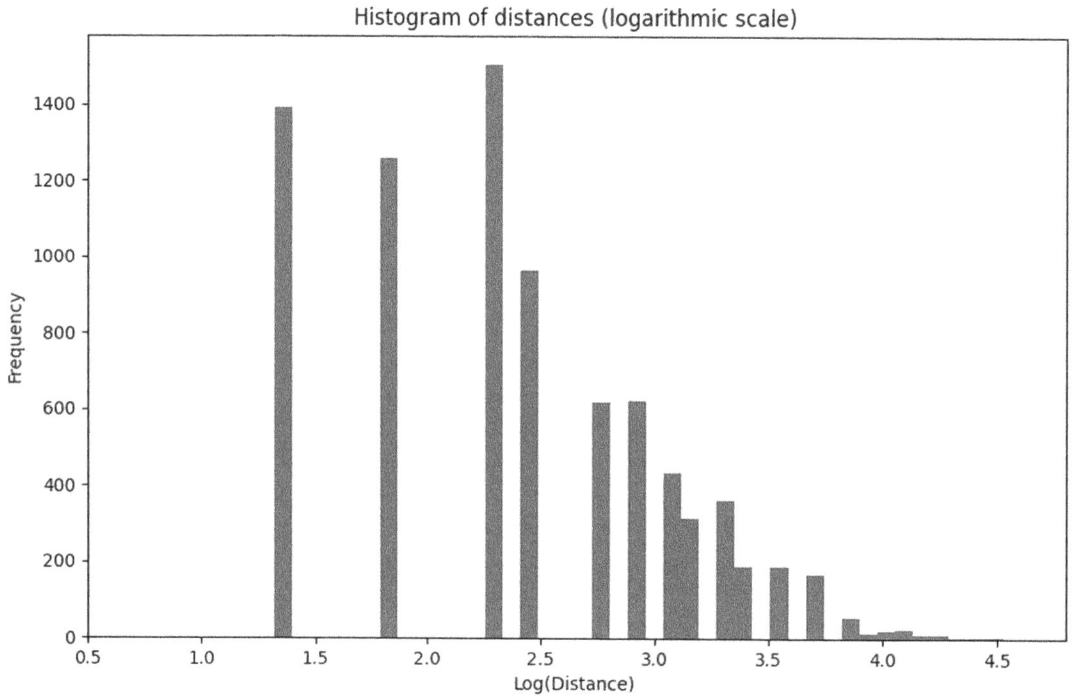

At this point, the researcher could explore the connection between the observed distribution and any combinatorial structures. Possible combinatorial probability models could refer to random permutations, whereby the distances between the first twin pairs are related to specific patterns or schemes in these permutations or to partitions in which the distribution of the distances of the first twins is related to the probabilities of obtaining specific partitions in a random process.

A curious graphical representation of prime numbers was created in 1963 by the Polish mathematician Stanislaw Ulam (1909-1986). During a conference, which was probably not particularly interesting, he found himself writing the natural numbers on a sheet of paper, arranging them along a spiral. When he marked the prime numbers, he noticed that they tended to be arranged on the diagonal lines of the number scheme. The Ulam spiral can be seen as combinatorial for several reasons: the arrangement of numbers in a spiral shape implies a well-defined and structured order typical of combinatorial configurations (da Fontoura Costa, 2023).

The formation of linear patterns of prime numbers within the spiral is a phenomenon that can be analysed using combinatorial techniques.

Figure 3. Ulam spiral

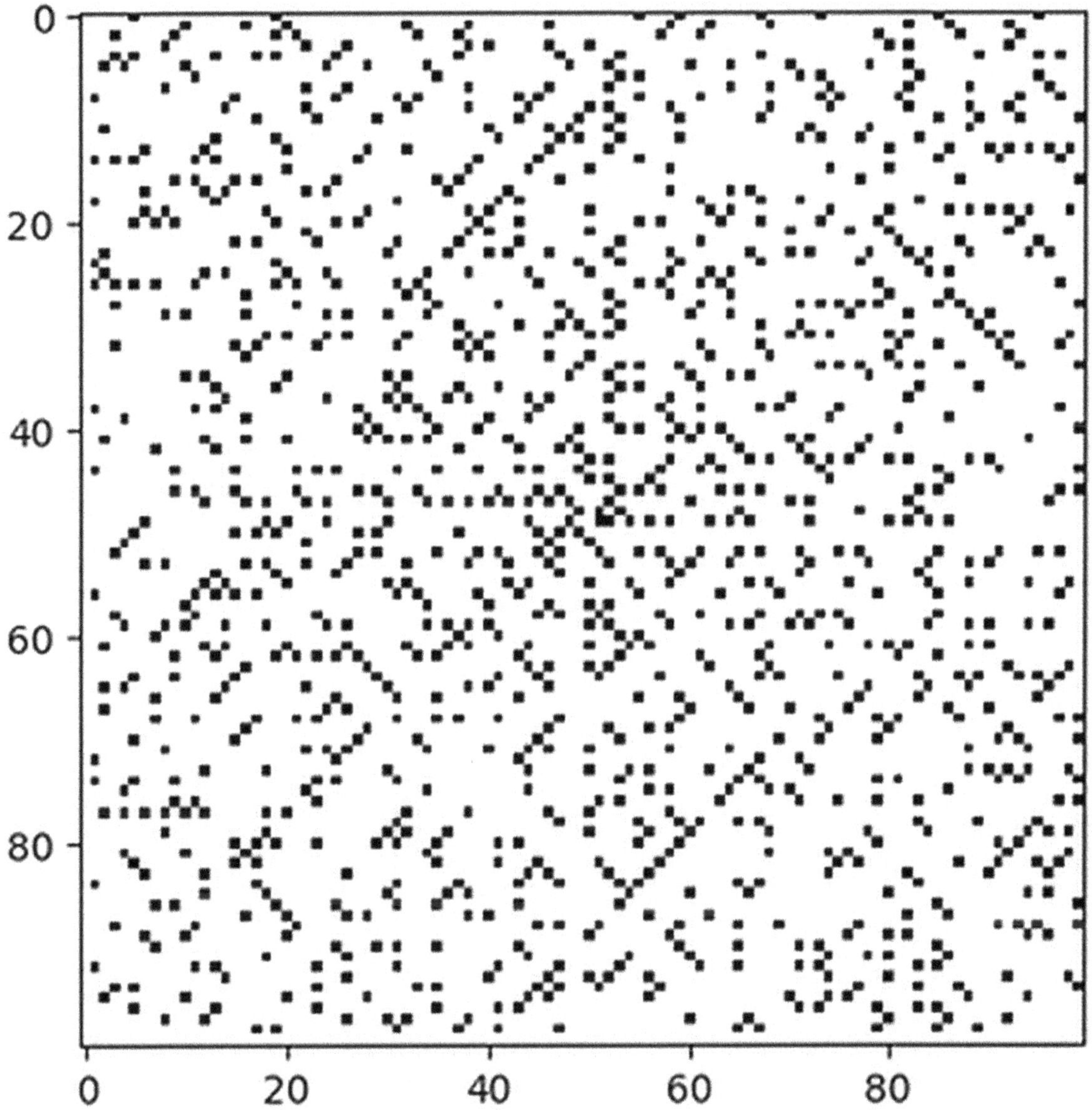

Program 17 Ulam Spiral

```
import numpy as np
import matplotlib.pyplot as plt
def ulam_spiral(n):
# Initialize directions for movement (dx, dy)
dx, dy = 0, -1
```

```
x, y = 0, 0
arr = np.zeros((n, n), dtype=int)
for i in range(1, n**2 + 1):
# Check if the position is within bounds
if (-n//2 < x <= n//2 - 1) and (-n//2 < y <= n//2 - 1):
arr[n//2 - y - 1, n//2 + x] = i
# Change direction when needed
if x == y or (x < 0 and x == -y) or (x > 0 and x == 1 - y):
dx, dy = -dy, dx
# Update the position
x, y = x + dx, y + dy
return arr
def is_prime(n):
if n < 2:
return False
if n == 2:
return True
if n % 2 == 0:
return False
sqrt_n = int(np.floor(np.sqrt(n)))
for i in range(3, sqrt_n + 1, 2):
if n % i == 0:
return False
return True
# Ask the user for the size of the spiral
n = int(input('Enter the size of the spiral (odd number): '))
# Generate the Ulam spiral
spiral = ulam_spiral(n)
# Apply the prime number check function to all elements in the matrix
prime_spiral = np.vectorize(is_prime)(spiral)
# Display the Ulam spiral with prime numbers highlighted
plt.imshow(prime_spiral, cmap='Greys', interpolation='nearest')
plt.title(f"Ulam Spiral of Prime Numbers ({n}x{n})")
plt.colorbar(label='Prime Numbers')
plt.show()
```

Many of the patterns in the Ulam spiral are primarily related to number theory, but some aspects can be analysed or explained using combinatorial concepts (Chmielewski et al., 2017) . Here are some working hypotheses:

1. The distribution of prime numbers in the spiral can be analysed in terms of residuals modulo 4 (since the primes, except for 2 and 3, are of the form 4k±1). This is an example of dividing a set (the natural numbers) into subsets, a combinatorial concept.
2. Diagonal patterns can be seen as sub-sequences of a more extensive sequence. The analysis of these sub-sequences has combinatorial elements, such as selection and sorting.
3. Let us consider prime numbers as vertices of a graph, with connections based on specific properties (such as the difference between two primes). Structures emerge that can be analysed using graph theory methods, a field closely related to combinatorics.
4. The count of primes in different regions of the spiral is a combinatorial problem. For example, count how many primes appear in a particular diagonal or circular spiral sector.

Program 18 Ulam Pattern

```
import numpy as np
import matplotlib.pyplot as plt
from sympy import isprime
def generate_ulam_spiral(dim):
spiral = np.zeros((dim, dim), dtype=int)
x, y = dim // 2, dim // 2
number = 1
spiral[x, y] = number
step = 1
while number < dim * dim:
for _ in range(step):
y += 1
number += 1
if number <= dim * dim:
spiral[x, y] = number
for _ in range(step):
x -= 1
number += 1
if number <= dim * dim:
spiral[x, y] = number
step += 1
for _ in range(step):
y -= 1
number += 1
if number <= dim * dim:
spiral[x, y] = number
for _ in range(step):
x += 1
number += 1
if number <= dim * dim:
spiral[x, y] = number
step += 1
return spiral
def find_diagonals(spiral):
dim = spiral.shape[0]
diagonals = []
for k in range(-dim + 1, dim):
diagonal = []
for i in range(max(0, -k), min(dim, dim - k)):
j = i + k
if isprime(spiral[i, j]):
diagonal.append(spiral[i, j])
if len(diagonal) > 1:
diagonals.append(diagonal)
for k in range(-dim + 1, dim):
diagonal = []
```

```
for i in range(max(0, k), min(dim, dim + k)):
j = i - k
if isprime(spiral[i, j]):
diagonal.append(spiral[i, j])
if len(diagonal) > 1:
diagonals.append(diagonal)
return diagonals
def display_ulam_spiral_and_diagonals(spiral, diagonals):
dim = spiral.shape[0]
image = np.zeros((dim, dim), dtype=int)
for i in range(dim):
for j in range(dim):
if isprime(spiral[i, j]):
image[i, j] = 1
fig, (ax1, ax2) = plt.subplots(2, 1, figsize=(10, 12))
ax1.imshow(image, cmap='binary', interpolation='none')
ax1.set_title("Ulam Spiral")
ax1.axis('off')
ax2.axis('off')
ax2.text(0.5, 0.5, f"Number of prime number diagonals found: {len(diagonals)}",
fontsize=12, ha='center', va='center')
plt.tight_layout()
plt.show()
dim = int(input("Spiral dimension (odd number): "))
spiral = generate_ulam_spiral(dim)
diagonals = find_diagonals(spiral)
display_ulam_spiral_and_diagonals(spiral, diagonals)
```

Another way to see the arrangement of the primes on a spiral might be the following:

Program 19 Spiral of Prime Numbers

```
import turtle
import math
def is_prime(n):
if n < 2:
return False
if n == 2:
return True
if n % 2 == 0:
return False
for i in range(3, int(math.sqrt(n)) + 1, 2):
if n % i == 0:
return False
return True
def draw_spiral(max_points, a=10, theta_step=0.1):
# Setup Turtle screen
screen = turtle.Screen()
```

```
screen.title("Spiral")
screen.bgcolor("white")
# Initialize Turtle
t = turtle.Turtle()
t.speed(0)
t.hideturtle()
# Colors
prime_color = "black"
non_prime_color = "grey"
num_points = 0
n = 1
while num_points < max_points:
# Calculate current point's x and y coordinates
r = a * math.sqrt(n)
theta = n * theta_step
x = r * math.cos(theta)
y = r * math.sin(theta)
# Draw the current point
t.penup()
t.goto(x, y)
t.pendown()
if is_prime(n):
t.dot(10, prime_color)
else:
t.dot(6, non_prime_color)
# Increment the current number and point counter
n += 1
num_points += 1
turtle.done() # <--- al posto di screen.mainloop()
# Run the function
draw_spiral(max_points=1000)
```

The visual result is a graphical representation of the distribution of prime numbers. Some interesting patterns can be observed: black dots (prime numbers) seem to be distributed unevenly but with some regularity. You may notice diagonal "lines" of grey dots, representing multiples of small numbers. The density of black dots decreases as you move away from the center, reflecting the fact that prime numbers become rarer among larger numbers.

Figure 4. Arrangement of prime numbers (in black) on a spiral

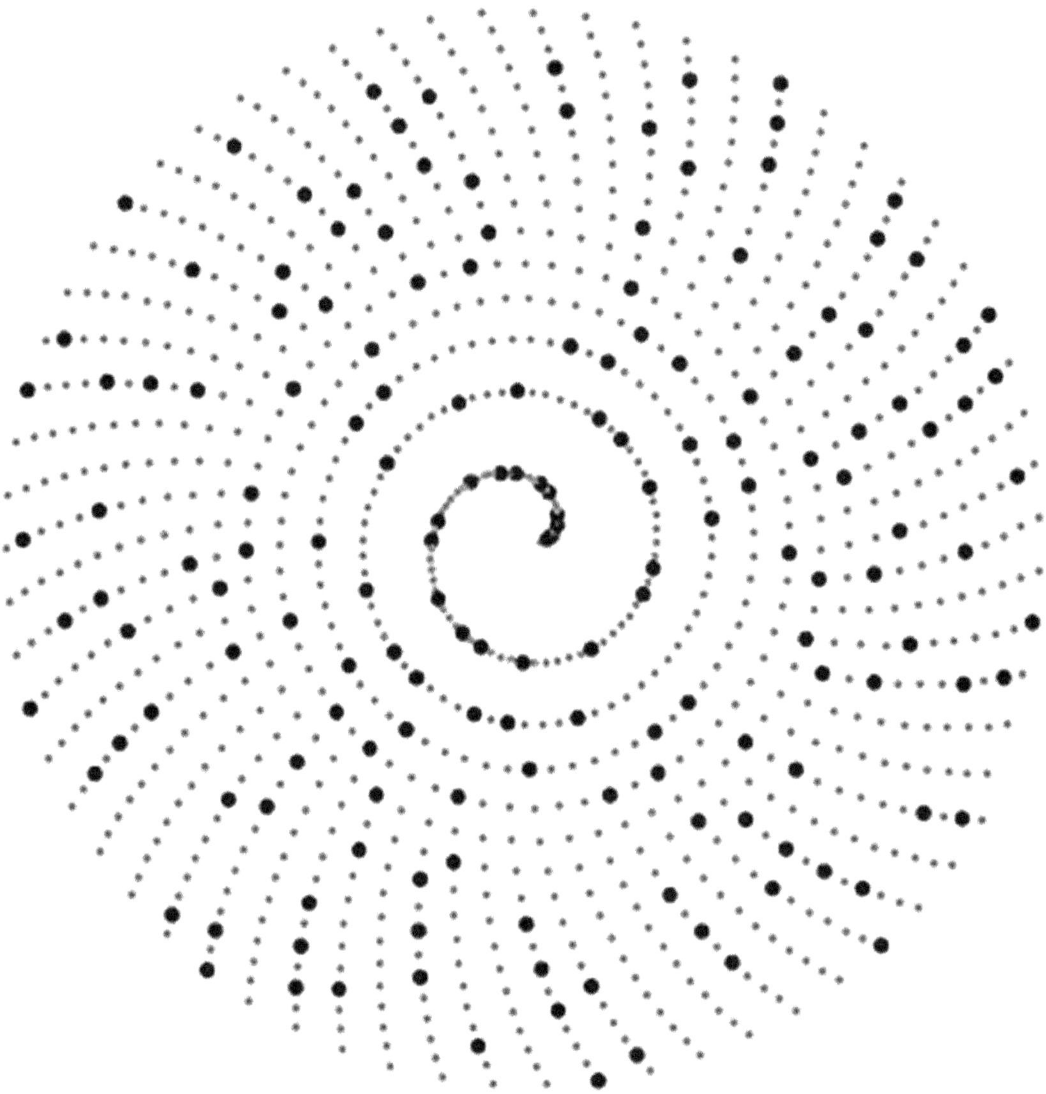

Although the program does not directly deal with classical combinatorial problems, combinatorics comes into play when considering how prime numbers are selected and distributed in relation to their position on the spiral. The arrangement of prime numbers is a practical case of a combinatorial problem that concerns the frequency and distribution of a subset of numbers (the prime numbers) within a larger set (the natural numbers). This visual representation of prime numbers can lead to several interesting combinatorial considerations. For example, the intervals between consecutive prime numbers are highlighted, allowing us to observe how these gaps vary and sometimes form sequences (Zhang, 2014; Holt, 2015; Vega, 2023); adjacent pairs can allow us to study the distribution and frequency of pairs of twin numbers; The spiral allows us to visualize the irregular but not random distribution of prime numbers and to identify some patterns and groupings.

As previously mentioned, specific optimisation challenges are typically characterised by their suitability for heuristic algorithms. These situations often involve a huge solution space, intricate or nonlinear constraints, and a limited timeframe in which to secure a satisfactory solution. We will delve into specific examples of these later; for now, we can focus on examining some classifying elements of heuristics and metaheuristics (Talbi, 2009; Desale et al., 2015). Each possesses unique characteristics and adapts well to different types of combinatorial problems (Lovász, 2007).

The cases referenced in the table will be explored in subsequent chapters.

Table 2. Comparison of combinatorial heuristic algorithm categories

Category	Exploration Strategy	Example
Constructive	Builds solution step-by-step using local decisions	Nearest Neighbor
Local Search	Improves a complete solution by exploring its neighborhood	Minimum Spanning Tree
Metaheuristics	Global search with adaptive mechanisms to escape local optima	Simulated Annealing
Hybrid Methods	Combines constructive and local/metaheuristic techniques	Genetic Algorithms
Randomized Algorithms	Explores solution space through stochastic sampling	Random Walk

THE CONJECTURES

By following a precise recipe, algorithms solve complex problems. What happens when we do not know the recipe yet? Conjectures come into play. Another topic that lends itself to deep reflection is that of conjecture. The interplay between combinatorics, algorithms, and conjecture is vast, representing a rich and ever-evolving area of study (Lovász & Prömel, 2004; Pólya et al., 2013; Vilenkin, 2014); Loehr, 2017; Mazur, 2022). Combinatorics often provides the ground on which conjectures arise. By studying discrete structures and recurring patterns, mathematicians formulate working hypotheses. These conjectures, in turn, stimulate the search for algorithms that can verify or refute them.

Algorithms, in turn, offer methods for solving combinatorial questions. This synergy between the three fields pushes the frontier of mathematical and computational knowledge, opening new avenues for solving complex problems and understanding fundamental structures. Many combinatorial conjectures, while seeming simple to enunciate, prove extremely difficult to prove or refute. Some remain open for decades or even centuries. The most interesting aspect is that attempting to prove conjectures often leads to the development of new mathematical techniques. False guesswork can also be helpful, leading to interesting counterexamples.

Let us take as an example the Erdős–Szekeres conjecture, proposed by the Hungarian mathematicians Paul Erdős and George Szekeres (1911-2005), one of the most famous in geometric combinatorics (Erdös & Szekeres, 1935). Conjecture is like a game: if enough points are scattered across a sheet, we can always find a few that form a perfect polygon, like when we connect the dots to form a figure. The statement of the conjecture is as follows:

For every integer n≥3, there exists an integer N(n) such that every set of N(n) non-collinear points in the plane contains n points that form a convex polygon.

After some checks with sufficiently small *n*, it can be shown that $N(n) = \binom{n-2}{2} + 1$ is sufficient to ensure that every set of N(n) points in general position contains *n* points capable of forming a convex polygon.

To better understand how these points are arranged in the plane and verify the polygon's convexity, it is helpful to visualize the points and the polygon. The following code uses Andrew's algorithm to find the convex hull of a set of randomly generated points and then visualizes these points and the convex envelope on a graph. It has a time complexity of O(n log n), where *n* is the number of points. This makes it very efficient.

The conjecture offers a theoretical basis for the existence of such polygons in sufficiently large sets of points, while the algorithm provides a practical and efficient method for finding them.

Program 20 Andrew's Algorithm

```
import random
import matplotlib.pyplot as plt
def generate_points(num_points, x_max=100, y_max=100):
points = set()
while len(points) < num_points:
x = random.randint(0, x_max)
y = random.randint(0, y_max)
points.add((x, y))
return list(points)
def andrew_monotone_chain(points):
points = sorted(points)
lower = []
for p in points:
while len(lower) >= 2 and cross_product(lower[-2], lower[-1], p) <= 0:
lower.pop()
lower.append(p)
upper = []
for p in reversed(points):
while len(upper) >= 2 and cross_product(upper[-2], upper[-1], p) <= 0:
upper.pop()
upper.append(p)
return lower[:-1] + upper[:-1]
def cross_product(p, q, r):
return (q[0] - p[0]) * (r[1] - p[1]) - (q[1] - p[1]) * (r[0] - p[0])
def plot_points_and_hull(total_points, hull):
total_x, total_y = zip(*total_points)
hull_x, hull_y = zip(*hull)
plt.scatter(total_x, total_y, c='blue', label='Generated Points')
plt.scatter(hull_x, hull_y, c='red', label='Convex Hull Points')
hull_x += (hull_x[0],) # Close the polygon
hull_y += (hull_y[0],) # Close the polygon
plt.plot(hull_x, hull_y, c='red')
plt.xlabel('X')
plt.ylabel('Y')
plt.title('Convex Hull of Generated Points')
plt.legend()
plt.grid(True)
plt.show()
def main():
n = 5 # Minimum number of vertices for the desired convex polygon
num_points = 9 # Total number of generated points
points = generate_points(num_points)
print("Generated points:", points)
hull = andrew_monotone_chain(points)
if len(hull) >= n:
print(f"Convex polygon with at least {n} points found:", hull)
```

```
plot_points_and_hull(points, hull)
else:
print(f"No convex polygon with at least {n} points found among the {num_points} generated
points.")
if __name__ == "__main__":
main()
```

So far, the following results are known:

The case n = 3 is trivial (each set of 3 points forms a triangle).
For n = 4, f(4) = 5 has been proved by Esther Klein.
For n = 5, f(5) = 9 has been proved by Erdős and Szekeres themselves.
For n = 6, f(6) = 17 has been proved by Tobias Gerken.
The known lower bound is $f(n) \geq 2^{(n-2)} + 1$ while the upper bound is $f(n) = C_{2n-5,\, n-2}$

Although considerable progress has been made in proving it for specific cases, research continues to find a general solution.

The use of a programming language is important for understanding specific concepts and verifying some results, as well as a tool for testing possible new conjectures.

Let us formulate a simple example that draws inspiration from the Erdős–Ginzburg–Ziv theorem. The conjecture does not precisely correspond to this theorem, as there are differences in the size of the set and the condition of the sum (Bonnetain et al., 2020); however, it is interesting to explore paths that can expand our knowledge of additive combinatorics. The conjecture is:

In any finite set of positive integers $\{a_1, a_2, \ldots, a_n\}$ with $n \geq 2$, there exists at least one pair (a_i, a_j) such that $a_i + a_j$ is a multiple of n.

A demonstrative route could refer to the analysis of the remains. If we consider the remainders of the numbers of the set modulo n, it is possible to determine whether at least one pair of remainders has a sum that is a multiple of *n*. The following program does what is requested. It verifies specific values and cannot arrive at the demonstration.

Program 21 Conjecture About Couples

```
from itertools import permutations
from math import sqrt
def is_prime(number):
if number < 2:
return False
if number in (2, 3):
return True
if number % 2 == 0 or number % 3 == 0:
return False
i = 5
while i * i <= number:
if number % i == 0 or number % (i + 2) == 0:
return False
i += 6
return True
def is_prime_jump_permutation(permutation):
n = len(permutation)
```

```
for i, element in enumerate(permutation):
jump = (element - i) % n
if jump == 0 or not is_prime(jump):
return False
return True
def count_prime_jump_permutations(n):
return sum(1 for perm in permutations(range(n)) if is_prime_jump_permutation(perm))
def verify_conjecture(max_limit):
print("n\tP(n)\tP(n) mod n\tConjecture Valid?")
print("-" * 50)
for n in range(3, max_limit + 1):
num_permutations = count_prime_jump_permutations(n)
remainder = num_permutations % n
conjecture_valid = remainder == 0
print(f"{n}\t{num_permutations}\t{remainder}\t{'Yes' if conjecture_valid else 'No'}")
# Verify the conjecture for n from 3 to 10
max_limit = 10
verify_conjecture(max_limit)
```

The principle of remainders must be used to prove the conjecture. First, look at the values of the numbers ai modulo n. The possible remainders are 0, 1, 2, ..., n−1, so we have n possible remainders. The pigeonhole principle can be applied since we have n numbers and only n possible remainders. If the numbers a_i and a_j have the same remainder modulo *n*, we can write: $r_i = r_j \bmod n$. This implies that $a_i \equiv a_j \bmod n$ and, consequently, the sum of these two numbers will be: $a_i + a_j \equiv r_i + r_j \equiv 2r_i \bmod n$. If the common remainder is zero, then $a_i + a_j$ is already a multiple of *n*.

If the remainder were non-zero, the sum $2r_i$ might not immediately be a multiple of *n*. However, considering non-zero remainders, for any pair of remainders r_i and r_j, there are combinations of remainders such that $r_i + r_j = n$, which would also lead to a sum that is a multiple of *n*.

Interestingly, one can determine if there is a counterexample that defeats the conjecture. For example, given the set of positive integers between 1 and 15, the conjecture is NOT verified for n = 31. The integers that test the conjecture are those equal to or less than 30. The conjecture is generally valid for sets with sufficient numbers and variations in the remainder modulo n. In general, for values of *n* that are large compared to the numbers in the given set, there may be no pairs that satisfy the condition of the conjecture. Thus, the conjecture is more robust when "sufficient" numbers are in the set than the value of *n*.

Conjecture has a strong relationship to combinatorics as it involves the analysis of all possible pairs (a_i, a_j) within a set of numbers. For a set of *n* elements, there are *n(n−1)/2* possible pairs. It is a classical combinatorial problem, where the combination of elements taken two by two is studied. The analysis of the remainder modulo *n* is also closely related to combinatorics since it concerns the distribution of numbers in a range of values). It is a problem of partitioning and combining the elements of the set according to their remainders.

Let us try another conjecture. First, we define a "prime-jump permutation" as a permutation in which each element moves from its original position by a number of positions that is a prime number (positive or negative), with the calculated displacements modulo n. The conjecture is:

For every n ≥ 3, the number of "prime-jump permutations" of n elements is NEVER divisible by n.

This conjecture has important implications in group theory and combinatorics, as it indicates some regularity in permutation structures. If true, it would mean that prime-jump permutations have intrinsic properties that prevent their divisibility by the number of elements *n*.

Program 22 Prime Jump Conjecture

```
from itertools import permutations
from math import sqrt
def is_prime(number):
if number < 2:
return False
if number in (2, 3):
return True
if number % 2 == 0 or number % 3 == 0:
return False
i = 5
while i * i <= number:
if number % i == 0 or number % (i + 2) == 0:
return False
i += 6
return True
def is_prime_jump_permutation(permutation):
n = len(permutation)
for i, element in enumerate(permutation):
jump = (element - i) % n
if jump == 0 or not is_prime(jump):
return False
return True
def count_prime_jump_permutations(n):
return sum(1 for perm in permutations(range(n)) if is_prime_jump_permutation(perm))
def verify_conjecture(max_limit):
print("{:<5} {:<10} {:<15} {:<20}".format("n", "P(n)", "P(n) mod n", "Conjecture Valid?"))
print("-" * 60)
for n in range(3, max_limit + 1):
num_permutations = count_prime_jump_permutations(n)
remainder = num_permutations % n
conjecture_valid = remainder == 0
print("{:<5} {:<10} {:<15} {:<20}".format(n, num_permutations, remainder, 'Yes' if conjec-
ture_valid else 'No'))
# Verify the conjecture for n from 3 to 10
max_limit = 10
verify_conjecture(max_limit)
```

Proving the conjecture is difficult. First, the definition of "first jump" introduces a complexity that makes it difficult to establish a precise and generalizable count. Upon evaluating the program, it becomes apparent that for n = 10, the verification is not complete. The breakdown of this value could have a decisive influence. New theories or approaches may be needed better to understand the relationships between permutations and their arithmetic properties.

Another example could refer to the aforementioned twin primes. It has been noted that they tend to distribute themselves with a certain regularity within the sequence, forming "groups" of twins close to each other, separated by gradually larger intervals. For example: (3,5), (11,13), (17,19), etc. A program can easily print a sequence of pairs and compare it with the function y = x +2.

Program 23 Twin Primes

```
import matplotlib.pyplot as plt
from sympy import isprime
def twin_primes(limit):
primes = []
for num in range(3, limit + 1, 2): # Start from 3 and check only odd numbers
if isprime(num):
primes.append(num)
twin_prime_pairs = []
for i in range(len(primes) - 1):
if primes[i + 1] - primes[i] == 2:
twin_prime_pairs.append((primes[i], primes[i + 1]))
return twin_prime_pairs
# Set the limit for the search
limit = 200 # You can increase this value to explore more numbers
twin_pairs = twin_primes(limit)
# Extract values for the plot
x_values = [pair[0] for pair in twin_pairs]
y_values = [pair[1] for pair in twin_pairs]
# Create the plot
plt.figure(figsize=(10, 6))
plt.scatter(x_values, y_values, color='blue', label='Twin Primes')
plt.title(f'Twin Prime Pairs up to {limit}')
plt.xlabel('Lower Prime')
plt.ylabel('Upper Prime')
plt.axline((0, 0), slope=1, color='red', linestyle='--', label='y = x + 2') # Reference
line
plt.legend()
plt.grid()
plt.show()
```

Of course, the question arises whether there is a regularity in the distribution of twin primes, in which it is observed that they tend to form "groups" of close pairs, separated by progressively increasing intervals. It might be interesting to connect this conjecture to the Hardy-Littlewood conjecture, which explores the density and distribution of twin primes. The conjecture formulated by G. H. Hardy and J. E. Littlewood in the 1920s concerns the distribution of prime numbers, particularly the behaviour of the sums of prime numbers under certain conditions. There are various versions of this conjecture; one of the most well-known versions states that:

For every integer k, there are infinitely many prime numbers that can be expressed in the form $n + k$, where n is a prime number.

This statement is a generalisation of the twin prime conjecture, explicitly addressing the case where $k = 2$.

In 2013, Yitang Zhang (1955-) published remarkable results that opened new avenues for the twin prime conjecture. He showed that there are infinitely many prime numbers separated by an interval of less than 70 million. After Zhang's work, other mathematicians, such as Terence Tao (1975-) and James Maynard (1987-), further refined this proof, reducing the limit to lower values. The limit is currently 246. Although the twin prime conjecture remains unsolved, progress represents significant steps toward a deeper understanding of the distribution of prime numbers and pairs of twins (Maynard, 2019).

A program can help us understand the complexity of the problem, even if only by exploring the values of *k* up to 20 and setting the limit for the generation of prime numbers at 100.

Program 24 Hardy-Littlewood Conjecture

```
def is_prime(num):
if num < 2:
return False
for i in range(2, int(num**0.5) + 1):
if num % i == 0:
return False
return True
def generate_primes(limit):
primes = []
for num in range(2, limit + 1):
if is_prime(num):
primes.append(num)
return primes
def explore_conjecture(k, limit):
primes = generate_primes(limit)
prime_set = set(primes) # For faster searches
valid_results = []
for prime in primes:
if (prime + k) in prime_set:
valid_results.append((prime, prime + k))
return valid_results
def count_infinite_pairs(limit):
for k in range(1, 21): # Explore values of k from 1 to 20
results = explore_conjecture(k, limit)
if len(results) > 0:
print(f"Found {len(results)} prime pairs of the form n + {k} up to {limit}:")
for n, n_plus_k in results:
print(f"{n} + {k} = {n_plus_k}")
print() # Empty line to separate results
else:
print(f"No pairs found of the form n + {k} up to {limit}.\n")
# Example parameters
limit = 1000 # Limit for generating prime numbers
# Run the exploration
count_infinite_pairs(limit)
```

Prime numbers have often been referred to as having an interesting and close connection with combinatorics (Wells, 2011). We can conclude with a digression that extends beyond the text's central theme but allows for an interesting connection between fundamental mathematical concepts. Let us discuss the fundamental theorem of arithmetic, the counting of divisors, and the structures underlying our numerical representation system. The starting point is 1729, made famous by the Indian mathematician Srinivasa Ramanujan (1887-1920). Like all natural numbers, this number can be written as the product of prime numbers; in this case, its prime factors are 7, 13, and 19. The divisors of 1729 are eight: 1, 7, 19, 91, 133, 247 and 1729.

The number 1729 can also be visualized as placed at the top of a cube, as suggested by Eugenia Cheng. This geometric representation makes the number more tangible and offers insights into the nature of numbers and their interconnectedness. Geometry and algebra are not separate disciplines but two sides of the same coin in mathematics, often intertwined in surprising ways.

Figure 5. Cube

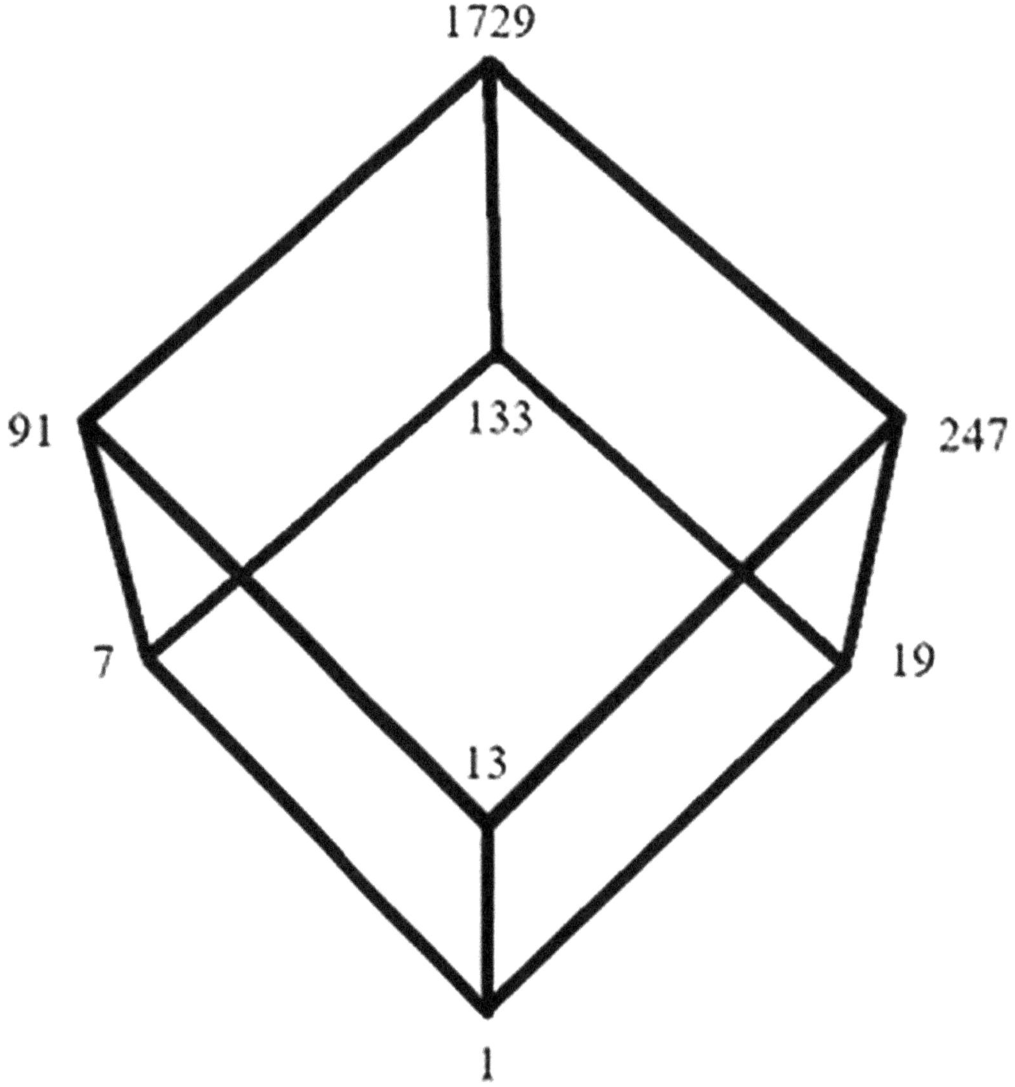

The fundamental theorem of arithmetic states that every integer greater than one can be written uniquely as a product of prime numbers, less than the order of factors. This principle emphasizes the importance of prime numbers as "building blocks" of mathematics and serves as the basis for many other areas of number theory and cryptography. Divisor counting is another area where prime numbers play a crucial role. If we consider a number n written in the factorial form: $n = p_1^{e_1} \cdot p_2^{e_2} \cdot \ldots \cdot p_k^{e_k}$, where p_i are the prime numbers and e^i are the corresponding exponents, the total number of divisors d(n) can be calculated with the formula:

$$d(n) = (e_1 + 1)(e_2 + 1)\ldots(e_k + 1)$$

In the case of the number 1729, having its prime factors as $7^1 \bullet 13^1 \bullet 19^1$, we can calculate the total number of divisors as: $d(1729) = (1+1)(1+1)(1+1) = 2 \bullet 2 \bullet 2 = 8$.

This shows the beauty of the structure of numbers and highlights the importance of combinatorics in the context of number theory (Greaves, 2013). Each combination of the exponents of the prime factors gives a unique divisor, and counting the possible combinations gives us the total number of divisors.

In conclusion, the number 1729 is not just a simple number but represents a meeting point between various branches of mathematics. Its connection to prime numbers, the fundamental theorem of arithmetic, and counting divisors gives us deeper insight into fundamental mathematical structures.

REFERENCES

Atkin, A., & Bernstein, D. (2004). Prime sieves using binary quadratic forms. *Mathematics of Computation, 73*(246), 1023–1030. DOI: 10.1090/S0025-5718-03-01501-1

Berman, G., & Fryer, K. D. (2014). *Introduction to combinatorics*. Elsevier.

Bonnetain, X., Bricout, R., Schrottenloher, A., & Shen, Y. (2020, December). Improved classical and quantum algorithms for subset-sum. In *International Conference on the Theory and Application of Cryptology and Information Security* (pp. 633-666). Cham: Springer International Publishing. DOI: 10.1007/978-3-030-64834-3_22

Cameron, P. J. (1994). *Combinatorics: topics, techniques, algorithms*. Cambridge University Press.

Chmielewski, L. J., Orłowski, A., & Gawdzik, G. (2017). Segment counting versus prime counting in the Ulam square. In *Intelligent Information and Database Systems: 9th Asian Conference, ACIIDS 2017, Kanazawa, Japan, April 3–5, 2017* [Springer International Publishing.]. *Proceedings, 9*(Part II), 227–236.

Colbourne, C., & Dinitz, J. (Eds.). (2007). *Handbook of combinatorial designs* (pp. 0018-9448). Boca Raton, FL: CRC press.

da Fontoura Costa, L. (2023). Prime Numbers Networks: Visualizing and Characterizing Relationships Between Prime Numbers.

Desale, S., Rasool, A., Andhale, S., & Rane, P. (2015). Heuristic and meta-heuristic algorithms and their relevance to the real world: A survey. *Int. J. Comput. Eng. Res. Trends, 351*(5), 2349–7084.

Erdös, P., & Szekeres, G. (1935). A combinatorial problem in geometry. *Compositio Mathematica, 2*, 463–470.

Ghidarcea, M., & Popescu, D. (2024). Sieve of Atkin Revisited. *Sci. Bull. Univ. Politeh. Buchar, 86*, 15–26.

Graham, R. L., Knuth, D. E., Patashnik, O., & Liu, S. (1989). Concrete mathematics: A foundation for computer science. *Computers in Physics, 3*(5), 106–107. DOI: 10.1063/1.4822863

Greaves, G. (2013). *Sieves in number theory* (Vol. 43). Springer Science & Business Media.

Hahn, H. K., & Sachs, R. (2008). The distribution of prime numbers on the square root spiral. *arXiv preprint arXiv:0801.1441*.

Holt, F. B. (2015). Combinatorics of the gaps between primes. *arXiv preprint arXiv:1510.00743*.

Loehr, N. (2017). *Combinatorics*. CRC Press.

Lovász, L. (2007). *Combinatorial problems and exercises* (Vol. 361). American Mathematical Soc.

Lovász, L., & Prömel, H. J. (2004). Combinatorics. *Oberwolfach Reports, 1*(1), 5–110. DOI: 10.4171/owr/2004/01

Maynard, J. (2019). The twin prime conjecture. *Japanese Journal of Mathematics, 14*(2), 175–206. DOI: 10.1007/s11537-019-1837-z

Mazur, D. R. (2022). *Combinatorics: a guided tour* (Vol. 55). American Mathematical Society.

Pólya, G., Tarjan, R. E., & Woods, D. R. (2013). *Notes on introductory combinatorics* (Vol. 4). Springer Science & Business Media.

Talbi, E. G. (2009). *Metaheuristics: from design to implementation*. John Wiley & Sons. DOI: 10.1002/9780470496916

Vega, F. (2023). *The smallest gap between primes*. Qeios.

Vilenkin, N. Y. (2014). *Combinatorics*. Academic Press.

Wells, D. (2011). *Prime numbers: the most mysterious figures in math*. Turner Publishing Company.

Zhang, Y. (2014). Bounded gaps between primes. *Annals of Mathematics*, *179*(3), 1121–1174. DOI: 10.4007/annals.2014.179.3.7

Chapter 5
Examples of Combinatorial Problems

ABSTRACT

Combinatorial problems are mathematical challenges that involve finding optimal solutions while satisfying specific constraints. The chapter explores various problems, including the Longest Common Subsequence (LCS), used in bioinformatics and text analysis, and clustering, which groups data based on similarities. It also covers the Minimum Spanning Tree (MST) problem, crucial for efficient networks, and the Traveling Salesman Problem (TSP), a classic optimization issue. Other problems discussed include the maximum flow in a graph, which is used for network and traffic optimization, and the maximum cut problem, which is applicable in social network analysis. The chapter also examines algorithms for maximum node independence in graphs, maximum coverage, stable matching (Gale-Shapley problem), and even soccer team formation strategies. Finally, it explores efficient algorithms for solving these problems, demonstrating their practical applications in various fields.

INTRODUCTION

Not everything that counts can be counted, and not everything that can be counted counts

Albert Einstein

Combinatorial challenges are, in essence, large-scale puzzles that require a successful outcome to adhere to a specific set of rules. Imagine, for instance, the task of assigning duties where each individual holds distinct preferences. Combinatorial methods facilitate the identification of the most effective solution. Practical instances abound; consider the intricacies of planning travel itineraries, assembling a football team's lineup, or curating a restaurant menu.

Numerous examples of complex combinatorial endeavours are currently active subjects of research within the mathematical community. Nonetheless, specific common characteristics stand out:

1. A finite set of elements can be people, objects, numbers, etc.
2. There are constraints or restrictions that the solutions must meet in order to satisfy certain conditions.
3. We try to find the best solution, which can be the one that maximizes or minimizes a specific function.

Here are some examples that refer to different areas:

The *Longest Common Sequence* (LCS) is a classic. It consists of finding the most extended sequence common to two (or more) strings while maintaining the relative order of the characters, but not necessarily the contiguity. The problem has several practical applications: in bioinformatics, it enables the comparison of DNA sequences and the identification of similarities between genes, while in text comparison, it is used to detect changes between different versions of a document.

The problem can be solved using several algorithmic techniques, including dynamic programming, which has a complexity of $O(n \quad m)$.

DOI: 10.4018/979-8-3373-3089-1.ch005

Program 25 longest common sequence

```
import random
import string
def generate_random_string(length):
# Generate a random string of uppercase letters
letters = string.ascii_uppercase
return ''.join(random.choice(letters) for _ in range(length))
def longest_common_subsequence(X, Y):
# Lengths of the two strings
n = len(X)
m = len(Y)
# Create a 2D table (n+1) x (m+1) initialized to zero for dynamic programming
dp = [[0] * (m + 1) for _ in range(n + 1)]
# Fill the dp table
for i in range(1, n + 1):
for j in range(1, m + 1):
# If the characters of the two strings match
if X[i - 1] == Y[j - 1]:
dp[i][j] = dp[i - 1][j - 1] + 1
else:
dp[i][j] = max(dp[i - 1][j], dp[i][j - 1])
# The length of the longest common subsequence is stored in dp[n][m]
length_lcs = dp[n][m]
# Reconstruct the longest common subsequence
lcs = []
i, j = n, m
while i > 0 and j > 0:
# If the characters match, they are part of the LCS
if X[i - 1] == Y[j - 1]:
lcs.append(X[i - 1])
i -= 1
j -= 1
# If they don't match, move in the direction with the higher value
elif dp[i - 1][j] >= dp[i][j - 1]:
i -= 1
else:
j -= 1
# Reverse the sequence since it was constructed backwards
lcs.reverse()
# Convert the list of characters to a string and return it along with the length
return length_lcs, ''.join(lcs)
# Generate two random strings
X = generate_random_string(20)
Y = generate_random_string(20)
# Compute the LCS
length, lcs = longest_common_subsequence(X, Y)
# Print the results
```

```
print(f"String 1: {X}")
print(f"String 2: {Y}")
print(f"The length of the longest common subsequence is: {length}")
print(f"The longest common subsequence is: {lcs}")
```

An interesting variant more suitable for comparing texts at the level of words and not of individual characters, returning a sequence of common words in the correct order is the following:

Program 26 longest common sequence for words

```
def longest_common_subsequence_words(X, Y):
# Split the texts into lists of words
words1 = X.split()
words2 = Y.split()
# Lengths of the word lists
n = len(words1)
m = len(words2)
# Create a 2D table (n+1) x (m+1) to store the results
dp = [[0] * (m + 1) for _ in range(n + 1)]
# Fill the dp table
for i in range(1, n + 1):
for j in range(1, m + 1):
# If the words are the same
if words1[i - 1] == words2[j - 1]:
dp[i][j] = dp[i - 1][j - 1] + 1
else:
dp[i][j] = max(dp[i - 1][j], dp[i][j - 1])
# The length of the longest common subsequence is in dp[n][m]
lcs_length = dp[n][m]
# Reconstruct the longest common subsequence
lcs_words = []
i, j = n, m
while i > 0 and j > 0:
if words1[i - 1] == words2[j - 1]:
lcs_words.append(words1[i - 1])
i -= 1
j -= 1
elif dp[i - 1][j] >= dp[i][j - 1]:
i -= 1
else:
j -= 1
# Reverse the sequence because it was constructed backwards
lcs_words.reverse()
# Convert the array of words into a string and return it along with the length
return lcs_length, ' '.join(lcs_words)
# Take input for two texts
text1 = input("Enter the first text: ")
text2 = input("Enter the second text: ")
```

```
# Compute the LCS on words
length, lcs = longest_common_subsequence_words(text1, text2)
# Print the results
print(f"\nText 1: {text1}")
print(f"Text 2: {text2}")
print(f"\nThe length of the longest common subsequence in terms of words is: {length}")
print(f"The longest common subsequence is: {lcs}")
```

LCS identification has many practical applications. In the academic and journalistic fields, comparing texts makes it possible to detect potential cases of plagiarism. Verification tools compare the content of a document with that of other existing texts, identifying the longest common sequences. When the number of words in common exceeds a certain threshold, the system can flag the document for further checks. In software version management, comparing code files helps you identify changes over time.

An interesting application is in bioinformatics, where genetic sequences are compared and similarities between genes are identified. In this case, we compare strings of letters representing nucleotide bases. Biologists then have a tool to identify common sequences between two DNA samples, suggesting possible evolutionary or functional relationships between species.

In philology, comparing ancient manuscripts and their later copies can provide valuable information on the transmission of texts. Scholars can thus compare variants and identify changes, thus reconstructing the original text more precisely.

The *Clustering Problem*: Python, as we have seen, is an exceptional tool for dealing with a wide range of combinatorial situations, thanks to the availability of numerous dedicated libraries. Let us see how one can group a set of data into subsets so that items within the same cluster are more similar to each other than items in other clusters. Imagine having a bag full of coloured balls of different sizes. Clustering is like taking this bag and splitting it up so that each subset contains balls that are similar to one another.

There are many clustering algorithms, each with their strengths and weaknesses. One of the most common is K-means. Let us look at a concrete example: a company wants to segment its customers into distinct groups based on two characteristics: income and age. The goal is to find similar customer groups to design targeted marketing strategies.

Program 27 clustering

```
import numpy as np
import matplotlib.pyplot as plt
from sklearn.cluster import KMeans
from sklearn.datasets import make_blobs
# Generate sample data: 300 customers with income and age
X, y_true = make_blobs(n_samples=300, centers=4, cluster_std=1.0, random_state=42)
# Create a KMeans model with 4 clusters (since we know there are 4 distinct groups)
kmeans = KMeans(n_clusters=4, n_init=10, random_state=42)
# Fit the model to the data
kmeans.fit(X)
# Get the cluster labels
y_kmeans = kmeans.predict(X)
# Visualize the results
plt.figure(figsize=(10, 8))
# Plot the data points, colored by cluster
plt.scatter(X[:, 0], X[:, 1], c=y_kmeans, s=50, cmap='viridis', label='Customer Points')
# Plot the cluster centers
centers = kmeans.cluster_centers_
```

```
plt.scatter(centers[:, 0], centers[:, 1], c='black', s=200, alpha=0.75, marker='X', la-
bel='Cluster Centers')
    # Add labels and a legend
    plt.title('Customer Segmentation with KMeans')
    plt.xlabel('Income')
    plt.ylabel('Age')
    plt.legend()
    plt.grid(True)
    plt.show()
```

The resulting graph shows customers coloured according to their cluster and cluster centres marked with black crosses. Each colour represents a group of customers with similar characteristics, while cluster centres indicate the midpoint of each group.

Figure 1. Clustering

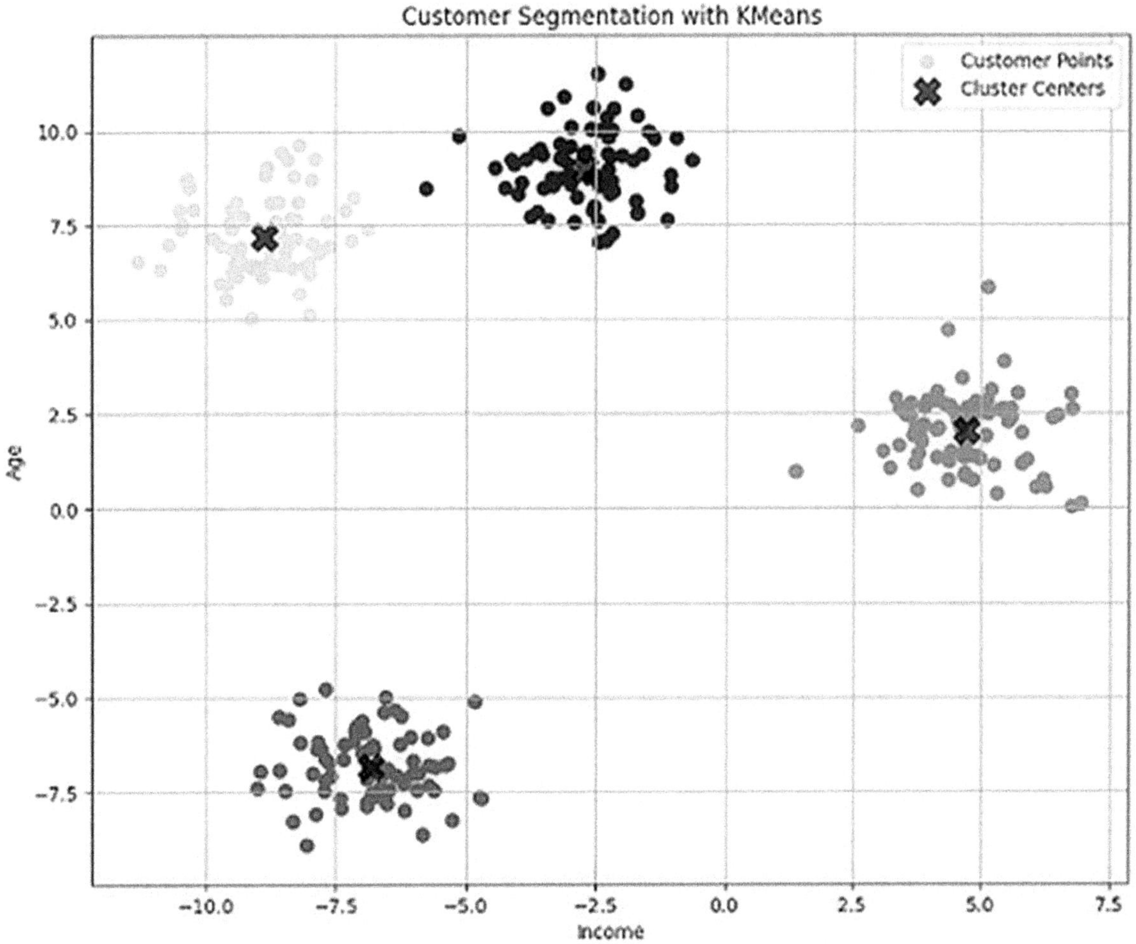

The *Minimum Spanning Tree* (MST) involves finding a tree, i.e., a subset of the edges of an undirected and weighted graph, that connects all the vertices while minimising the overall weight of the edges used. This problem is fundamental in computer science and graph theory (Diestel, 2024), as it has numerous practical applications, including designing efficient networks, reducing transportation costs, and optimising resource management (Ali & Sik, 2012).

Several combinatorial algorithms have been developed to solve the Minimum Spanning Tree (MST) problem (Wang et al., 2021). Among the most well-known and widely used is Kruskal's algorithm (1928–2010). The algorithm begins by sorting all the edges of the graph in ascending order according to their weights. It then incrementally adds edges to the spanning tree, ensuring that no cycles are formed in the process (Kruskal, 1956). Specifically, an edge is added only if it connects two distinct components of the graph—that is, vertices that are not yet part of the same connected subgraph. This process continues until all vertices are connected, guaranteeing that the resulting spanning tree has minimum total weight.

Program 28 MST

```
import numpy as np
import networkx as nx
import matplotlib.pyplot as plt
def generate_random_adjacency_matrix(num_nodes, max_weight=10):
# Create a random adjacency matrix with float values
matrix = np.random.uniform(1, max_weight + 1, size=(num_nodes, num_nodes))
# Make the matrix symmetric (undirected graph)
matrix = np.triu(matrix)
matrix += matrix.T
# Replace zeros with infinity to represent the absence of edges
matrix[matrix == 0] = np.inf
return matrix
def create_graph_from_matrix(matrix):
G = nx.Graph()
num_nodes = matrix.shape[0]
for i in range(num_nodes):
for j in range(i + 1, num_nodes):
weight = matrix[i, j]
if weight < np.inf:
G.add_edge(i, j, weight=weight)
return G
def draw_graph_with_mst(G, mst):
pos = nx.spring_layout(G) # Layout for graph drawing
plt.figure(figsize=(10, 8))
# Draw the original graph
nx.draw(G, pos, with_labels=True, node_color='grey', node_size=500, edge_color='gray',
font_size=16, font_weight='bold')
# Draw the minimum spanning tree
nx.draw_networkx_edges(G, pos, edgelist=mst.edges(), edge_color='grey', width=4)
# Display edge weights with two decimal points
edge_labels = nx.get_edge_attributes(G, 'weight')
formatted_edge_labels = {k: f"{v:.2f}" for k, v in edge_labels.items()}
nx.draw_networkx_edge_labels(G, pos, edge_labels=formatted_edge_labels, font_col-
or='black')
plt.title('Minimum Spanning Tree')
plt.show()
def main():
```

```
# Number of nodes in the graph
num_nodes = 5
# Generate a random adjacency matrix
matrix = generate_random_adjacency_matrix(num_nodes)
# Create the graph from the adjacency matrix
G = create_graph_from_matrix(matrix)
# Compute the minimum spanning tree using Kruskal's algorithm
mst = nx.minimum_spanning_tree(G)
# Print the adjacency matrix
matrix_rounded = np.round(matrix, 2)
print("Adjacency matrix with weights rounded to two decimal places:")
print(matrix_rounded)
# Print the weights of the edges in the minimum spanning tree
mst_weights = nx.get_edge_attributes(G, 'weight')
print("\nWeights of the edges in the minimum spanning tree:")
mst_weights_rounded = {k: round(v, 2) for k, v in mst_weights.items()}
print(mst_weights_rounded)
# Draw the graph and the minimum spanning tree
draw_graph_with_mst(G, mst)
if __name__ == "__main__":
main()
```

Figure 2. Minimum spanning tree

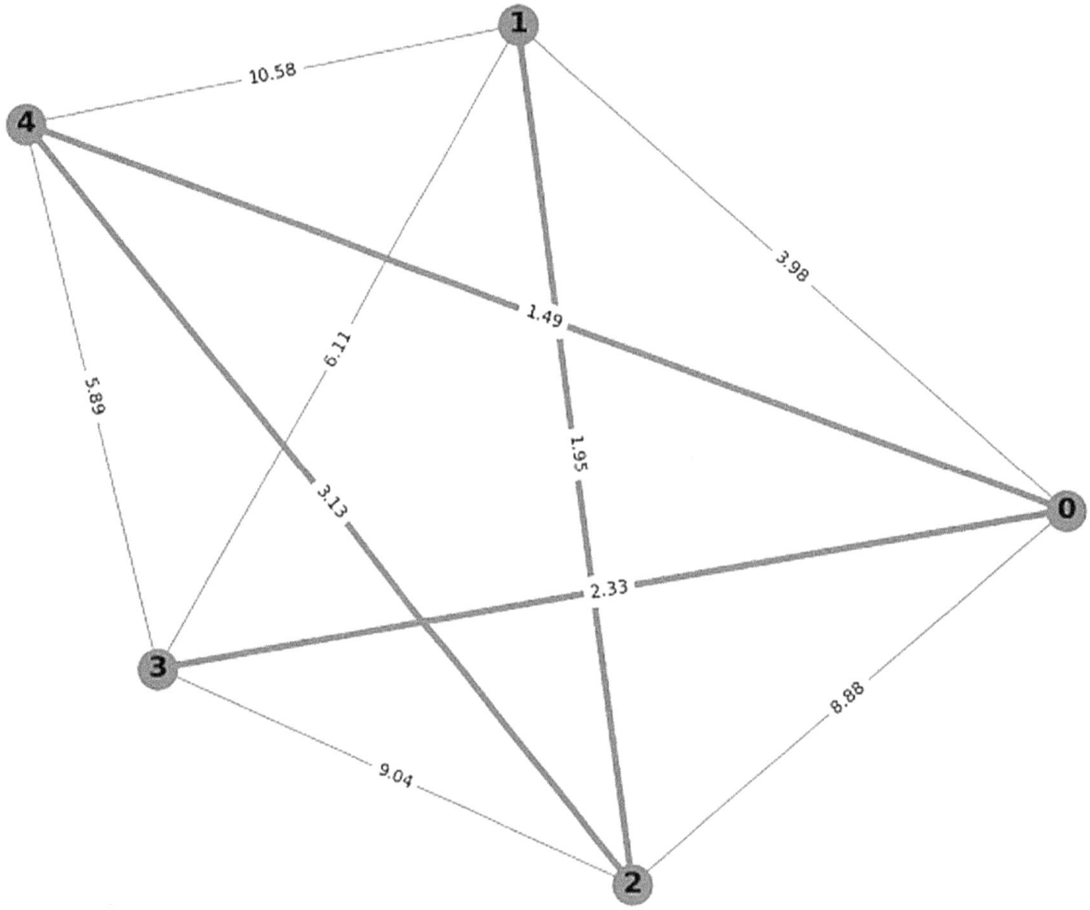

A possible interpretation of the graph is that it is necessary to select a minimum number of emergency stations to cover as many areas of the city as possible.

Imagine that the graph represents a city, where the nodes represent areas, and the edges represent connections between them, with the weights indicating the cost of each connection.

In addition to Kruskal's algorithm, another widely used procedure to solve the least spanning tree problem is Prim's algorithm. This algorithm builds the spanning tree incrementally, starting from a single vertex and subsequently adding the lightest edges that connect the vertices already included in the tree to those not yet included. Both are efficient and find application in different contexts, depending on the specific needs of the scenario.

Both Kruskal and Prim are efficient algorithms for finding a graph's MST. The former is preferable for sparse graphs, while the latter is generally more efficient for dense graphs, particularly when the number of edges is significantly greater than the number of vertices.

Program 29 Prim's algorithm

```
import networkx as nx
```

```
import matplotlib.pyplot as plt
import heapq
import time
def prim_step_by_step(graph):
num_vertices = len(graph)
in_mst = [False] * num_vertices
min_edge = [float('inf')] * num_vertices
parent = [-1] * num_vertices
min_edge[0] = 0
min_heap = [(0, 0)]
mst_edges = []
# Create the graph using NetworkX
G = nx.Graph()
# Add nodes and edges to the graph
for u in graph:
for v, w in graph[u]:
G.add_edge(u, v, weight=w)
# Position of the nodes
pos = nx.spring_layout(G)
# Initialize the figure
plt.figure(figsize=(8, 6))
nx.draw(G, pos, with_labels=True, node_color='lightblue', node_size=700, font_size=15)
labels = nx.get_edge_attributes(G, 'weight')
nx.draw_networkx_edge_labels(G, pos, edge_labels=labels)
plt.title("Building the Minimum Spanning Tree (MST)")
plt.ion() # Enable interactive mode
plt.show()
# Step-by-step MST construction process
while min_heap:
weight, u = heapq.heappop(min_heap)
if in_mst[u]:
continue
in_mst[u] = True
if parent[u] != -1:
mst_edges.append((parent[u], u))
# Update the graph
plt.clf() # Clear the current figure
nx.draw(G, pos, with_labels=True, node_color='lightblue', node_size=700, font_size=15)
labels = nx.get_edge_attributes(G, 'weight')
nx.draw_networkx_edge_labels(G, pos, edge_labels=labels)
nx.draw_networkx_edges(G, pos, edgelist=mst_edges, edge_color='black', width=2)
plt.title(f"Step {len(mst_edges)}: Add edge ({parent[u]}, {u})")
plt.draw()
plt.pause(1) # Wait 1 second before the next step
for v, w in graph[u]:
if not in_mst[v] and w < min_edge[v]:
min_edge[v] = w
parent[v] = u
heapq.heappush(min_heap, (w, v))
```

```
# Keep the window open after the last step
plt.ioff() # Disable interactive mode
plt.show()
# Example graph represented as an adjacency list
graph = {
0: [(1, 2), (3, 6)],
1: [(0, 2), (2, 3), (3, 8), (4, 5)],
2: [(1, 3), (4, 7)],
3: [(0, 6), (1, 8), (4, 9)],
4: [(1, 5), (2, 7), (3, 9)]
}
# Compute the minimum spanning tree using Prim's algorithm with step-by-step visualization
prim_step_by_step(graph)
```

The *Travelling Salesman Problem* (TSP) is a classic combinatorial optimisation problem (Vesselinova et al., 2020; Kool et al., 2022), which, despite its apparent simplicity, has captured the attention of numerous researchers over the years. It can be defined as the task of finding the least-cost Hamiltonian path in a complete and weighted graph. In other words, given a set of cities represented by nodes and the distances between them represented by the weights on the arcs, we try to determine the optimal order to visit each city only once and return to the starting point, minimising the total distance travelled.

TSP represents a paradigmatic case of combinatorial optimisation (Yang & Whinston, 2023), where the goal is to find a solution among a finite set of possible solutions that grows enormously with the increase in the number of cities. With four cities, there are 24 possible combinations, but with 40 cities, the number of possible routes increases to approximately 8.16×10^{47}.

The applications of TSP are numerous and transversal to different sectors, from logistics and route planning to integrated circuit design and bioinformatics. This issue is significant in logistics and information technology, with applications in various fields, including vehicle route planning and integrated circuit design. TSP can also be used to determine the optimal sequence of gene reads in a DNA sequencing experiment.

The TSP is an NP-hard problem, which means that no efficient algorithm can solve large instances in polynomial time. This computational intractability has prompted the development of a wide range of exact and approximate procedures. Exact algorithms, such as branch-and-bound programming and dynamic programming, are guaranteed to reach the optimal solution; however, they are often prohibitively slow for large instances. In contrast, approximate algorithms, such as the nearest neighbour heuristic, 2-opt, and metaheuristics, including simulated annealing and genetic algorithms, provide suboptimal solutions in a reasonable amount of time, making them more suitable for practical applications. The so-called "nearest neighbour" algorithm captures the problem well and is used to find an approximate optimal route. The algorithm is a simple heuristic that starts from any city and, at each step, chooses to visit the nearest unvisited city. While it does not guarantee the achievement of an optimal solution, it often produces high-quality results relatively quickly, making it a valuable algorithm for large-scale challenges.

Program 30 TSP

```
import numpy as np
import matplotlib.pyplot as plt
def distance(city1, city2):
return np.linalg.norm(city1 - city2)
def nearest_neighbor(nodes):
num_nodes = len(nodes)
unvisited_nodes = set(range(1, num_nodes))
```

```
tour = [0] # The tour starts at city 0
total_distance = 0.0
labels = {}
while unvisited_nodes:
current_city = tour[-1]
closest_city = min(unvisited_nodes, key=lambda city: distance(nodes[current_city],
nodes[city]))
total_distance += distance(nodes[current_city], nodes[closest_city])
labels[(tour[-1], closest_city)] = f"{distance(nodes[current_city], nodes[closest_
city]):.2f}"
tour.append(closest_city)
unvisited_nodes.remove(closest_city)
# Return to the starting point
tour.append(tour[0])
total_distance += distance(nodes[tour[-2]], nodes[tour[-1]])
labels[(tour[-2], tour[-1])] = f"{distance(nodes[tour[-2]], nodes[tour[-1]]):.2f}"
return tour, total_distance, labels
def plot_nodes_and_tour(nodes, tour, total_distance, labels):
x = nodes[:, 0]
y = nodes[:, 1]
plt.figure(figsize=(8, 6))
# Show city points
plt.scatter(x, y, color='blue', label='Cities')
# Draw all edges in gray
num_nodes = len(nodes)
for i in range(num_nodes):
for j in range(i + 1, num_nodes):
plt.plot([x[i], x[j]], [y[i], y[j]], color='gray', linestyle='--', alpha=0.5)
label = f"{distance(nodes[i], nodes[j]):.2f}"
plt.text((x[i] + x[j]) / 2, (y[i] + y[j]) / 2, label, color='gray')
# Draw the optimal path in black
tour_x = [x[i] for i in tour]
tour_y = [y[i] for i in tour]
plt.plot(tour_x, tour_y, color='black', linestyle='-', marker='o', label='Optimal Path')
# Add labels to cities
for i, txt in enumerate(tour[:-1]):
plt.annotate(f'{txt}', (x[txt], y[txt]), textcoords="offset points", xytext=(0, 5),
ha='center')
# Add labels to edges in the optimal path
for edge, label in labels.items():
plt.text((x[edge[0]] + x[edge[1]]) / 2, (y[edge[0]] + y[edge[1]]) / 2, label, col-
or='green')
plt.title('Traveling Salesman Problem')
plt.xlabel('Coord. X')
plt.ylabel('Coord. Y')
plt.legend()
plt.grid(True)
plt.show()
# Input: Number of cities
```

```
while True:
try:
num_nodes = int(input("Enter the number of cities: "))
if num_nodes <= 0:
print("Please enter a valid number of cities.")
else:
break
except ValueError:
print("Please enter a valid number of cities.")
# Generate cities with random coordinates
nodes = np.random.rand(num_nodes, 2)
# Compute the optimal path using the Nearest Neighbor algorithm
tour, total_distance, labels = nearest_neighbor(nodes)
# Print the optimal path and the total distance
print("Optimal path:", tour)
print("Total path distance:", total_distance)
# Draw the set of cities and the optimal path with distances and gray edges
plot_nodes_and_tour(nodes, tour, total_distance, labels)
```

Randomly generated distances can lead to a result like this:

Figure 3. Optimal path

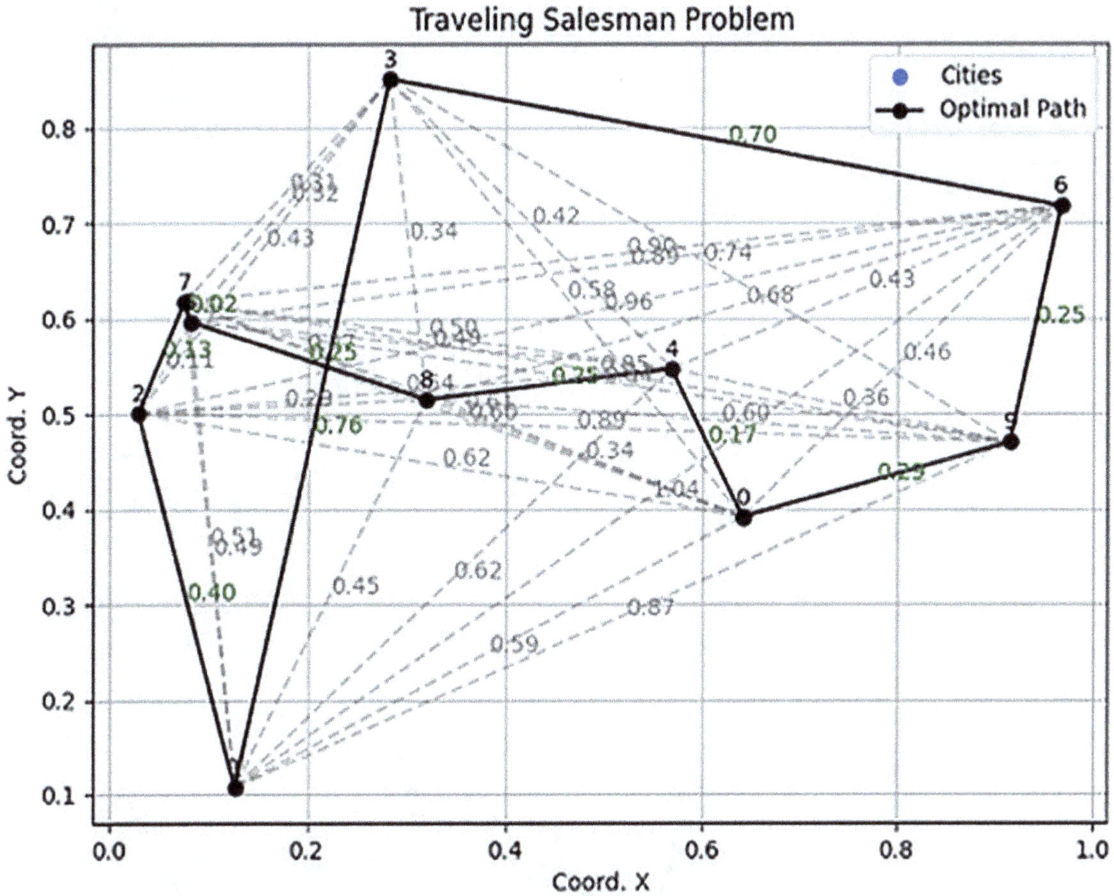

An intriguing possibility is to use a program like GeoGebra (Falcón Ganfornina & Rios Collantes de Terán, 2015). The software enables the creation of points on the Cartesian plane to represent cities and immediately visualises the calculated route. It is more intuitive to understand the problem and the solutions obtained. Additionally, we can adjust the position of the points (cities) and observe how the optimal route changes. This interactivity makes it easy to explore different configurations. GeoGebra has useful commands, such as TravelerSalesperson (List of Points). The procedure is simple:

1. Insert the map
2. Each city is associated with a point, and a List of points is created
3. In the Input bar, type TravelerSalesperson(l1)
4. GeoGebra provides the optimal route
5. Moving points recalculate the optimal route

In the figure, the result of a simulation in which the minimum route between different European capitals is:

Figure 4. The salesman problem

The *Maximum Flow Problem* (MFP): Given a directed graph with capacitance on its edges, find the maximum flow from the source node to the destination node. The goal is to find the flow that maximizes the total flow from the source to the well in a network, given a set of constraints dictated by the capacities of the arcs. Exploring the set of all possible admissible flows naturally leads to a combinatorial problem.

This particular challenge is widely used in graph theory (Wilson, 1979), with numerous applications in linear programming and operations research (Dantzig, 2002; Hillier & Lieberman, 2015), e.g., designing telecommunications networks (Prim, 1957), traffic management (land and air), material management in an industrial plant, and many other applications. There are several algorithms, for example, the Edmonds-Karp algorithm:

Program 31 Edmonds-Karp algorithm

```
import numpy as np
from collections import deque
def bfs(capacity, flow, source, sink, n):
queue = deque([source])
paths = {source: []}
while queue:
u = queue.popleft()
for v in range(n):
if capacity[u][v] - flow[u][v] > 0 and v not in paths:
paths[v] = paths[u] + [(u, v)]
if v == sink:
return paths[v]
queue.append(v)
return None
def max_flow(capacity):
num_nodes = len(capacity)
flow = np.zeros((num_nodes, num_nodes), dtype=int)
source = 0
sink = num_nodes - 1
while True:
path = bfs(capacity, flow, source, sink, num_nodes)
if not path:
break
flow_increment = min(capacity[u][v] - flow[u][v] for u, v in path)
for u, v in path:
flow[u][v] += flow_increment
flow[v][u] -= flow_increment
return np.sum(flow[source])
# Input: Number of nodes
while True:
try:
num_nodes = int(input("Enter the number of nodes: "))
if num_nodes <= 0:
print("Please enter a valid number of nodes.")
else:
break
except ValueError:
```

```
print("Please enter a valid number of nodes.")
# Generate capacity matrix with random values
capacity = np.random.randint(0, 21, size=(num_nodes, num_nodes))
# Set the diagonal to zero (no cycles between nodes)
np.fill_diagonal(capacity, 0)
# Print the capacity matrix
print("Capacity Matrix:")
print(capacity)
# Compute the maximum flow
max_flow_value = max_flow(capacity)
print("The maximum flow is", max_flow_value)
```

This algorithm is a more efficient implementation of the Ford-Fulkerson algorithm. It uses breadth-first search (BFS) to find the increment path with the maximum remaining capacity, which ensures that each increment is as large as possible for each iteration. In practice, it finds the minimum capacity along the path, increases the flow along the entire path by this amount, and finally updates the residual capacities of the arcs along the path and the reverse arcs. It has a time complexity of $O(VE^2)$, where V is the number of vertices and E is the number of edges in the graph. It ensures that the maximum flow is found in polynomial time and is best suited for practical applications.

In a computer network, the algorithm can be used to determine the maximum amount of data that can be transmitted between two network nodes to optimize bandwidth usage and prevent congestion. In social networks, where nodes represent people and arcs represent social connections, can be used to analyse the maximum flow of information or influences from one individual or group of individuals to another. This can help to understand how information spreads within a social network. The following program shows an example of the algorithm applied to this situation.

Program 32 social network

```
import random
import networkx as nx
import matplotlib.pyplot as plt
from collections import deque
def bfs(capacity, source, sink, parent):
visited = [False] * len(capacity)
queue = deque([source])
visited[source] = True
while queue:
u = queue.popleft()
for v, cap in enumerate(capacity[u]):
if not visited[v] and cap > 0:
queue.append(v)
visited[v] = True
parent[v] = u
if v == sink:
return True
return False
def edmonds_karp(capacity, source, sink):
parent = [-1] * len(capacity)
max_flow = 0
while bfs(capacity, source, sink, parent):
```

```
path_flow = float('Inf')
s = sink
while s != source:
path_flow = min(path_flow, capacity[parent[s]][s])
s = parent[s]
v = sink
while v != source:
u = parent[v]
capacity[u][v] -= path_flow
capacity[v][u] += path_flow
v = parent[v]
max_flow += path_flow
return max_flow
def generate_adjacency_matrix(num_nodes, max_capacity):
matrix = [[0] * num_nodes for _ in range(num_nodes)]
for i in range(num_nodes):
for j in range(i + 1, num_nodes):
capacity = random.randint(0, max_capacity)
matrix[i][j] = capacity
matrix[j][i] = 0
return matrix
def visualize_network(capacity, nodes, max_flow):
G = nx.DiGraph()
for i in range(len(capacity)):
for j in range(len(capacity[i])):
if capacity[i][j] > 0:
G.add_edge(nodes[i], nodes[j], capacity=capacity[i][j])
pos = nx.spring_layout(G)
edges = G.edges(data=True)
capacities = [d['capacity'] for (u, v, d) in edges]
# Draw the nodes and edges
nx.draw(G, pos, with_labels=True, node_color='lightblue', node_size=3000, font_size=15,
font_weight='bold')
nx.draw_networkx_edges(G, pos, edgelist=edges, width=capacities, edge_color='grey')
# Add labels to edges
edge_labels = {(u, v): f"{d['capacity']}" for u, v, d in edges}
nx.draw_networkx_edge_labels(G, pos, edge_labels=edge_labels, label_pos=0.3, font_col-
or='black', font_size=10)
plt.title(f"Social Network - Maximum Flow: {max_flow}")
plt.show()
# Social network parameters
num_nodes = 6
max_capacity = 20
capacity = generate_adjacency_matrix(num_nodes, max_capacity)
nodes = ["A", "B", "C", "D", "E", "F"]
source = 0
sink = 5
max_flow = edmonds_karp(capacity, source, sink)
# Print the adjacency matrix
```

```
print("Adjacency Matrix (capacity):")
for row in capacity:
print(row)
print(f"\nThe maximum flow of information from {nodes[source]} to {nodes[sink]} is: {max_
flow}")
# Visualize the social network and the maximum flow
visualize_network(capacity, nodes, max_flow)
```

The *Maximum Cut Problem* (Max-Cut): Can be used to detect distinct communities or groups in social network theory. By separating a graph into subgroups with the highest share, one gets groups with strong connections within and weak connections between them. Since polynomial time-solving algorithms are unknown, the main difficulties lie in the explosive growth of possible partitions. The number of ways to partition n vertices into two sets is 2^n, but this count includes both trivial partitions (where one set is empty) and considers swapping the sets as different partitions. To adjust for this, we subtract 2 (for the trivial partitions) and divide by 2 to account for the symmetry (i.e., swapping Set 1 with Set 2 should not result in a new partition). Thus, the number of distinct ways to partition the n vertices into two non-empty sets is $2^{n-1}-1$.

Therefore, it is necessary to resort to approximation algorithms for large graphs. A program can simulate the maximum cut problem using a randomly generated partition approach. The idea is to randomly divide the graph's vertices into two sets and calculate the sum of the weights of the edges that cross the cut (Pemantle, 2012). Repeating this process several times, the program approximates the optimal solution.

Program 33 the maximum cut problem

```
import random
import networkx as nx
import matplotlib.pyplot as plt
# Function to generate a random graph with weighted edges
def generate_random_graph(num_nodes, edge_prob, min_weight, max_weight):
# Use NetworkX to create a random graph
G = nx.gnp_random_graph(num_nodes, edge_prob)
# Add random weights to edges
graph = {}
for u, v in G.edges():
weight = random.randint(min_weight, max_weight)
graph[(u, v)] = weight
return graph
# Function to calculate the weight of the cut given a set of nodes S
def calculate_cut_weight(graph, set_s):
cut_weight = 0
for (u, v), weight in graph.items():
# If one node is in S and the other is outside, it is a cut edge
if (u in set_s and v not in set_s) or (u not in set_s and v in set_s):
cut_weight += weight
return cut_weight
# Function to find an approximate solution to the maximum cut problem
def maximum_cut(graph, num_iterations=1000):
nodes = set()
for edge in graph.keys():
```

```
nodes.update(edge)
best_cut_weight = 0
best_cut = None
# Perform the calculation for a defined number of iterations
for _ in range(num_iterations):
# Random partition of nodes into two sets
set_s = set(node for node in nodes if random.choice([True, False]))
# Calculate the cut weight for this partition
current_cut_weight = calculate_cut_weight(graph, set_s)
# If the solution is better, update it
if current_cut_weight > best_cut_weight:
best_cut_weight = current_cut_weight
best_cut = set_s
return best_cut, best_cut_weight
# Visualization of the cut
def visualize_cut(graph, set_s, best_cut_weight):
# Create the graph using NetworkX
G = nx.Graph()
for (u, v), weight in graph.items():
G.add_edge(u, v, weight=weight)
# Set the layout of the graph
pos = nx.spring_layout(G)
# Create a figure with larger dimensions
plt.figure(figsize=(10, 8)) # Size of the plot window
# Determine node colors based on the partition
node_colors = ['lightgrey' if node in set_s else 'darkgrey' for node in G.nodes()]
# Draw the nodes
nx.draw_networkx_nodes(G, pos, node_color=node_colors, node_size=500)
nx.draw_networkx_labels(G, pos, font_size=10, font_color='black')
# Draw the edges
edge_colors = []
for (u, v, d) in G.edges(data=True):
# Color cut edges gray and other edges black
if (u in set_s and v not in set_s) or (v in set_s and u not in set_s):
edge_colors.append('grey') # Cut edges
else:
edge_colors.append('black') # Non-cut edges
# Draw the edges and add weight labels
nx.draw_networkx_edges(G, pos, edge_color=edge_colors, width=2)
nx.draw_networkx_edge_labels(G, pos, edge_labels={(u, v): d['weight'] for u, v, d in
G.edges(data=True)})
# Add the legend
handles = [
plt.Line2D([0], [0], marker='o', color='w', markerfacecolor='lightgrey', markersize=10,
label='Nodes in maximum cut'),
plt.Line2D([0], [0], marker='o', color='w', markerfacecolor='darkgrey', markersize=10, la-
bel='Nodes not in maximum cut'),
plt.Line2D([0], [0], color='grey', lw=4, label='Cut edges'),
plt.Line2D([0], [0], color='black', lw=4, label='Non-cut edges'),
```

```
]
plt.legend(handles=handles, loc='lower left')
# Show the graph
plt.title(f"Maximum Cut: {best_cut_weight}")
plt.axis('off') # Hide axes
plt.show()
# Parameters for the random graph
num_nodes = 10 # Number of nodes
edge_prob = 0.3 # Probability of an edge between two nodes
min_weight = 1 # Minimum edge weight
max_weight = 10 # Maximum edge weight
# Generate the random graph
graph = generate_random_graph(num_nodes, edge_prob, min_weight, max_weight)
# Find the maximum cut
best_cut, best_cut_weight = maximum_cut(graph)
# Visualize the maximum cut found
visualize_cut(graph, best_cut, best_cut_weight)
```

The graphical representation of the solution can certainly help the understanding of the problem.

Figure 5. Maximum cut

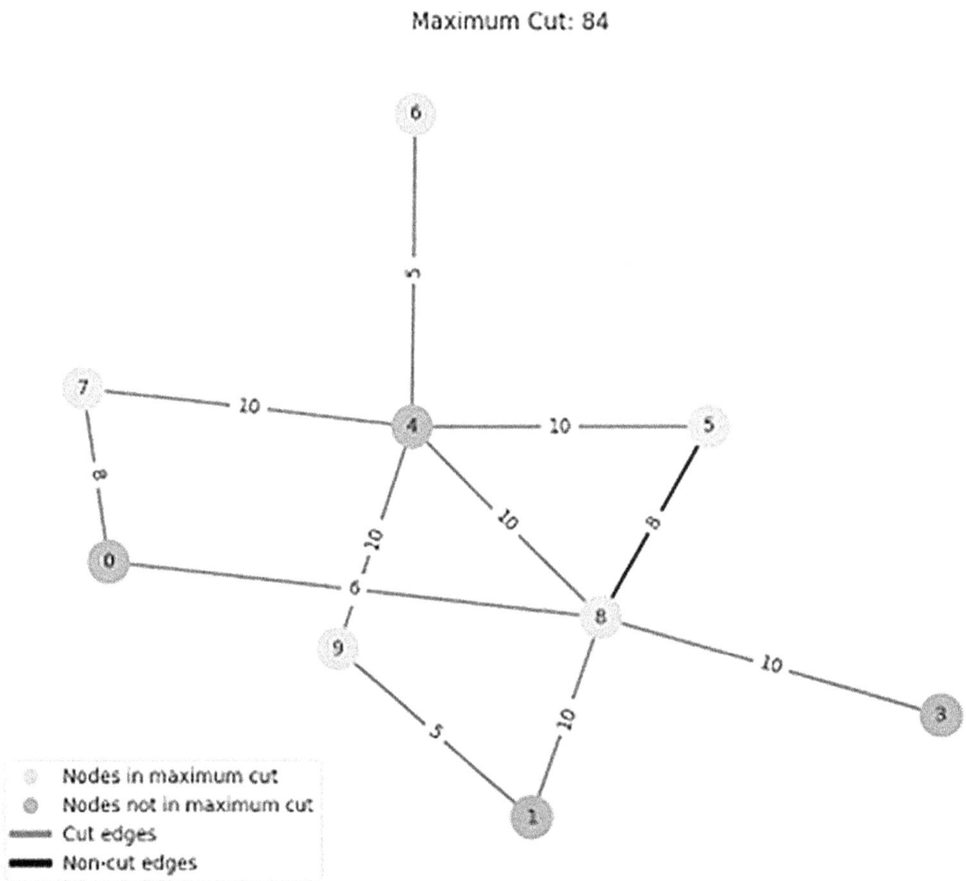

One can explain the maximum cut as finding the optimal way to divide a network into isolated groups.

For the following subsequence problem, one can think of a concrete situation, such as finding the most interesting 5-minute part in a 10-minute video.

The *problem of the sum of subsequences*: Given a set of numbers, we want to find the subsequence of maximum length whose sum is maximum. Of course, there are many possible cases, continuous or not. Suppose we have the sequence of the first ten natural numbers taken with alternating signs [1, -2, 3, -4, 5, -6, 7, -8, 9, -10]; the combinations (as we will see later) are 1023, while the possible subsequences are $10 \bullet (10 + 1)/2 = 55$.

The key difference between the two concepts is that consecutive elements of the array must form continuous sub-vectors, while combinations do not have this restriction. Manually calculating all the subsequences would be a very laborious process. For the solution, once the length of the desired subsequence has been established, an approach based on a variant of Kadane's algorithm is used. Given the length 5, we obtain the array [5, -6, 7, -8, 9] with sum 7 in the proposed example. This problem has important applications in computer science, particularly in designing algorithms (Skiena, 1998) for natural language recognition and computer vision. In some machine learning applications can be used to extract patterns from complex datasets.

Program 34 Kadane algorithm

```
def kadane_algorithm_fixed_length():
# Take the length of the subarray as input
while True:
try:
k = int(input("Enter the length of the subarray: "))
if k <= 0:
print("The length of the subarray must be positive.")
else:
break
except ValueError:
print("Enter a valid number for the subarray length.")
# Take the array components as input
while True:
try:
array = list(map(int, input("Enter the array elements separated by space: ").split()))
if len(array) < k:
print("The array must contain at least 'k' elements.")
else:
break
except ValueError:
print("Enter only valid numbers for the array elements.")
# Calculate the sum of the first subarray of length k
current_max = sum(array[:k])
global_max = current_max
start_index = 0
# Traverse the array to find the maximum sum subarray of length k
for i in range(k, len(array)):
current_max = current_max - array[i - k] + array[i]
if current_max > global_max:
global_max = current_max
start_index = i - k + 1
```

```
# Return the maximum sum and the subarray with the maximum sum
max_subarray = array[start_index:start_index + k]
return f"The maximum sum is {global_max} and the subarray with the maximum sum is {max_
subarray}."
print(kadane_algorithm_fixed_length())
```

The algorithm has a time complexity of *O(n)*, where *n* is the length of the array. A practical problem might be the following: considering a random array of profits and losses, we have to find the continuous sequence of days that provides the maximum total profit over a month. Initially, the maximum value coincides with the first element, then you iterate through the array and update the total maximum value if the current value exceeds it.

Program 35 profits

```
import random
def kadane(profits):
current_max = profits[0]
total_max = profits[0]
for i in range(1, len(profits)):
current_max = max(profits[i], current_max + profits[i])
total_max = max(total_max, current_max)
return total_max
def generate_profits(days, min_value=-10, max_value=20):
return [random.randint(min_value, max_value) for _ in range(days)]
def main():
days = 31 # Number of days in a month
profits = generate_profits(days)
# Print the daily profits and losses array
print("Daily profits/losses for a month:")
print(profits)
# Calculate and print the maximum achievable profit
max_profit = kadane(profits)
print(f"\nThe maximum achievable profit for this month is: {max_profit}")
if __name__ == "__main__":
main()
```

The *Zero-Sum Subset Problem* (ZSSP) is a subset of a given set of numbers whose total value is zero. In other words, given a set $S = \{a_1, a_2, \ldots, a_n\}$, we look for a non-empty subset of elements $S' \subseteq S$ such that the sum of the elements in S' is equal to zero. The goal of finding a zero-sum subset can be solved using several approaches, including exhaustive search (brute-force) or optimised algorithms that utilise partial sums and hashing. As it iterates through the array, the algorithm calculates a cumulative sum, and the dictionary stores each partial sum associated with the index in which it first appears. The zero-sum subset problem is a combinatorial challenge that involves finding a subset of numbers that meets a specific condition: the sum of the numbers equals zero (Hardy & Wright, 1979).

It is a classic example of a combination search task. Since the possible count of subsets grows exponentially with the number of elements, simply checking every combination becomes unfeasible very quickly. To efficiently solve this, an effective strategy involves combining partial sums with hashing, enabling a significantly faster approach than exhaustive enumeration.

Program 36 zero-sum subset problem

```
def subset_sum_zero(arr):
# Dictionary to store partial sums and indices
partial_sums = {}
current_sum = 0
# Iterate through each element of the array
for i, num in enumerate(arr):
# Add the current number to the cumulative sum
current_sum += num
# If the current sum is zero, print the subset from position 0 to i
if current_sum == 0:
subset = arr[:i+1]
print(f"Subset with sum zero found: {' + '.join(map(str, subset))} = 0")
return True
# If the current sum is already in the dictionary, print the subset with sum zero
if current_sum in partial_sums:
# Find the corresponding subset
start_index = partial_sums[current_sum] + 1
subset = arr[start_index:i+1]
print(f"Subset with sum zero found: {' + '.join(map(str, subset))} = 0")
return True
# Store the current sum in the dictionary
partial_sums[current_sum] = i
# If no subset with sum zero is found
print("No subset with sum zero found")
return False
# Input the array from user input
array = list(map(int, input("Enter the elements of the array separated by spaces:
").split()))
# Call the function
subset_sum_zero(array)
```

The zero-sum subset problem has applications in several fields. For example, it could be used by a company trying to locate suspicious financial transactions or balance items in a balance sheet. Business accounting requires that various entries and expenses be recorded, and the goal is to identify groups of transactions that cancel each other out, i.e., whose sum is zero. This procedure is essential for identifying posting errors or discrepancies that indicate unauthorised movements or suspicious transactions.

Another example would be the following: in energy analysis models, balancing different sources and consumption can be helpful to ensure that the sum of the input and output energies is zero. Suppose a system has various energy components, and we want to know if there are entirely balanced configurations. Using the zero-sum subset model, we can identify groups of energy inputs and outputs that balance each other, thereby allowing us to optimise energy flows.

The *Maximum Independent Set Problem* (MIS): Given a graph, one must find the maximum subset of nodes not connected by edges. In other words, given a graph G = (V, E) where V is the set of vertices and E is the set of edges, it is necessary to determine the largest subset of vertices $V' \subseteq V$ such that no pair of vertices in V' is connected by an arc in E. This problem has many applications in graph theory, game theory, and theoretical computer science (Braverman et al., 2024).

A simple example clarifies the concept: suppose we have an unoriented graph with nodes 0, 1, 2, 3 and edges (0, 1), (0, 2), (1, 2), (2, 3).

Figure 6. Graph

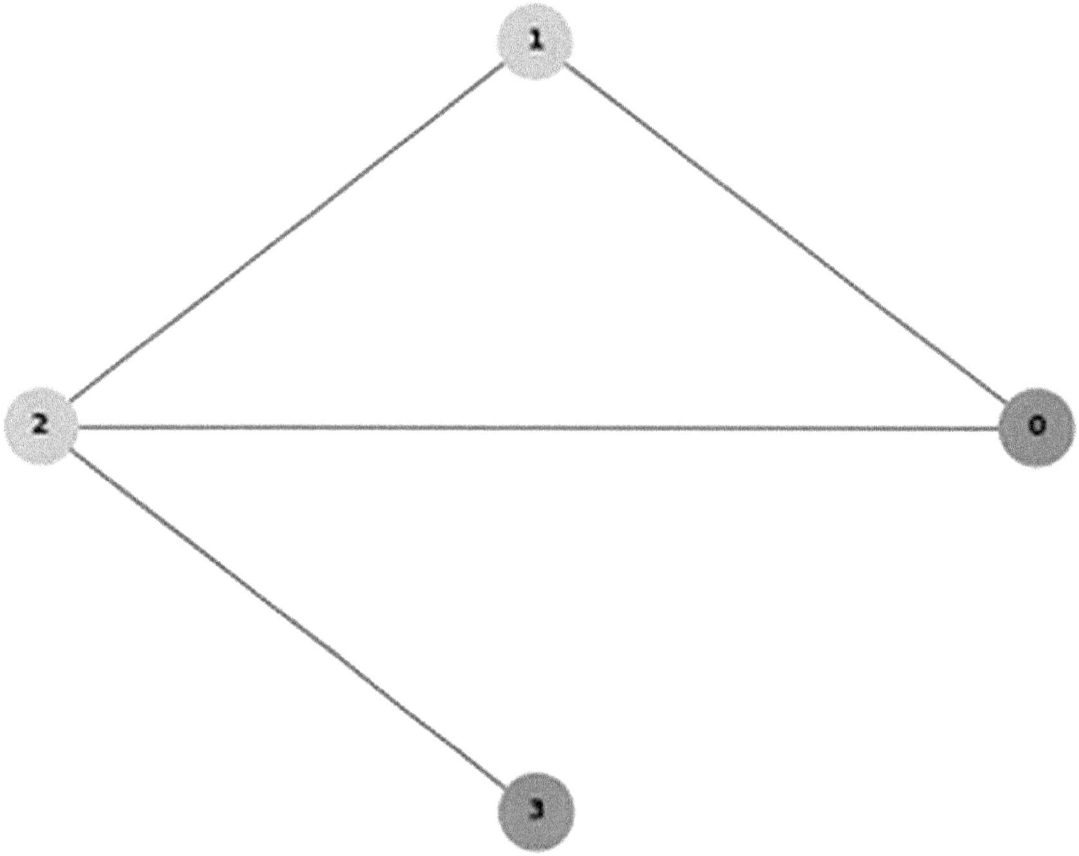

This corresponds to the matrix

0, 1, 1, 0

1, 0, 1, 0

1, 1, 0, 1

0, 0, 1, 0

The goal is to find the maximum independent set in this graph. In other words, we want to find the most extensive set of nodes where an arc connects no pairs of nodes. The solution is {1, 4}. In this set, there are no arcs between any pair of nodes, so it is a combination of maximum independence. This is a case of NP-hard. However, there are approximation algorithms that can provide reasonable solutions. The following program draws the graph and determines the nodes with the highest approximate independence from a randomly generated adjacency matrix.

Program 37 maximum independence

```
import numpy as np
import networkx as nx
import matplotlib.pyplot as plt
from networkx.algorithms.approximation import maximum_independent_set
def create_graph_from_matrix(matrix):
graph = nx.Graph()
num_nodes = len(matrix)
for i in range(num_nodes):
for j in range(i + 1, num_nodes):
if matrix[i][j] == 1:
graph.add_edge(i, j)
return graph
def visualize_graph(graph, independent_set):
pos = nx.circular_layout(graph)
plt.figure(figsize=(10, 8))
# Draw the graph
nx.draw(graph, pos, with_labels=True, font_weight='bold', node_color='grey', node_
size=800, font_size=10)
# Highlight the independent set
nx.draw_networkx_nodes(graph, pos, nodelist=independent_set, node_color='red', node_
size=800)
plt.title('Graph with Approximate Independent Set')
plt.show()
def main():
# Input: Number of nodes
while True:
try:
num_nodes = int(input("Enter the number of nodes: "))
if num_nodes <= 0:
print("Please enter a positive number of nodes.")
else:
break
except ValueError:
print("Please enter a valid number of nodes.")
# Generate the adjacency matrix with random 0 or 1 values
adjacency_matrix = np.random.randint(2, size=(num_nodes, num_nodes))
# Ensure the adjacency matrix is symmetric (undirected graph)
adjacency_matrix = np.triu(adjacency_matrix, k=1)
adjacency_matrix += adjacency_matrix.T
# Print the adjacency matrix
print("Adjacency Matrix:")
print(adjacency_matrix)
# Create the graph
graph = create_graph_from_matrix(adjacency_matrix)
# Find an approximate independent set
independent_set = maximum_independent_set(graph)
# Print the independent set
print("Approximate Independent Set:", independent_set)
# Visualize the graph
```

```
visualize_graph(graph, independent_set)
if __name__ == "__main__":
main()
```

The *Maximum Coverage Problem* (MCP): This combinatorial optimization problem arises in several contexts, such as allocating resources or selecting elements to maximize a more extensive set's coverage (McGregor et al., 2021). Given a set and a subsets of these elements, one must find the maximum subset of subsets covering all elements. Maximum coverage is critical in many real-world applications where decisions must be made about how to allocate limited resources.

It is an NP-hard problem, and there is no polynomial algorithm that solves it exactly; however, a "greedy" algorithm exists that guarantees an approximation of the optimal solution. To better understand the concept, consider an example: suppose we have a set {1,2,3,4,5} and a collection of subsets {{1,2},{2,3},{3,4},{4,5}} with associated costs {3,2,4,1}. The goal is to select subsets that maximise the coverage of the set at the lowest overall cost. The solution is [[1, 2], [3, 4], [4, 5]] with cost 8. This problem finds applications in many fields.

In experimental design, it helps choose the most effective set of experiments to cover a range of conditions. In genome sequencing, it applies to selecting DNA sequences, often called primers. Here, the subsets correspond to short primer sequences, and the challenge is to select a set of primers that covers as many target DNA sequences of interest as possible, considering that each primer incurs a cost.

A program that implements a greedy algorithm:

Program 38 maximum coverage

```
def maximum_coverage(elements, subsets, costs):
covered_elements = set()
selected_subsets = []
total_cost = 0
while covered_elements < elements:
# Find the subset with the maximum number of uncovered elements per unit of cost.
best_subset, best_cost = max(
zip(subsets, costs),
key=lambda x: len(set(x[0]) - covered_elements) / x[1]
)
# Update the covered elements
covered_elements.update(best_subset)
# Add the selected subset and its cost to the total
selected_subsets.append(best_subset)
total_cost += best_cost
# Remove the selected subset to avoid duplicate selections
subsets.remove(best_subset)
costs.remove(best_cost)
# Check to avoid infinite loops in case of duplicate subsets or zero costs
if not costs:
break
return selected_subsets, total_cost
# Test Program
elements = {1, 2, 3, 4, 5}
subsets = [{1, 2}, {2, 3}, {3, 4}, {4, 5}]
costs = [1, 2, 3, 4] # Add costs for each subset
selected_subsets, total_cost = maximum_coverage(elements, subsets, costs)
```

```
print("Elements:", elements)
print("Subsets:", subsets)
print("Costs:", costs)
print("Selected subsets:", selected_subsets)
print("Total cost:", total_cost)
```

The *Stable Marriage Problems* (SMP): This is also known as the "Gale-Shapley Problem". It concerns the stable matching between two groups of people in which everyone has preferences over those of the other group. Given a set of *a*'s and *b*'s, finding a configuration of couples in which no one prefers another partner over the current one is necessary. There is an algorithm that solves the problem of stable mates efficiently (Gokhale et al., 2024). There are two groups, and each person has a list of preferences over the other group; the goal is to make everyone in a "stable" way in the sense that a pairing is considered as such if there are no two people who would prefer to be paired with each other over their current partners. It starts with each element of the set a that proposes element b at the top of its list; if the latter is free, it accepts (temporarily); otherwise, it chooses between the new proposer and the current partner based on its preferences.

The process continues as long as everyone is paired. The algorithm is optimal for group *a* and has a complexity of $O(n^2)$ in the worst case. Lloyd Shapley (1923-2016) and Alvin Roth (1951-) won the Nobel Prize in Economics in 2012 for their work on the theory of stable allocations.

Program 39 Gale-Shapley

```
def gale_shapley(preferences_a, preferences_b):
a_free = list(preferences_a.keys())
pairs = {}
while a_free:
a = a_free.pop(0)
preferences = preferences_a[a]
for b in preferences:
current_preferences_b = preferences_b[b]
current_pair = pairs.get(b)
if not current_pair:
pairs[b] = a
break
elif current_preferences_b.index(current_pair) > current_preferences_b.index(a):
pairs[b] = a
a_free.append(current_pair)
break
return pairs
def main():
n_a = int(input("Enter the number of a: "))
n_b = int(input("Enter the number of b: "))
preferences_a = {}
preferences_b = {}
# Enter preferences
for i in range(n_a):
preferences = input(f"Enter the preferences for a{i+1} (e.g., b1 b2 ...): ").split()
preferences_a[f'a{i+1}'] = preferences
for i in range(n_b):
preferences = input(f"Enter the preferences for b{i+1} (e.g., a1 a2 ...): ").split()
```

```
preferences_b[f'b{i+1}'] = preferences
result = gale_shapley(preferences_a, preferences_b)
# Print the result
print("The result of the Gale-Shapley algorithm is: ")
for b, a in result.items():
print(f"{b} is paired with {a}")
if __name__ == "__main__":
main()
```

The *Problem of Football Formations*: How many ways are there to field eleven players in a football team? Of course, it depends on the coach's strategy, but how can we calculate all the options? The formation is the arrangement of a team's players on the field. Excluding the goalkeeper, there are ten players to be placed on the pitch. In general, we can distinguish between defenders, midfielders (in turn, it is possible to make the distinction between defensive and offensive), and forwards. The question has two aspects: one tactical and the other combinatorial.

How many solutions are there, especially in the practice of the game? The problem is not trivial, as historically, we have transitioned from the 1-0-9 module of the 1850-1860 era to the more recent 5-3-2 (or from 2-3-2-3 to 3-2-4-1 in the case of quadripartition). If we dwell on the scheme that does not distinguish between midfielders, it is possible to say that there are 58,024 ways in which players can be arranged. In the other case, the modalities become 989,526. Here are the two programs:

Program 40 football modules 1

```
from math import comb
def formations():
total = 0
# Iterate over the number of defenders possible
for d in range(1, 10): # Number of defenders, from 1 to 9
# Iterate over the number of midfielders possible
for c in range(0, 10 - d + 1): # Number of midfielders, remaining spots
# Calculate the number of forwards left
a = 10 - d - c
# Check that the number of forwards is valid
if a >= 0:
# Calculate the number of combinations for each role
defender_combinations = comb(10, d)
midfielder_combinations = comb(10 - d, c)
forward_combinations = comb(10 - d - c, a)
# Add the product of the combinations to the total
total += defender_combinations * midfielder_combinations * forward_combinations
return total
# Print the total number of possible formations
print(f"The total number of formations is {formations()}.")
```

Program 41 football modules 2

```
from math import comb
```

```
def formations():
total = 0
# Iterate over the number of defenders
for d in range(1, 10): # Number of defenders, from 1 to 9
# Iterate over the number of defensive midfielders
for cd in range(0, 10 - d + 1): # Number of defensive midfielders
# Iterate over the number of offensive midfielders
for co in range(0, 10 - d - cd + 1): # Number of offensive midfielders
# Calculate the number of forwards left
a = 10 - d - cd - co
# Check that the number of forwards is valid
if a >= 0:
# Calculate the number of combinations for each role
defender_combinations = comb(10, d)
defensive_midfielder_combinations = comb(10 - d, cd)
offensive_midfielder_combinations = comb(10 - d - cd, co)
forward_combinations = comb(10 - d - cd - co, a)
# Add the product of the combinations to the total
total += defender_combinations * defensive_midfielder_combinations * offensive_midfield-
er_combinations * forward_combinations
return total
# Print the total number of possible formations
print(f"The total number of formations is {formations()}.")
```

To know the distributions, one needs to be able to print all the formations, at least those most used over time, as shown in the table:

Table 1. Formation with a midfield line

Form	Year of first use
1-0-9	1850
1-1-8	1860
1-2-7	1870
2-2-6	1870
2-3-5	1880
3-2-5	1920
3-4-3	1920
3-3-4	1950
4-2-4	1950
4-3-3	1960
4-4-2	1960
4-5-1	1960
5-3-2	1960
5-4-1	1970
3-5-2	1980
3-6-1	2000

Program 42 midfield module 1

```
from math import comb
# Considered formations
module = [
(1, 0, 9),
(1, 1, 8),
(1, 2, 7),
(2, 2, 6),
(2, 3, 5),
(3, 2, 5),
(3, 3, 4),
(3, 4, 3),
(3, 5, 2),
(3, 6, 1),
(4, 2, 4),
(4, 3, 3),
(4, 4, 2),
(4, 5, 1),
(5, 3, 2),
(5, 4, 1),
]
for formation in module:
d, m, a = formation
combinations = comb(10, d) * comb(10 - d, m) * comb(10 - d - m, a)
print(f"Defenders: {d}, Midfielders: {m}, Forwards: {a}, Combinations: {combinations}")
```

A subdivision that includes two types of midfielders leads to the table:

Table 2. Formation with two midfield lines

Form	Year of first use
3-2-3-2	1950
4-1-3-2	1960
4-3-2-1	1970
3-5-1-1	1980
5-3-1-1	1980
3-3-1-3	1990
3-4-1-2	2000
4-2-3-1	2000
4-3-1-2	2000
3-4-2-1	2010
4-1-4-1	2020

Consequently, the program becomes:

Program 43 midfield module 2

```
from math import comb
# Considered formations
formations = [
(3, 2, 3, 2),
(3, 3, 1, 3),
(3, 4, 1, 2),
(3, 4, 2, 1),
(3, 5, 1, 1),
(4, 1, 3, 2),
(4, 1, 4, 1),
(4, 2, 3, 1),
(4, 3, 2, 1),
(4, 3, 1, 2),
(5, 3, 1, 1),
]
def calculate_combinations():
for d, cd, co, a in formations:
total_combinations = comb(10, d) * comb(10 - d, cd) * comb(10 - d - cd, co) * comb(10 - d
- cd - co, a)
print(f"Formation: ({d}, {cd}, {co}, {a}) - Number of ways: {total_combinations}")
calculate_combinations()
```

For example, the formation (4-2-3-1) indicates that there are four defenders (D), two defensive midfielders (DM), three attacking midfielders (AM), and one striker (S). There are C(10, 4) ways to place the four defenders in 10 available positions. After placing the defenders, we have six positions for the defensive midfield players, so there are C(6, 2) ways to fill them. We now have four positions for attacking midfielders, so there are C(4, 3) ways. Finally, there is only one way to position the attacker.

The total number of ways to form this formation is the product of these combinations:

C(10, 4) • C(6, 2) • C(4, 3) • 1 = 210 • 15 • 4 • 1 = 12,600.

REFERENCES

Ali, M. A., & Sik, Y. H. (2012). Transportation problem: A special case for linear programing problems in mining engineering. *International Journal of Mining Science and Technology*, *22*(3), 371–377. DOI: 10.1016/j.ijmst.2012.04.015

Braverman, V., Dharangutte, P., Shah, V., & Wang, C. (2024). Learning-augmented maximum independent set. *arXiv preprint arXiv:2407.11364.*

Dantzig, G. B. (2002). Linear programming. *Operations Research*, *50*(1), 42–47. DOI: 10.1287/opre.50.1.42.17798

Diestel, R. (2024). Extremal graph theory. In *Graph theory* (pp. 179–226). Springer Berlin Heidelberg.

Falcón Ganfornina, R. M., & Rios Collantes de Terán, R. (2015). The use of GeoGebra in Discrete Mathematics. *GeoGebra International Journal of Romania*, *4*(1), 39–50.

Gokhale, K., Mallik, A. K., Misra, A. K., & Nath, S. (2024). A Gale-Shapley View of Unique Stable Marriages. In *ECAI 2024* (pp. 3652–3659). IOS Press. DOI: 10.3233/FAIA240922

Hardy, G. H., & Wright, E. M. (1979). *An introduction to the theory of numbers*. Oxford university press.

Hillier, F. S., & Lieberman, G. J. (2015). *Introduction to operations research*. McGraw-Hill.

Kool, W., van Hoof, H., Gromicho, J., & Welling, M. (2022, June). Deep policy dynamic programming for vehicle routing problems. In *International conference on integration of constraint programming, artificial intelligence, and operations research* (pp. 190-213). Cham: Springer International Publishing. DOI: 10.1007/978-3-031-08011-1_14

Kruskal, J. B. (1956). On the shortest spanning subtree of a graph and the traveling salesman problem. *Proceedings of the American Mathematical Society*, *7*(1), 48–50. DOI: 10.1090/S0002-9939-1956-0078686-7

McGregor, A., Tench, D., & Vu, H. T. (2021). Maximum coverage in the data stream model: Parameterized and generalized. *arXiv preprint arXiv:2102.08476.*

Pemantle, R. (2012). A Survey of Random Processes with Reinforcement. *Probability Surveys*, *9*, 116–158.

Prim, R. C. (1957). Shortest connection networks and some generalizations. *The Bell System Technical Journal*, *36*(6), 1389–1401. DOI: 10.1002/j.1538-7305.1957.tb01515.x

Skiena, S. S. (1998). *The algorithm design manual* (Vol. 2). New York: springer.

Vesselinova, N., Steinert, R., Perez-Ramirez, D. F., & Boman, M. (2020). Learning combinatorial optimization on graphs: A survey with applications to networking. *IEEE Access : Practical Innovations, Open Solutions*, *8*, 120388–120416. DOI: 10.1109/ACCESS.2020.3004964

Wang, Y., Yu, S., Gu, Y., & Shun, J. (2021, June). Fast parallel algorithms for euclidean minimum spanning tree and hierarchical spatial clustering. In *Proceedings of the 2021 international conference on management of data* (pp. 1982-1995). DOI: 10.1145/3448016.3457296

Wilson, R. J. (1979). *Introduction to graph theory*. Pearson Education India.

Yang, Y., & Whinston, A. (2023, July). A survey on reinforcement learning for combinatorial optimization. In *2023 IEEE World Conference on Applied Intelligence and Computing (AIC)* (pp. 131-136). IEEE DOI: 10.1109/AIC57670.2023.10263956

Chapter 6
The Sub-Disciplines

ABSTRACT

This chapter dives into the rich diversity of combinatorics, emphasizing how it goes far beyond simple combinatorial calculation. Six important subdisciplines are presented: combinatorial design theory, focused on the construction of structures with particular properties (such as block drawings); code theory, which uses combinatorics to design and analyse codes for reliable data transmission, with examples like Hamming and Golay codes; geometric combinatorics, which investigates the combinatorial properties of geometric objects; algebraic combinatorics, which links combinatorics to algebra to solve problems through algebraic tools and includes algorithms like Jarvis for convex hulls; topological combinatorics, which creates a bridge between combinatorics and the topological properties of objects and examines discrete structures using topological methods, such as Sperner's theorem and Delaunay triangulation; and finally, experimental design, which explores the link between combinatorics and the structuring of experiments to obtain meaningful results.

INTRODUCTION

Combinatorics is the mathematics of choices

David Gale

The discipline is vast and diverse and is not limited to the better-known enumerative combinatorics (Bóna, 2015), which is closely related to combinatorics, computational combinatorics, or combinatorial game theory. It encompasses numerous subfields and has practical applications in various scientific and technological fields, including the design of web search algorithms and DNA analysis.

The sub-disciplines are numerous; those covered are represented in the figure:

DOI: 10.4018/979-8-3373-3089-1.ch006

Figure 1. Map of some sub-disciplines

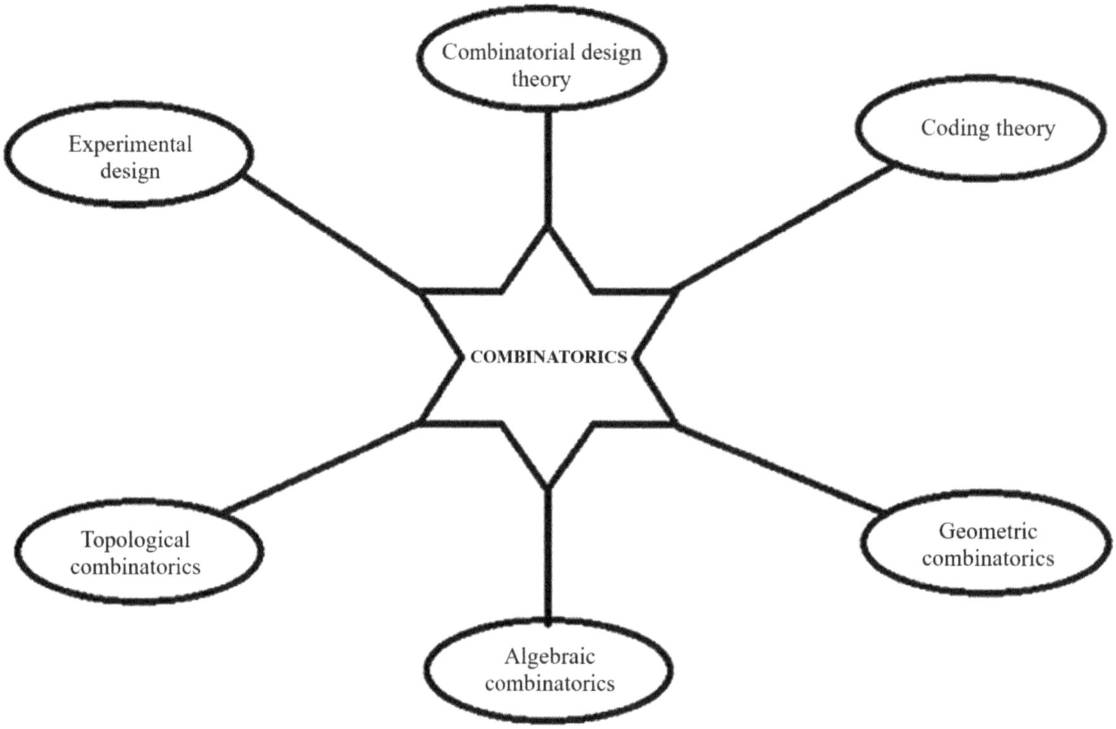

Six branches extend from the central node, each representing a specific area of application or study within combinatorics:

1. *Combinatorial design theory* focuses on constructing combinatorial structures with specific properties, such as block designs.
2. *Coding theory* applies combinatorics to the design and analysis of codes for reliable information transmission.
3. *Geometric combinatorics* studies the combinatorial properties of geometric objects.
4. *Algebraic combinatorics* bridges the gap between combinatorics and algebra, utilising algebraic tools to solve complex combinatorial problems.
5. *Topological combinatorics* explores the relationship between combinatorics and the topological properties of the studied objects.
6. *Experimental design* highlights the connection between combinatorics and the planning of experiments, where multiple factors are combined to obtain meaningful and statistically valid results.

Let us examine these six sub-disciplines.

Combinatorial design theory: involves constructing statistical experiments, known as "experimental designs," that satisfy specific properties (Dukes et al., 2008; Colbourn & Mathon, 2011). Combinatorial designs allow efficient experiments that systematically evaluate the effect of different variables with a minimum number of observations. This is useful in statistics and probability theory but also in the field of experimental psychology.

In experimental psychology, combinatorial designs serve to arrange experimental setups that test multiple independent variables.

In physics, these designs are applied to examine the configuration of fractals and to produce fractal constructs with defined characteristics. Fractals are geometric structures that show self-similarity at different scales.

A combinatorial design, in its most general form, is a pair (X, B) where X is a finite set of elements (called “points”), and B is a collection of subsets of X (called “blocks”) that satisfy certain symmetry and balancing properties. A simple mathematical example of a combinatorial design is an (e, b, n)-design, where *e* denotes the number of elements, *b* the number of blocks, and each b-subset of the *e* elements appears in exactly *n* blocks. Suppose we have six people who want to play any game. One way to do this is to use a (6, 2, 1)-design. In this case, the design blocks represent the pairs of players who must play together. One way to create a (6, 2, 1)-design is to generate six pairs of players. The number of possible designs can be determined by calculating combinations and is equal to 15.

The blocks of a project, generated according to the binomial coefficient, are printed by the following program:

Program 44 Combinatorial Design

```
from itertools import combinations
def generate_project(n, b):
elements = set(range(n))
project = [set(subset) for subset in combinations(elements, b)]
return project
def print_project(project):
for i, subset in enumerate(project, 1):
print(f"Subset {i}: {subset}")
def main():
n = int(input("Enter the number of elements in the project: "))
b = int(input("Enter the number of elements per subset: "))
e = int(input("Enter the number of common elements between each pair of subsets: "))
if b <= n and e < b:
project = generate_project(n, b)
print(f"Generated a valid ({n}, {b}, {e})-project:")
print_project(project)
else:
print("It is not possible to generate a valid project with the provided parameters.")
if __name__ == "__main__":
main()
```

Coding theory concerns the design and analysis of codes for transmitting information through potentially incorrect communication channels, has applications in many areas, including cryptography, telecommunications, and information theory (Golay, 1954; Ho & Lun, 2008; Huffman et al., 2021; Vermani, 2022). It is used to transmit digital data reliably through communication channels such as the Internet, cellular networks and coaxial cables. This theory is based on several fundamental concepts: a code is a set of strings constructed from a finite alphabet; the number of positions in which two codewords differ is called the Hamming distance, while the Hamming weight indicates the number of non-zero symbols in a codeword. Combinatorics is widely used to generate sets of valid codes.

For example, when people design error-correcting code, they must determine how to combine symbols or digits so that the code can detect or correct errors. There are two main categories of codes: error-correcting codes, designed to detect and correct errors during data transmission, and compression codes, designed to compress data efficiently without losing information. *Hamming code* is an error-correcting code that utilises combinatorial concepts in its design. This code is designed to detect and correct errors in bit sequences during transmission.

We can verify the speech using a program that implements the Hamming code (7,4), an error correction code that allows for the detection and correction of errors in binary data transmission. This code can correct a 7-bit error corresponding to a 2-(7,3,1) design. The Generating array G is used to encode the 4-bit message into a 7-bit code word. The H parity check matrix decodes the code word and detects and corrects errors. The encoding function takes a 4-bit message and encodes it, via an array-vector product, into a 7-bit code word. Finally, the transmission is simulated by adding a possible error to the code word and the received word is decoded, correcting a possible error of a single bit.

Program 45 Hamming Code

```
import numpy as np
import random
# Generator matrix for the Hamming (7,4) code
G = np.array([
[1, 0, 0, 0, 1, 1, 1],
[0, 1, 0, 0, 1, 1, 0],
[0, 0, 1, 0, 1, 0, 1],
[0, 0, 0, 1, 0, 1, 1]
])
# Parity-check matrix
H = np.array([
[1, 1, 1, 0, 1, 0, 0],
[1, 1, 0, 1, 0, 1, 0],
[1, 0, 1, 1, 0, 0, 1]
])
def encode(message):
return np.dot(message, G) % 2
def add_error(codeword, error_probability=0.1):
if random.random() < error_probability:
bit_to_flip = random.randint(0, 6)
codeword[bit_to_flip] ^= 1
return codeword
def decode(received):
syndrome = np.dot(H, received) % 2
error_position = int(''.join(map(str, syndrome[::-1])), 2) - 1
if error_position >= 0:
received[error_position] ^= 1
return received[:4] # Returns the first 4 bits (the original message)
def simulate_transmission(message):
print(f"Original message: {message}")
# Encoding
codeword = encode(message)
print(f"Codeword: {codeword}")
# Simulate transmission (possible error)
received = add_error(codeword.copy())
print(f"Received word: {received}")
# Decoding
decoded = decode(received)
print(f"Decoded message: {decoded}")
print(f"Correction successful: {np.array_equal(message, decoded)}")
# Example usage
```

```
message = np.array([1, 0, 1, 1])
simulate_transmission(message)
```

Another example is *Golay codes*. They are error corrector codes used to ensure accurate data transmission in the presence of disturbances. They are particularly appreciated for their efficiency in detecting and correcting errors, making them ideal for applications where reliability is critical, such as space communications or telecommunications networks. The mathematical structure of Golay's codes is extraordinarily rich and has interesting connections with other areas of mathematics, such as finite group theory and combinatorial geometry. Golay code (23,12) is a binary code that encodes 12 bits of information into 23 bits, corrects up to 3 errors in a 23-bit code, and detects up to 6 errors.

Marcel Golay (1902-1989) was a Swiss mathematician and engineer known for his contributions to the field of code theory, particularly for the development of the code of the same name but is also known for his work in algebraic geometry.

Program 46 Golay Code

```
import numpy as np
from itertools import combinations
class GolayCode23:
def __init__(self):
# Generator matrix for the Golay (23,12) code
self.generator_matrix = self._create_generator_matrix()
self.parity_check_matrix = self._create_parity_check_matrix()
self.syndrome_table = self._precompute_syndrome_table()
def _create_generator_matrix(self):
B = np.array([
[0,1,1,1,1,1,1,1,1,1,1],
[1,1,1,0,1,1,1,0,0,0,1],
[1,1,0,1,1,1,0,0,0,1,0],
[1,0,1,1,1,0,0,0,1,0,1],
[1,1,1,1,0,0,0,1,0,1,1],
[1,1,1,0,0,0,1,0,1,1,0],
[1,1,0,0,0,1,0,1,1,0,1],
[1,0,0,0,1,0,1,1,0,1,1],
[1,0,0,1,0,1,1,0,1,1,1],
[1,0,1,0,1,1,0,1,1,1,0],
[1,1,0,1,1,0,1,1,1,0,0],
[1,0,1,1,0,1,1,1,0,0,0]
])
return np.column_stack((np.eye(12), B))
def _create_parity_check_matrix(self):
B = self.generator_matrix[:, 12:]
return np.column_stack((B.T, np.eye(11)))
def _precompute_syndrome_table(self):
"""Precompute the syndrome table for all single, double, and triple errors."""
table = {}
n = 23
# Generate all single, double, and triple error patterns
for error_size in range(1, 4): # Single, double, and triple errors
for error_positions in combinations(range(n), error_size):
```

```
error = np.zeros(n, dtype=int)
for pos in error_positions:
error[pos] = 1
syndrome = tuple(np.dot(error, self.parity_check_matrix.T) % 2)
table[syndrome] = error
return table
def encode(self, message):
if len(message) != 12:
raise ValueError("The message must be 12 bits long")
return np.dot(message, self.generator_matrix) % 2
def decode(self, received):
if len(received) != 23:
raise ValueError("The encoded message must be 23 bits long")
# Compute the syndrome
syndrome = tuple(np.dot(received, self.parity_check_matrix.T) % 2)
if syndrome in self.syndrome_table:
# Correct the error
error = self.syndrome_table[syndrome]
corrected = (received + error) % 2
return corrected[:12]
# If syndrome is zero, no error
if all(bit == 0 for bit in syndrome):
return received[:12]
raise ValueError("Too many errors to correct")
def add_noise(self, codeword, error_rate):
noise = np.random.random(codeword.shape) < error_rate
return (codeword + noise) % 2
def binary_string_to_array(s):
return np.array([int(bit) for bit in s])
def array_to_binary_string(arr):
return ''.join(str(int(bit)) for bit in arr)
def get_valid_input():
while True:
message = input("Enter a 12-bit binary message (only 0s and 1s): ")
if len(message) == 12 and all(bit in '01' for bit in message):
return message
print("Invalid input. Make sure to enter exactly 12 bits (0 or 1).")
# Main function
def main():
golay = GolayCode23()
# Get the message from the user
original_message = get_valid_input()
message_array = binary_string_to_array(original_message)
# Encode
encoded = golay.encode(message_array)
print(f"\nOriginal message: {original_message}")
print(f"Encoded message: {array_to_binary_string(encoded)}")
# Simulate noise in the channel
error_rate = 0.1
```

```
received = golay.add_noise(encoded, error_rate)
print(f"Received message with errors: {array_to_binary_string(received)}")
# Decode
try:
decoded = golay.decode(received)
print(f"Decoded message: {array_to_binary_string(decoded)}")
if np.array_equal(message_array, decoded):
print("Decoding was successful: the original message was recovered.")
else:
print("Warning: the decoded message does not match the original.")
except ValueError as e:
print(f"Decoding error: {str(e)}")
if __name__ == "__main__":
main()
```

Geometric combinatorics: Merges geometry with combinatorics, tackling challenges that involve the spatial arrangement of objects, such as points, line segments, or polygons (Miller & Reiner, 2004).

Consider the classic problem of determining the maximum number of regions into which n lines can divide a plane. We can easily apply this to a concrete scenario, such as cutting a cake. With three straight cuts, one can slice a cake into a maximum of 7 pieces (not necessarily equal). With four cuts, one can achieve 11 slices. And with n cuts? The formula gives the solution: This result combines geometric concepts (intersection of lines) with a combinatorial counting problem.

$$F = \frac{n^2}{2} + \frac{n}{2} + 1$$

A particular enumerative aspect, in which the calculation of possible combinations depends on geometric reasoning, considers an important object, even if used less than in the past: the stamp. It is a notched rectangle of paper, printed on one side and gummed on the back, used for postage and correspondence services. Stamps are divided into two broad categories: ordinary and commemorative.

Suppose we have bought a whole sheet of 12 stamps (three lines and four columns), and you want to detach four of them so that they remain united on at least one side. In how many ways can this division be achieved?

According to the English mathematician Henry Dudeney (1857-1930), author of many logic games, there are five possible cases:

Figure 2. Five separation modes

The next enumerative problem requires knowing how many ways the five figures can be realized. Supposing twelve tiles are numbered progressively, the first shape, square, can be made in six ways:

Table 1. The square shape

1	2		2	3		3	4
5	6		6	7		7	8
5	6		6	7		7	8
9	10		10	11		11	12

The other figures can be made in 14, 28, 14, and 3 ways, so there are 65 possible cases.

Interestingly, the pieces are those of Tetris, a video game of logic, skill and reasoning invented in 1985 by Russian researcher Alexey Pazhitnov. In truth, there are seven pieces, but two more can be obtained by symmetry.

The convex envelope can also be considered within geometric combinatorics, although it is not one of the central topics of this discipline. Discrete geometry, which includes the study of polyhedra and polytopes, is a branch of combinatorics that intersects with the theory of convex envelopes. The convex envelope of a set of points can be seen as a convex polytope, an object studied in discrete geometry. Studies of convex polytopes, including the analysis of their faces, vertices, and edges, as well as their combinatorial representation, are typical topics in geometric combinatorics.

The Jarvis algorithm, also known as *gift wrapping*, finds the convex envelope of a finite set of points in a plane. The algorithm is based on "wrapping" the dots as if you were wrapping a gift. We start from a known point belonging to the convex envelope (often the leftmost point) and proceed iteratively by finding the successive vertices of the convex envelope. We start with the point with the least x coordinate, as it will undoubtedly be a vertex of the convex envelope. Then, for each point not yet included in the envelope, we determine the angle formed by the current line and the line connecting the current point to that point and select the point that forms the largest angle (i.e., the furthest clockwise) with respect to the current line.

Program 47 Jarvis Algorithm

```
import matplotlib.pyplot as plt
import random
def generate_random_points(n, min_range=0, max_range=10):
return [(random.uniform(min_range, max_range), random.uniform(min_range, max_range)) for _
in range(n)]
def leftmost_point(points):
return min(points, key=lambda p: (p[0], p[1]))
def orientation(p, q, r):
value = (q[1] - p[1]) * (r[0] - q[0]) - (q[0] - p[0]) * (r[1] - q[1])
if value == 0:
return 0
return 1 if value > 0 else 2
def jarvis_algorithm(points):
if len(points) < 3:
return points
hull = []
start = leftmost_point(points)
point = start
while True:
hull.append(point)
next_point = points[0]
for candidate in points:
orient = orientation(point, next_point, candidate)
if next_point == point or orient == 2 or (orient == 0 and distance(point, candidate) >
distance(point, next_point)):
next_point = candidate
point = next_point
if point == start:
break
return hull
def distance(p1, p2):
return (p1[0] - p2[0])**2 + (p1[1] - p2[1])**2
def plot_convex_hull(points, hull):
plt.figure(figsize=(8, 8))
plt.scatter(*zip(*points), label='Points', color='blue', alpha=0.6)
for i in range(len(hull)):
p1 = hull[i]
p2 = hull[(i + 1) % len(hull)]
```

```
plt.plot([p1[0], p2[0]], [p1[1], p2[1]], 'r')
plt.scatter(*zip(*hull), color='red', label='Convex Hull', zorder=5)
plt.title("Convex Hull with Jarvis Algorithm")
plt.xlabel("X-Axis")
plt.ylabel("Y-Axis")
plt.grid(True)
plt.legend()
plt.show()
points = generate_random_points(20)
convex_hull = jarvis_algorithm(points)
print("Convex Hull Points:")
for point in convex_hull:
print(point)
plot_convex_hull(points, convex_hull)
```

In GeoGebra, somebody can easily apply the Convex Envelope (List of Points) command to a concrete problem. Let us consider the distribution of Auchan supermarkets in Paris and find the envelope through the following steps:

1. Insert the map of the city with the location of the supermarkets
2. A dot is placed on each supermarket
3. Once the points have been selected, a List of points is created (l1)
4. In the Input bar, type Convex(l1) Envelope
5. GeoGebra draws the outline

Figure 3. Auchan supermarkets in Paris

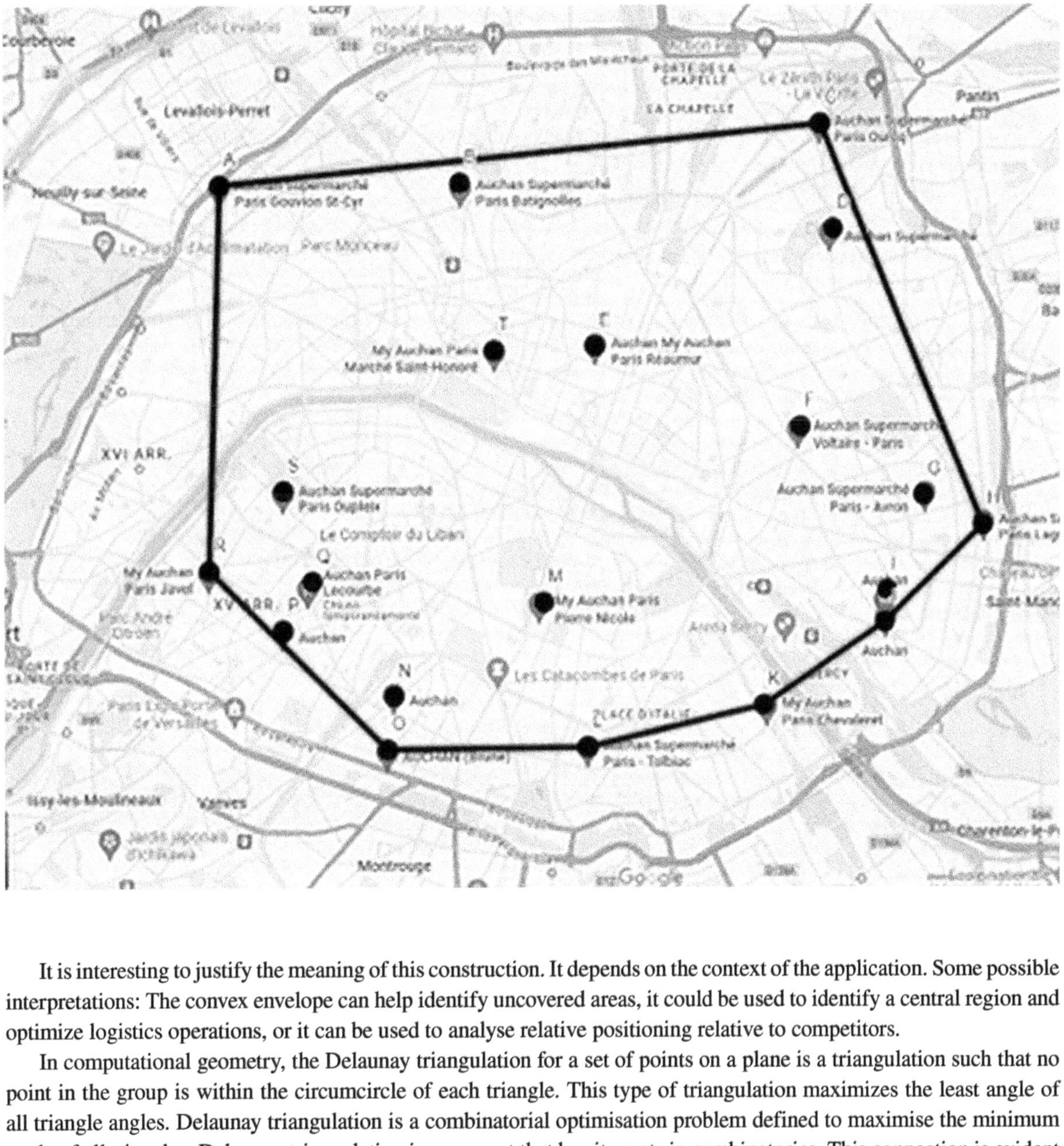

It is interesting to justify the meaning of this construction. It depends on the context of the application. Some possible interpretations: The convex envelope can help identify uncovered areas, it could be used to identify a central region and optimize logistics operations, or it can be used to analyse relative positioning relative to competitors.

In computational geometry, the Delaunay triangulation for a set of points on a plane is a triangulation such that no point in the group is within the circumcircle of each triangle. This type of triangulation maximizes the least angle of all triangle angles. Delaunay triangulation is a combinatorial optimisation problem defined to maximise the minimum angle of all triangles. Delaunay triangulation is a concept that has its roots in combinatorics. This connection is evident in several aspects, for example, in the enumeration of the number of triangulations and the count of the elements present. Euler already studied the determination of the number of triangulations of a convex polygon. Calculating the number of triangulations of a non-convex polygon, however, is a complex problem that requires the use of advanced techniques, and there is no general and straightforward formula for all cases. This number grows exponentially with *n*. The current upper limit is $O(30^n)$, while the lower limit is $\Omega(2.33^n)$.

In summary, Delaunay triangulation is a concept that combines geometry with combinatorics. Its importance lies in its ability to represent sets of points efficiently and accurately, as well as in the numerous applications that derive from it, including computer graphics, cartography, crystallography, and data analysis. The algorithms for constructing Delaunay triangulations are based on "divide and conquer" techniques, a typical approach in algorithmic combinatorics..

Some algorithms can display these triangulations, such as the Bowyer-Watson triangulation, but we will refer to GeoGebra in this case. Among the commands related to Discrete Mathematics is TriangulationDelaunay(List of points). Also, in this case, the procedure is simple:

1. When the Spreadsheet is displayed, *n* pairs of random points are created with RandomBetween(min, max)
2. Once the points have been selected, a List of points is created (l1)
3. In the Insertion bar, type TriangulationDelaunay(l1)
4. GeoGebra provides the optimal route
5. With the F9 key, new random points and new routes are generated

Thanks to the geometric possibilities of the program, it is easy to draw the circles passing through the vertices of the triangles and verify that no point is inside the circumcircle of each triangle.

Figure 4. Delaunay triangulations and circumcircles

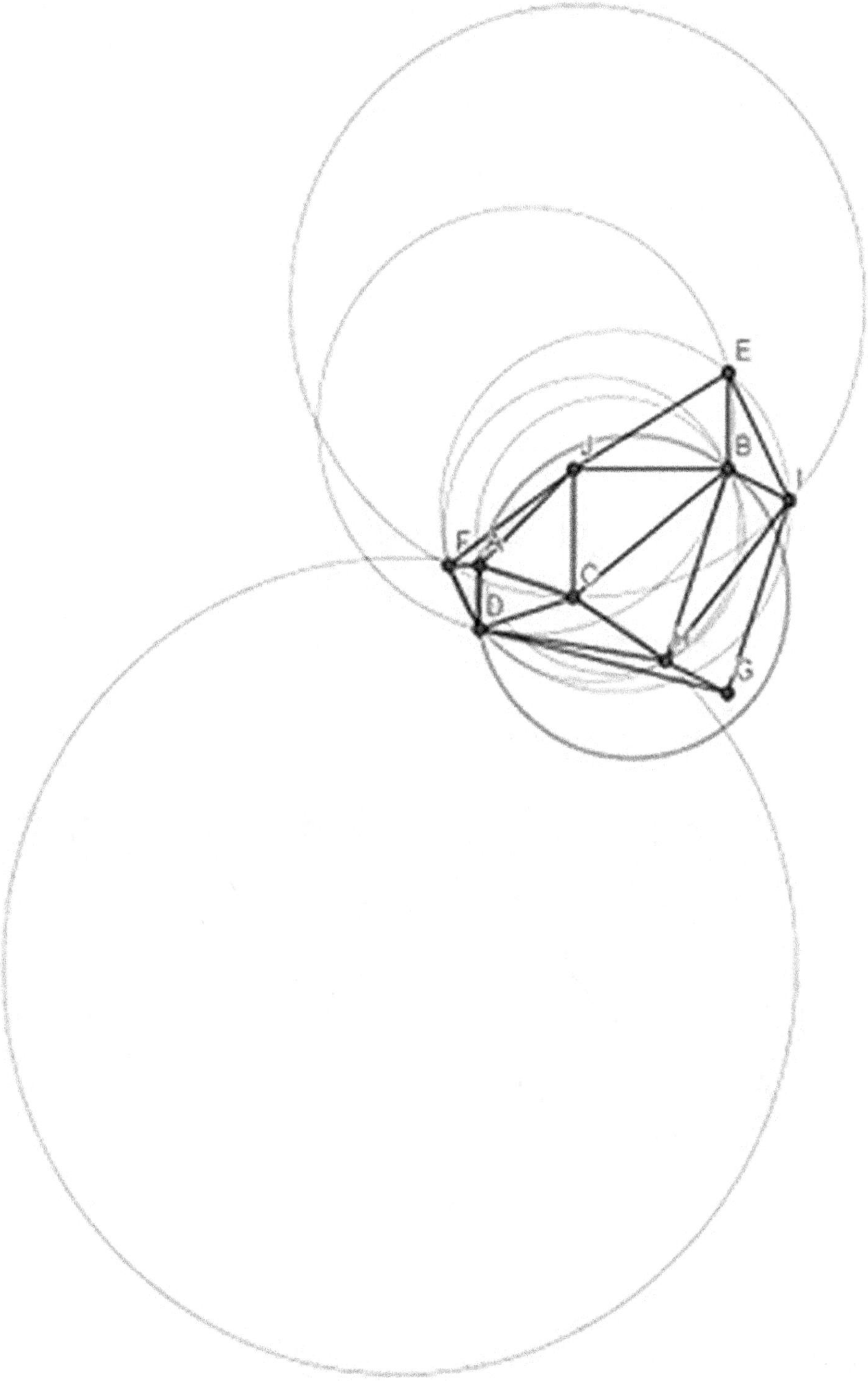

It is important to note that each triangle is associated with a single vertex of the corresponding Voronoi diagram, of which it is the dual. If we consider a set of points in the plane, the Voronoi diagram associates each point with a polygon that represents the area closest to that point compared to the others (Aurenhammer et al., 2013; Miller & Sheehy, 2014). In GeoGebra, there is the Voronoi(List of points) command, and the procedure is similar to the one seen above:

1. The map of the city is inserted with the subdivisions into neighbourhoods
2. At the center of the neighbourhoods is placed a point
3. Once the points have been selected, a List of points is created (l1)
4. In the Insertion bar, type Voronoi (l1)
5. GeoGebra draws the diagram
6. By moving the points, one can try to adapt the model to the real subdivision

Figure 5. Voronoi diagram

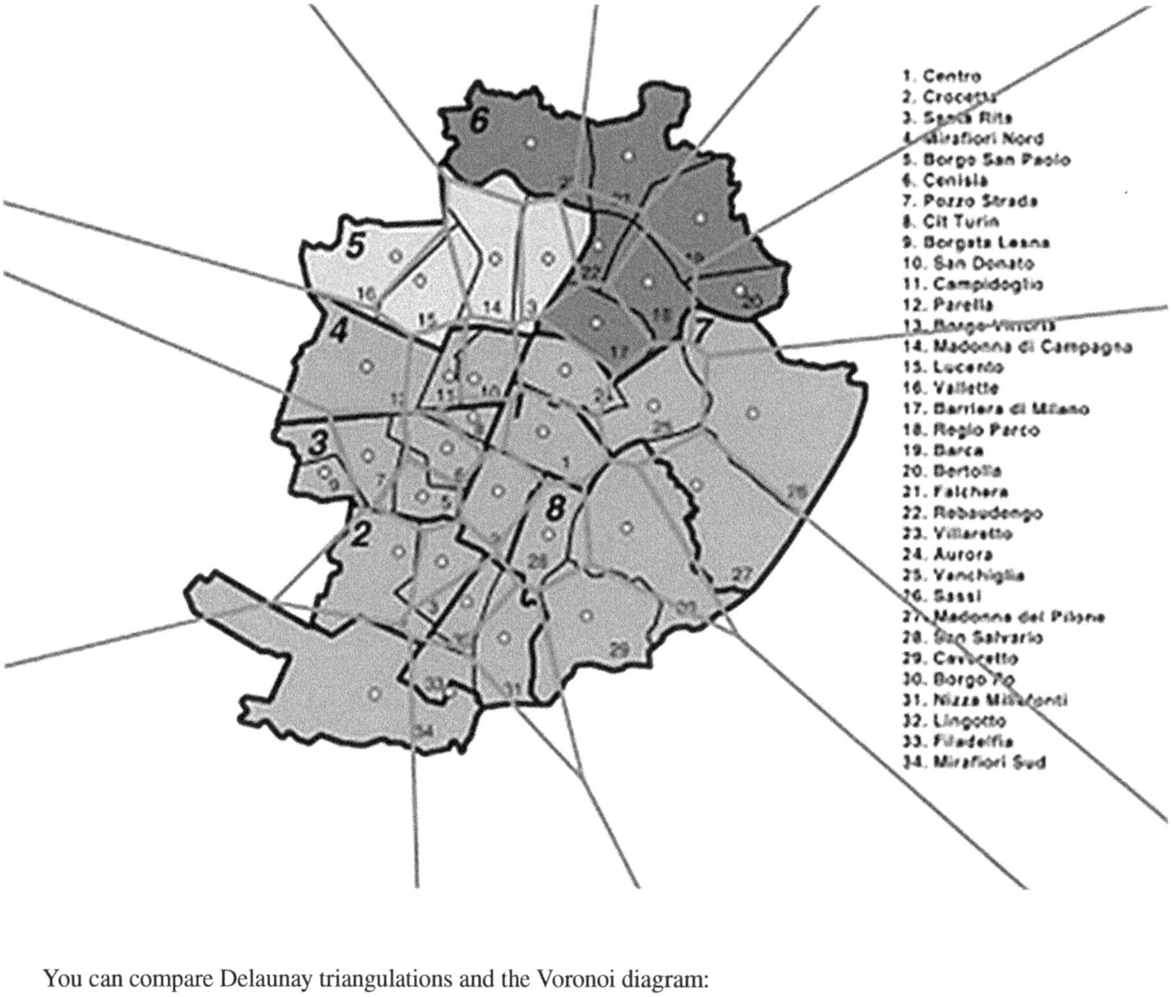

You can compare Delaunay triangulations and the Voronoi diagram:

Figure 6. Delaunay and Voronoi

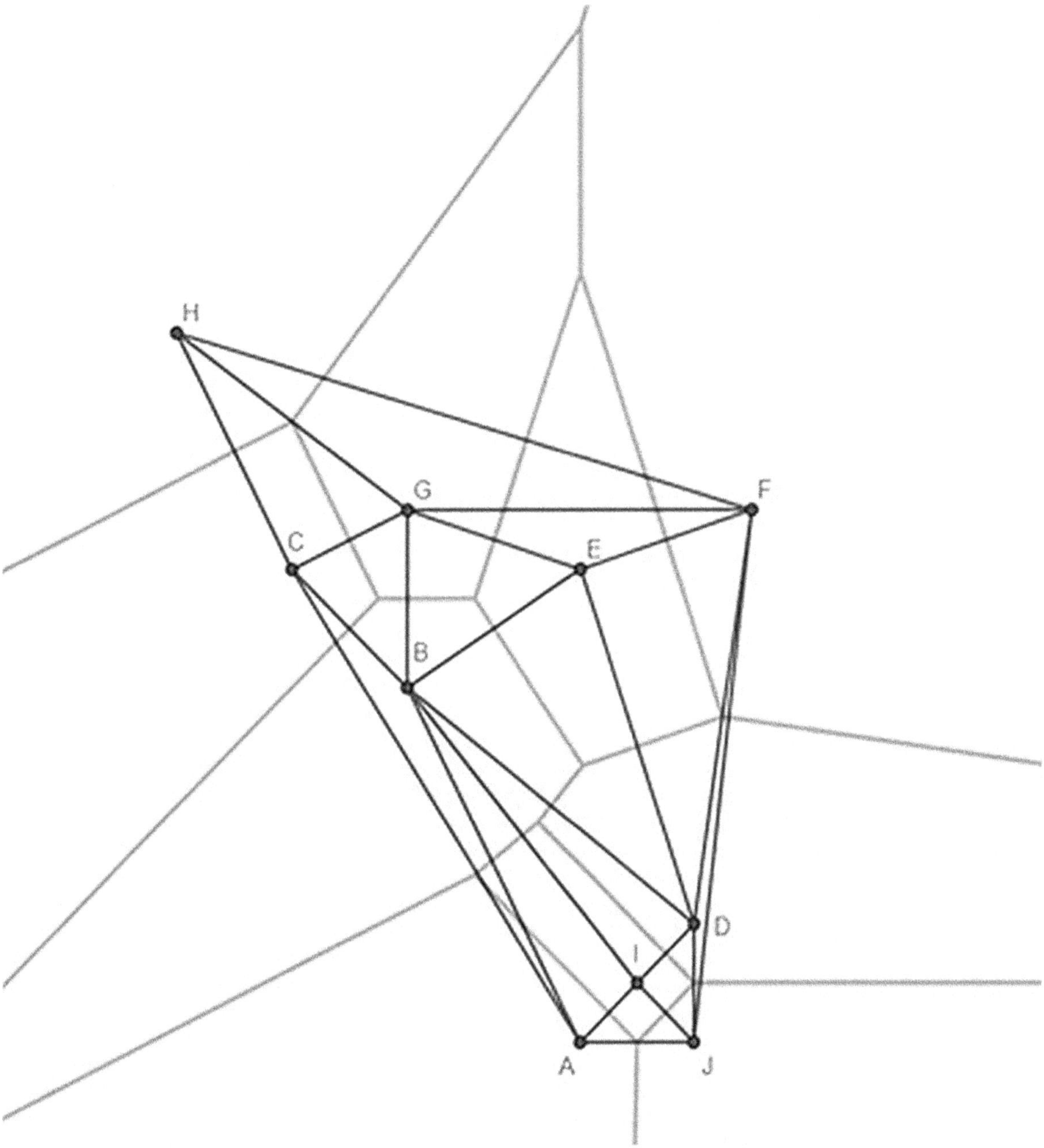

The problem of Erdős–Szekeres points is also a classical question in geometric combinatorics. This problem concerns the arrangement of points in a plane and the convex polygon they can form. The statement can be formulated as follows: for every positive integer *n*, there exists a minimum number N(n) such that any set of non-collinear points contains at least *n* points that form a convex polygon. The Erdős-Szekeres conjecture states that: $N(n) = 2^{n-2} + 1$.

For example, for n = 3 we get value 3; in this case the set of three non-collinear points forms a triangle. Conversely, with 9 points, the figure is that of a pentagon as they are obtained for n = 5.

A simple program can solve both cases.

Program 48 Geometric Combinatorics

```
import math
while True:
choice = input("\nDo you want to calculate N given n (enter '1'), n given N (enter '2'),
or exit the program (enter '3')? ")
if choice == '1':
try:
n = int(input("Enter the value of n: "))
if n < 2:
print("n must be at least 2.")
continue
N = 2**(n-2) + 1
print(f"N = {N}")
except ValueError:
print("Enter a valid number for n.")
elif choice == '2':
try:
N = int(input("Enter the value of N: "))
if N < 3:
print("N must be at least 3.")
continue
n = math.log2(N - 1) + 2
if n.is_integer():
print(f"n = {int(n)}")
else:
print("There is no integer value for n.")
except ValueError:
print("Enter a valid number for N.")
elif choice == '3':
print("You chose to exit the program.")
break
else:
print("Invalid choice. Enter '1', '2', or '3'.")
```

Geometric combinatorics solve practical problems thanks to algorithms that are often used in applications such as pattern recognition, computational graphics and computer-aided design. It encompasses various sub-areas, including the study of the faces of convex polyhedra, the investigation of the combinatorics of convex set intersections, and discrete geometry.

In engineering, geometric combinatorics are used to design components and systems. Geometric combinatorics determine the number of ways in which circuits can be arranged in an electronic board. In the field of graphics, it can be used to determine the number of pixels and edges of an image or to generate natural shapes, such as landscapes, plants, and animals. Geometric combinatorics can be useful to determine how many ways and how leaves can be arranged on a branch or petals on a flower.

We can see how we can draw a tree:

Program 49 Tree Drawing

```
from turtle import Turtle, colormode, tracer, mainloop
from random import randrange
from time import perf_counter as clock
def simRandom(n: int) -> int:
"""Generates a random integer between -n and n."""
return randrange(-n, n + 1)
def randomize(branchList: list, angleVariation: int, sizeVariation: int) -> list:
"""Randomly modifies the list of branches."""
return [(angle + simRandom(angleVariation), sizeFactor * 1.01 ** simRandom(sizeVariation))
for angle, sizeFactor in branchList]
def randomfd(t: Turtle, distance: float, parts: int, angleVariation: int):
"""Draws a line with random angles and advances."""
for _ in range(parts):
t.left(simRandom(angleVariation))
t.forward(distance / parts)
def tree(tList: list, size: float, level: int, widthFactor: float, branchList: list, ang-
leVariation=10, sizeVariation=5):
"""Draws a tree recursively."""
if level > 0:
lst = []
brs = []
for t, branches in zip(tList, branchList):
t.pensize(size * widthFactor)
t.pencolor(255 - (180 - 11 * level + simRandom(15)),
180 - 11 * level + simRandom(15),
0)
t.pendown()
randomfd(t, size, level, angleVariation)
yield 1
for angle, sizeFactor in branches:
t.left(angle)
lst.append(t.clone())
brs.append(randomize(branches, angleVariation, sizeVariation))
t.right(angle)
for x in tree(lst, size * sizeFactor, level - 1, widthFactor, brs, angleVariation, size-
Variation):
yield None
def initialize(t: Turtle, x: int, y: int):
"""Initializes the turtle for drawing the tree."""
colormode(255)
t.reset()
t.speed(0)
t.hideturtle()
t.left(90)
t.penup()
t.setpos(x, y)
```

```
t.pendown()
def drawSingleTree(level: int, x: int, y: int):
"""Draws a single tree with a given branching level."""
pen = Turtle(undobuffersize=1)
pen.hideturtle()
initialize(pen, x, y)
t = tree([pen], 80, level, 0.1, [[(45, 0.69), (0, 0.65), (-45, 0.71)]])
list(t) # Consumes the generator to actually execute the drawing
def main():
"""Main function to start drawing the tree."""
tracer(75, 0)
drawSingleTree(6, 20, -208)
mainloop()
if __name__ == '__main__':
main()
```

Algebraic combinatorics is a branch of mathematics that studies the relationships between combinatorial functions and algebraic structures (Panova, 2023). For example, it is used to study the relationships between the permutations, combinations, and partitions of a set, as well as algebraic operations such as addition, multiplication, and division.

Symmetric functions are fundamental objects in algebraic combinatorics. These are multivariable polynomials that remain invariant under any permutation of their variables. A notable subclass is the Schur polynomials, which play a pivotal role in the representation theory of symmetric groups and have significant connections to algebraic geometry (Cheung & Crapo, 1976) and mathematical physics. These polynomials, generally denoted by $S_\lambda(x_1,x_2,...,x_n)$), are associated with a partition λ of an integer n, where a partition λ is a sequence of non-negative numbers ordered in a non-ascending manner. For example, the Schur polynomial associated with the partition $\lambda = (2,1)$ in three variables is:

$$s_{(2,1)}(x_1,x_2,x_3) = x_1^2x_2 + x_1x_2x_3 + x_2^2x_3 + x_3^2x_2$$

Each term corresponds to a permutation of the variables.

Algebraic combinatorics are based on several algebraic concepts and techniques, including generating functions and incidence algebras (Stanley, 2013). An incidence algebra is an algebraic structure that represents the relationships between the subsets of a set. In its versatility, it deals with the study of graphs, i.e., sets of nodes connected by edges.

Graph theory is an interdisciplinary discipline that overlaps with combinatorics, discrete mathematics (Matousek, 2013), and data science. Graph theory and combinatorics overlap when analysing combinatorial structures within graphs and solving problems involving both fields.

Graph enumeration is the task of determining the count of graphs that possess a specified number of vertices and edges. It is a complex, combinatorial undertaking, yet one with broad applications in areas such as network design and statistics.

Consider, for instance, the challenge of counting bipartite graphs with a given count of vertices in each part. A bipartite graph is one where its vertices can be divided into two distinct subsets. Importantly, each vertex in one subset connects exclusively to vertices in the other. This particular structure makes it an excellent model for work assignment scenarios. The fascinating aspect, from a combinatorial perspective, is how the variety of possible bipartite graphs (3,2) changes in response to imposed conditions. If no conditions are applied, there are 64 such graphs. However, if we stipulate that each vertex of the first part must link to precisely one vertex of the second part, the array of possible bipartite graphs (3,2) shrinks to just 6.

One can easily appreciate the reduced solution with the following program:

Program 50 Algebraic Combinatorics

```
import itertools
import networkx as nx
import matplotlib.pyplot as plt
def generate_bipartite_graph(nodes1, nodes2):
graph = nx.Graph()
graph.add_nodes_from(nodes1, bipartite=0)
graph.add_nodes_from(nodes2, bipartite=1)
edges = itertools.product(nodes1, nodes2)
graph.add_edges_from(edges)
return graph
def enumeration_and_visualization_of_combinations(graph):
nodes = graph.nodes(data=True)
bipartite_sets = [set(n for n, d in nodes if d['bipartite'] == i) for i in range(2)]
all_combinations = list(itertools.product(*bipartite_sets))
plt.figure(figsize=(8, 6))
for i, combination in enumerate(all_combinations):
current_graph = graph.subgraph(combination)
# Manually assign positions
pos = {'A1': (0, 1), 'A2': (0, 0), 'A3': (0, -1), 'B1': (2, 0), 'B2': (2, -1)}
nx.draw(current_graph, pos, with_labels=True, font_weight='bold', node_size=700, node_col-
or='grey', font_size=8, edge_color='gray', linewidths=0.5, alpha=0.8)
plt.title(f"All Combinations")
plt.show()
if __name__ == "__main__":
nodes1 = ["A1", "A2", "A3"]
nodes2 = ["B1", "B2"]
bipartite_graph = generate_bipartite_graph(nodes1, nodes2)
enumeration_and_visualization_of_combinations(bipartite_graph)
```

Topological combinatorics is a branch of mathematics that combines the principles of combinatorics and topology (De Longueville, 2013; Moscovich, 2015). This field examines how discrete configurations emerge within topological spaces, offering a distinctive perspective on the form and organisation of mathematical entities. While topology studies the properties of spaces that are preserved under continuous deformation, combinatorics focuses on discrete patterns and the relationships among their components. Together, they offer powerful tools for understanding the interplay between continuity and discreteness in mathematical systems. Combining these two fields creates a powerful tool for analysing complex problems in mathematics and beyond.

Topological Combinatorics finds applications in numerous fields, from theoretical physics to computational biology. One of the key concepts relating to simplicial complexes is a collection of simplexes (points, lines, triangles, etc.) that come together coherently. Topological graph theory, on the other hand, explores the properties of graphs embedded in surfaces or spaces of higher dimensions. Finally, knot theory studies the properties of closed curves in three-dimensional space to classify and distinguish nodes. Several techniques are employed, ranging from triangulations to calculating topological invariants and various counting methods.

As a non-trivial example of a combinatorial application, we can turn to Betti's numbers, a sequence that describes a topological space's connectivity. These invariants are named after the Italian mathematician Enrico Betti (1823-1892). Specifically, we denote by bn the count of connected components (n=0), holes (n=1), cavities (n=2), and so forth.

Consider a torus (the classic doughnut shape):

$b_0 = 1$ signifies that the torus forms a single connected piece.

$b_1 = 2$ represents two fundamental cycles: one wrapping around the central "hole" (the inner circumference) and another encircling the "section" of the torus (the outer circumference).

$b_2 = 1$ accounts for the singular cavity enclosed by the torus's surface.

To compute these values using topological combinatorics, we can view them as the ranks of their respective homology groups, such that $b_n = \text{rank}(H_n(T))$. For instance, if $H_1 = \mathbb{Z}^2$ the rank is 2, then $b_1 = 2$. This approach elegantly connects the abstract concept of connectivity to concrete; computable quantities derived from combinatorial structures.

A discrete representation of continuous space is often used to calculate Betti's numbers. In the case of the torus, one can use a triangulation of the rectangle from which one can start to build the torus. This grid represents a torus if we identify the vertices, the edges, and the triangular faces. At this point, combinatorial counting comes into play:

Vertices (V): Reduce to 1 as all vertices are identified after winding

Edges (E): Reduced to 3: one horizontal, one vertical, one diagonal

Faces (F): Reduced to 2, one for each orientation

We use the generalized Euler formula $\chi = V - E + F$, so $\chi = 1 - 3 + 2 = 0$. For surfaces, the Euler characteristic χ can also be expressed in terms of Betti's numbers, from which $\chi = b_0 - b_1 + b_2 = 0$. Knowing that $b_0 = 1$ (connected), we can deduce b_1 and b_2.

The transformation of the rectangle into a torus can be visualized with an animation.

Program 51 From Rectangle to Torus

```
import numpy as np
import matplotlib.pyplot as plt
import matplotlib.animation as animation
from mpl_toolkits.mplot3d import Axes3D
# Torus parameters
R = 1 # Radius of the central circle
r = 0.3 # Radius of the secondary circle
def rectangle_to_torus(t, n=10):
"""Gradually transforms a triangulated rectangle into a torus."""
u = np.linspace(0, 2 * np.pi, n)
v = np.linspace(0, 2 * np.pi, n)
u, v = np.meshgrid(u, v)
x = (R + r * np.cos(v)) * np.cos(u)
y = (R + r * np.cos(v)) * np.sin(u)
z = r * np.sin(v)
# Transition effect
x_rect = np.linspace(-1, 1, n)
y_rect = np.linspace(-1, 1, n)
x_rect, y_rect = np.meshgrid(x_rect, y_rect)
z_rect = np.zeros_like(x_rect)
# Linear interpolation between rectangle and torus
x = (1 - t) * x_rect + t * x
y = (1 - t) * y_rect + t * y
z = (1 - t) * z_rect + t * z
return x, y, z
def update(num, ax, n=10):
ax.clear()
```

```
t = num / 100
x, y, z = rectangle_to_torus(t, n)
# Creating the torus mesh
ax.plot_surface(x, y, z, color='lightskyblue', edgecolor='k')
ax.set_xlabel('X')
ax.set_ylabel('Y')
ax.set_zlabel('Z')
ax.set_title('Transformation from Rectangle to Torus')
fig = plt.figure()
ax = fig.add_subplot(111, projection='3d')
n = 10
# Increase interval to slow down the animation
ani = animation.FuncAnimation(fig, update, frames=100, fargs=(ax, n), interval=100) # interval in milliseconds
plt.show()
```

Turning a rectangle into a torus has direct applications in interconnection networks for parallel and distributed computing. A two-dimensional toroidal topology is widely used in supercomputers and communication networks. Many supercomputers use toroidal networks because they allow high connectivity and reduce communication latency between nodes.

Another example related to topological combinatorics is Sperner's lemma. Given a family of subsets of a set of n elements, if no subset of the family is contained in another, then the largest possible family will be composed of subsets of size int(n/2) = 2. The maximum number of subsets in this family is $\binom{n}{\text{int}(\frac{n}{2})}$.

Suppose we have a set of 4 numbers {1,2,3,4}. The lemma tells us that there are exactly = 6 subsets of dimension 2, and none of these are contained in another. If we consider the previous set, its subsets are: $\binom{n}{\text{int}(\frac{n}{2})}\binom{4}{2}$.

1. {∅}
2. {1} {2} {3} {4}
3. {1,2} {1,3} {1,4} {2,3} {2,4} {3,4}
4. {1,2,3} {1,2,4} {1,3,4} {2,3,4}
5. {1,2,3,4}

If we choose subsets of dimension 2, we can observe that no set of that dimension can contain another set of dimension 2.

Program 52 Sperner's Lemma

```
from itertools import combinations
def sperner_subsets(elements):
n = len(elements)
max_size = n // 2 # Calculate the maximum subset size
subsets = list(combinations(elements, max_size)) # Generate subsets of size max_size
return subsets
# Input of set elements
input_elements = input("Enter the elements of the set separated by spaces: ")
elements = input_elements.split() # Split the string by spaces
# Find and print the subsets according to Sperner's lemma
result = sperner_subsets(elements)
print(f"Subsets of maximum size {len(elements) // 2}:")
```

```
for subset in result:
print(subset)
```

However, in what sense does the lemma fit into topological combinatorics? The lemma of Emanuel Sperner (1905-1980) originally applied to a subdivision of an object called "simplex". A simplex is a geometric shape that generalizes the concept of a triangle to multiple dimensions. For example, a two-dimensional simplex is a triangle, while a three-dimensional simplex is a pyramid with a triangular base. The subdivision of a simplex is an important concept in combinatorial topology. Simplicial complexes approximate continuous spaces with discrete structures, allowing combinatorial methods (using counts and combinations) to be applied to topological problems (involving the shape and structure of spaces). The lemma can also prove Brouwer's fixed point theorem in a combinatorial way.

The theorem is due to Luitzen Brouwer (1881-1996) and states that if we take a continuous function that maps a specific type of space to itself, then at least one point remains fixed, i.e., that is mapped onto itself. In summary, Sperner's lemma bridges combinatorics and topology as it shows how discrete structures (vertex colourations) can capture topological properties (such as the existence of fixed points).

Figure 7. Sperner and topology

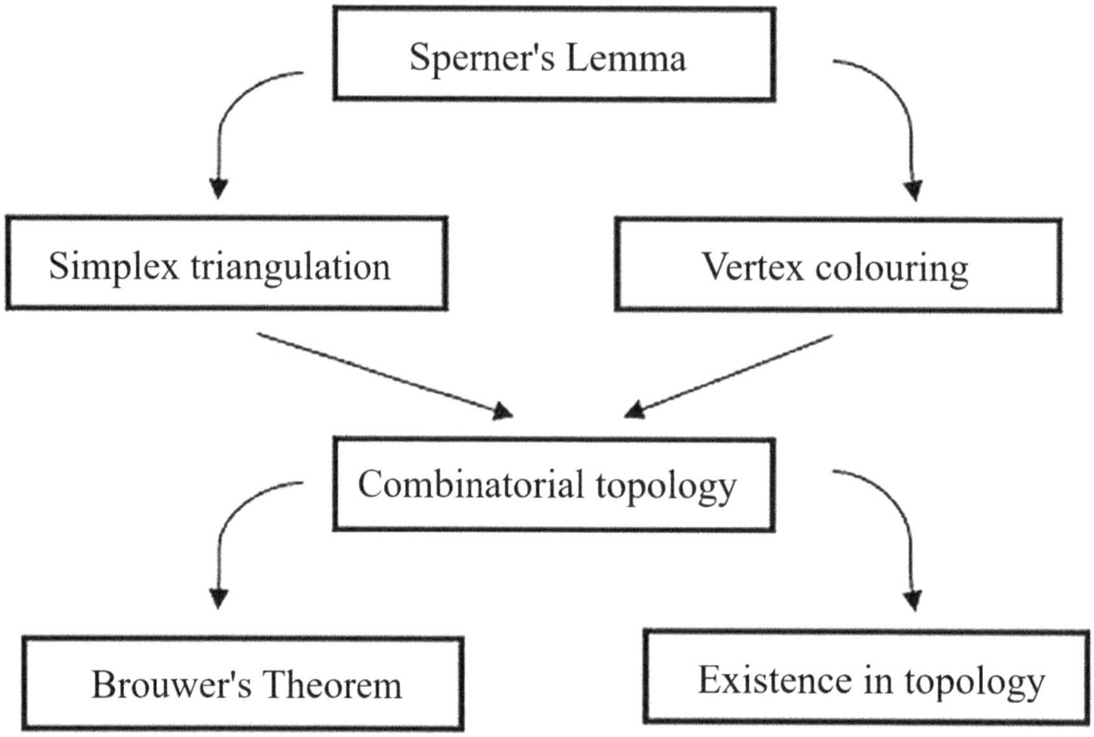

To further develop the argument, we can use another theorem, that is due to Adolf Kneser (1862-1930). The theorem states that if we have a set of n elements and we want to form subsets of dimension k and m with $k + m \leq n$, then the maximum number of pairs of possible disjoint subsets is given by:

$$Max = \binom{n}{k}\binom{n-k}{m}$$

If we reconsider the set formed by the four elements {1, 2, 3, 4}, we have seen that there are ten disjoint subsets of dimensions 1 and 2. If we consider {1} and {2, 3}, we see that they are disjointed; they do not share any element. According to Kneser's theorem, the maximum number of disjoint pairs we can form is given by:

$$Max = \binom{4}{1}\binom{3}{2} = 12$$

Program 53 Kneser's Theorem

```
import math
import itertools
def binom(n, k):
if k > n or k < 0:
return 0
return math.comb(n, k)
def kneser_theorem(n, k, m):
if k + m > n:
raise ValueError("The sum of k and m must be less than or equal to n.")
return binom(n, k) * binom(n - k, m)
def generate_subsets(elements, size):
return list(itertools.combinations(elements, size))
# User input
n = int(input("Enter the total number of elements (n): "))
k = int(input("Enter the size of the first subset (k): "))
m = int(input("Enter the size of the second subset (m): "))
# Generate the set of elements
elements = list(range(1, n + 1))
# Generate subsets of size k and m
subsets_k = generate_subsets(elements, k)
subsets_m = generate_subsets(elements, m)
# Calculate the maximum number of disjoint pairs
max_pairs = kneser_theorem(n, k, m)
# Print the results
print(f"Subsets of size {k}: {subsets_k}")
print(f"Subsets of size {m}: {subsets_m}")
print(f"The maximum number of disjoint pairs of subsets of size {k} and {m} in a set of
{n} elements is: {max_pairs}")
```

Forming pairs of disjoint subsets reflects the ability to find points or configurations in a topological space that satisfy certain conditions.

Experimental design involves the planning, execution, and analysis of experiments (Montgomery, 2017; Stinson, 2021; Kenett, 2023). It may include the use of combinatorial techniques to optimize the process. A combinatorial experimental design can assess all possible combinations of variables and create control groups to minimize the effects of confounding factors. The applications of experimental design are diverse, spanning fields such as medicine, economics, engineering, and biology.

Consider the following example: We aim to investigate the impact of two different fertilisers on plant growth. We can create several treatment groups: one where a single fertilizer is applied, a control group with no fertilizer, and groups where both fertilizers are combined. Combinatorics plays a role in determining the different treatment combinations to be tested. Furthermore, the order in which the treatments are applied can influence the results. Therefore, deciding whether to use one fertilizer first and then the other is essential, or vice versa. This introduces various combinations of treatment orders that one must consider.

Numerous types of combinatorial experimental designs exist. These range from randomized block designs, where participants are divided into homogeneous blocks to which randomized treatments are applied, to factorial designs, in which multiple factors are examined simultaneously. Other designs include paired-pair designs, where one of the two treatments serves as a control, and Balanced Incomplete Block Designs (BIBDs), which allow a set of treatments to be tested while ensuring that each treatment appears in a balanced number of blocks.

Once the data are collected, statistical analysis is performed. Here, we can use permutation and combination techniques to evaluate the null hypothesis and assess its significance.

The model is complex, but it can be understood thanks to the following program:

Program 54 Experimental Design 1

```
import math
import itertools
def binom(n, k):
if k > n or k < 0:
return 0
return math.comb(n, k)
def kneser_theorem(n, k, m):
if k + m > n:
raise ValueError("The sum of k and m must be less than or equal to n.")
return binom(n, k) * binom(n - k, m)
def generate_subsets(elements, size):
return list(itertools.combinations(elements, size))
# User input
n = int(input("Enter the total number of elements (n): "))
k = int(input("Enter the size of the first subset (k): "))
m = int(input("Enter the size of the second subset (m): "))
# Generate the set of elements
elements = list(range(1, n + 1))
# Generate subsets of size k and m
subsets_k = generate_subsets(elements, k)
subsets_m = generate_subsets(elements, m)
# Calculate the maximum number of disjoint pairs
max_pairs = kneser_theorem(n, k, m)
# Print the results
print(f"Subsets of size {k}: {subsets_k}")
print(f"Subsets of size {m}: {subsets_m}")
print(f"The maximum number of disjoint pairs of subsets of size {k} and {m} in a set of
{n} elements is: {max_pairs}")
```

One can add data analysis to extend this example to a more realistic context. For example, calculate the mean and variance or perform a statistical analysis to determine the significant effect of treatments. Five replications are performed for each combination of treatments to simulate natural variability.

Extended code provides a more complete view of experimental design and data analysis, helping us better understand how combinatorics and experimental design combine to produce valid and meaningful results. The program provides, in essence, a practical example of how to manage and analyse an experiment's data.

Program 55 Experimental Design 2

```
import itertools
import random
import numpy as np
from scipy.stats import f_oneway
import matplotlib.pyplot as plt
# Setting the seed for the random number generator
random.seed(42)
# Definition of treatments
treatments = {
'fertilizer A': [1, 2, 3],
'fertilizer B': [0.5, 1, 1.5]
}
# Creating all possible combinations of treatments
treatment_combinations = list(itertools.product(*treatments.values()))
# Initializing results
results = []
# Number of replications for each combination
replications = 5
# Experiment simulation
for i, combination in enumerate(treatment_combinations, start=1):
print(f"Experiment {i}: Treatments {combination}")
combination_results = []
for _ in range(replications):
# Simulating the effect of natural variability
simulated_result = random.uniform(0.8, 1.2)
simulated_result *= sum(combination)
combination_results.append(simulated_result)
# Adding the results of the combination to the overall list
results.append((combination, combination_results))
# Calculating descriptive statistics
mean = np.mean(combination_results)
variance = np.var(combination_results)
print(f"Simulated results: {combination_results}")
print(f"Mean: {mean:.2f}")
print(f"Variance: {variance:.2f}")
print()
# Overall data analysis
print("Overall data analysis:")
for combination, combination_results in results:
mean = np.mean(combination_results)
variance = np.var(combination_results)
print(f"Combination {combination}: Mean = {mean:.2f}, Variance = {variance:.2f}")
# Example of statistical analysis (ANOVA)
# Preparing data for ANOVA
anova_data = [combination_results for _, combination_results in results]
# Performing the ANOVA
```

```
anova_result = f_oneway(*anova_data)
print("\nANOVA results:")
print(f"F-statistic: {anova_result.statistic:.2f}")
print(f"P-value: {anova_result.pvalue:.4f}")
# Visualization of results
def visualize_results(results):
combinations, values = zip(*results)
mean = [np.mean(v) for v in values]
variance = [np.var(v) for v in values]
plt.figure(figsize=(10, 5))
plt.subplot(1, 2, 1)
plt.bar(range(len(mean)), mean, tick_label=[str(c) for c in combinations])
plt.title('Mean Results per Combination')
plt.xticks(rotation=90)
plt.subplot(1, 2, 2)
plt.bar(range(len(variance)), variance, tick_label=[str(c) for c in combinations])
plt.title('Variance of Results per Combination')
plt.xticks(rotation=90)
plt.tight_layout()
plt.show()
# Visualizing the results
visualize_results(results)
```

An interesting example of drawing refers to *Hadamard's design* (Hedayat & Wallis, 1978). It is a set of blocks (or treatments), each associated with an orthogonal Hadamard matrix row. Projects have the property that blocks (rows of the matrix) are orthogonal to each other, i.e., the scalar product between two distinct rows is zero. Projects are closely tied to a matrix structure, in which each row of the matrix represents a "block", and the rows are orthogonal.

An example of practical application is in biomedicine. Imagine that we want to conduct a clinical trial to evaluate the efficacy of different drugs on various patient groups. A Hadamard design can distribute treatments so that each combination of dosages and patient groups is tested orthogonally, minimizing the possibility of results being confounded by factors not considered. Suppose we have a clinical trial with four drugs and four patient groups. Each group receives a combination of drugs, but each combination must be assigned to ensure that treatments are distributed evenly among patient groups. In the context of the following program, the health score is a numerical simulation that represents the effectiveness of treatments (such as medications) on patients. It is a numerical value that simulates the effect of treatment on the health of participants in a clinical trial. The health score is randomly generated to simulate a patient's response to a treatment. It is a numerical value that varies between 50 and 100, with 50 representing a low health condition and 100 representing an optimal health value. Of course, in an actual clinical trial, the health score would be determined by concrete measurements based on specific clinical parameters, such as improvements in symptoms, changes in biological parameters, and recovery times.

Program 56 Clinical Trial

```
import numpy as np
import random
# Function to construct a Hadamard matrix using Sylvester's algorithm
def hadamard(n):
if n == 1:
return np.array([[1]])
```

```
elif n == 2:
return np.array([[1, 1], [1, -1]])
else:
H = hadamard(n // 2)
top = np.hstack((H, H)) # Upper part of the matrix
bottom = np.hstack((H, -H)) # Lower part of the matrix
return np.vstack((top, bottom))
# Function to verify the orthogonality of the matrix
def verify_orthogonality(H):
n = len(H)
product = np.dot(H, H.T) # Product of the matrix with its transpose
I = np.eye(n) # Identity matrix of order n
return np.allclose(product, n * I) # Check if the product is identical to n*I
# Function to create a Hadamard design for a clinical study
def clinical_study(n, treatments):
if (n & (n - 1)) != 0:
print(f"Error: n = {n} is not a power of 2.")
return None
# Create the Hadamard matrix
H = hadamard(n)
# Verify orthogonality
if verify_orthogonality(H):
print(f"Clinical study for n = {n} with treatments {treatments}:")
print("Hadamard matrix (orthogonal combinations of treatments):")
print(H)
# Assign treatments to the rows of the Hadamard matrix
for i in range(n):
combination = H[i,:]
treatment_assignment = [treatments[j] if combination[j] == 1 else f"Placebo_{treat-
ments[j]}" for j in range(len(treatments))]
print(f"Group {i + 1}: {treatment_assignment}")
# Simulate the results for each group
print("\nSimulating results for each group:")
results = {}
for i in range(n):
combination = H[i,:]
treatment_assignment = [treatments[j] if combination[j] == 1 else f"Placebo_{treat-
ments[j]}" for j in range(len(treatments))]
# Simulate health results for the group
health = random.uniform(50, 100) # Simulated health score (from 50 to 100)
if "Placebo" in str(treatment_assignment):
health -= random.uniform(5, 15) # Groups with placebo will have a lower health score
results[f"Group {i + 1}"] = {
"Treatments": treatment_assignment,
"Health": health
}
# Display the results
for group, result in results.items():
```

```
    print(f"{group}: Treatments = {result['Treatments']}, Health Score = {re-
sult['Health']:.2f}")
    else:
    print("The Hadamard matrix is not orthogonal.")
    return None
    # Define the treatments (medications)
    treatments = ["Drug_A", "Drug_B", "Drug_C", "Drug_D"]
    # Run the program for n = 4 (number of groups and treatments)
    n = 4 # Number of groups (also the number of treatments is 4 in this case)
    clinical_study(n, treatments)
```

The program generates an experimental Hadamard design for a clinical trial. This design allows for orthogonal combinations of treatments, ensuring that the effects of each treatment can be separated and analysed independently. The program simulates a clinical trial based on this Hadamard design. By using orthogonal matrices, the treatments are combined so that their effects can be evaluated separately. The simulated results suggest that placebo-treated groups generally exhibit lower health scores, which reflects the potential impact of the medicine or sham treatment.

Let us examine a potential simulation outcome for a clinical study with n = 4 treatments: ['Drug_A', 'Drug_B', 'Drug_C', 'Drug_D'].

Hadamard matrix (orthogonal combinations of treatments):

```
[[ 1 1 1 1]
 [ 1 -1 1 -1]
 [ 1 1 -1 -1]
 [ 1 -1 -1 1]]
```

Treatment Groups:

Group 1: ['Drug_A', 'Drug_B', 'Drug_C', 'Drug_D']
Group 2: ['Drug_A', 'Placebo_Drug_B', 'Drug_C', 'Placebo_Drug_D']
Group 3: ['Drug_A', 'Drug_B', 'Placebo_Drug_C', 'Placebo_Drug_D']
Group 4: ['Drug_A', 'Placebo_Drug_B', 'Placebo_Drug_C', 'Drug_D']

Simulating results for each group:

Group 1: Treatments = ['Drug_A', 'Drug_B', 'Drug_C', 'Drug_D'], Health Score = 50.29
Group 2: Treatments = ['Drug_A', 'Placebo_Drug_B', 'Drug_C', 'Placebo_Drug_D'], Health Score = 62.39
Group 3: Treatments = ['Drug_A', 'Drug_B', 'Placebo_Drug_C', 'Placebo_Drug_D'], Health Score = 82.30
Group 4: Treatments = ['Drug_A', 'Placebo_Drug_B', 'Placebo_Drug_C', 'Drug_D'], Health Score = 77.51

The simulation is based on a random value between 50 and 100, with a reduction in the health score for the placebo-treated groups.

Group 1 received the lowest health score (50.29), indicating that the entire treatment (without any placebos) had a relatively modest effect in this simulation. This result could be due to random variability or the simulated effectiveness of the medications.

Group 2, including placebo treatments, has a higher health score (62.39) than Group 1. The treatments have a positive effect, as indicated by the improved health score, despite including a placebo.

Group 3 achieved the highest health score (82.30), implying that the actual treatments, especially Drug_A and Drug_B, significantly improved the subjects' health. The placebo treatments did not significantly lower the treatment's effectiveness in this group.

Group 4, despite having placebos, also shows a relatively high health score (77.51). While the placebo treatments did not drastically reduce the health score, the overall effectiveness of the drugs remains strong, though slightly lower than in Group 3.

The program efficiently implements a combinatorial design using the Hadamard matrix to ensure orthogonality and separability of treatment effects in a clinical trial. The combinatorial properties of the Hadamard matrix reduce the number of experiments required while still allowing for precise identification of treatment effects. This method ensures that any interactions or confounding variables are minimized, providing a robust framework for analysing the effectiveness of treatments in clinical trials.

REFERENCES

Aurenhammer, F., Klein, R., & Lee, D. T. (2013). *Voronoi diagrams and Delaunay triangulations*. World Scientific Publishing Company. DOI: 10.1142/8685

Bóna, M. (2015). *Introduction to enumerative and analytic combinatorics*. CRC Press. DOI: 10.1201/b19267

Cheung, A., & Crapo, H. (1976). A combinatorial perspective on algebraic geometry. *Advances in Mathematics*, *20*(3), 388–414. DOI: 10.1016/0001-8708(76)90203-6

Colbourn, C. J., & Mathon, R. (Eds.). (2011). *Combinatorial design theory* (Vol. 34). Elsevier.

De Longueville, M. (2013). *A course in topological combinatorics*. Springer Science & Business Media. DOI: 10.1007/978-1-4419-7910-0

Dukes, P., Lamken, E., & Wilson, R. (2008). *Combinatorial design theory*. Banff International Research Station for Mathematical Innovation and Discovery.

Golay, M. J. E. E. (1954). Binary coding. *Transactions of the IRE Professional Group on Information Theory*, *4*(4), 23-28.

Hedayat, A., & Wallis, W. D. (1978). Hadamard matrices and their applications. *Annals of Statistics*, •••, 1184–1238.

Ho, T., & Lun, D. (2008). *Network coding: an introduction*. Cambridge University Press. DOI: 10.1017/CBO9780511754623

Huffman, W. C., Kim, J. L., & Solé, P. (2021). *Concise encyclopedia of coding theory*. Chapman and Hall/CRC. DOI: 10.1201/9781315147901

Kenett, R. S., Zacks, S., & Gedeck, P. (2023). Classical design and analysis of experiments. In *Industrial Statistics: A Computer-Based Approach with Python* (pp. 141–224). Springer International Publishing. DOI: 10.1007/978-3-031-28482-3_5

Matousek, J. (2013). *Lectures on discrete geometry* (Vol. 212). Springer Science & Business Media.

Miller, E., & Reiner, V. (2004). What is Geometric Combinatorics?–An Overview of the Graduate Summer School. *Ezra Miller and Vic Reiner What is Geometric Combinatorics? An Overview of the Graduate Summer School*, 1.

Miller, G. L., & Sheehy, D. R. (2014). A new approach to output-sensitive construction of Voronoi diagrams and Delaunay triangulations. *Discrete & Computational Geometry*, *52*(3), 476–491. DOI: 10.1007/s00454-014-9629-y

Montgomery, D. C. (2017). *Design and analysis of experiments*. John Wiley & sons.

Moscovich, I. (2015). The puzzle universe: A history of mathematics in 315 puzzles. Firefly Books.

Panova, G. (2023). Computational complexity in algebraic combinatorics. *arXiv preprint arXiv:2306.17511*.

Stanley, R. P. (2013). Algebraic combinatorics. *Springer*, *20*(22), 4.

Stinson, D. R. (2021). *Techniques for Designing and Analyzing Algorithms*. Chapman and Hall/CRC. DOI: 10.1201/9780429277412

Vermani, L. R. (2022). *Elements of algebraic coding theory*. Routledge. DOI: 10.1201/9780203758533

Chapter 7
Examples of the Main Sub-Disciplines

ABSTRACT

The chapter explores the significant subdisciplines of combinatorics through concrete examples. Problems such as combinatorial design applied to smartphones and restaurant menus, Kirkman's fifteen-student puzzle, and the use of Steiner systems in agriculture are presented. The chapter also discusses Gardner's square packing problem, restaurant code assignment using coding theory, and dining room layout planning using geometric combinatorics. The chapter includes examples of graph coloring for radio frequency assignment, the art gallery problem, and the use of Stirling numbers in algebraic combinatorics. Finally, graph theory examples are provided for scheduling home deliveries and using Hasse diagrams in project management and partition representation. It is noted that Hasse diagrams are widely used in discrete mathematics to visualize ordering relationships between elements without explicitly showing all redundant connections. These examples highlight how combinatorics offers powerful tools for solving optimization and organization problems in diverse contexts.

EXAMPLES OF THE MAIN SUB-DISCIPLINES

Design is the central core of human creation

Steve Jobs

Exploring all the sub-disciplines of *Combinatorial Design* means delving into a field of vast complexity. However, it's possible to illustrate the topic's scope by limiting the discussion to a few concrete examples.

Combinatorial Design is a valuable methodology for designing and optimizing systems and products (Colbourn & Van Oorschot, 1989; Stinson, 2008). A practical example is the design of a new smartphone. First, we need to identify the main variables that influence the product's features, such as processor type, RAM size, camera resolution, and battery life. Using Combinatorial Design, we can explore different combinations of these variables. In a lab setting, performance can be tested, and based on the goals, the most promising combination can be selected.

To calculate the total of possible combinations, we multiply the number of options for each variable.

For example, if the variables are:

Processor: (A, B, C)

RAM: (4GB, 8GB, 16GB)

Camera: (12MP, 24MP)

Battery: (3000mAh, 4000mAh)

The total number of possible combinations is 36. The final evaluation can then be done using a formula that weights the various components, considering factors such as cost.

Program 57 smartphone design

DOI: 10.4018/979-8-3373-3089-1.ch007

```
import itertools
# Definition of variables and costs
processors = {
'A': 55, # Cost in monetary units
'B': 60,
'C': 50
}
ram_memory = {
4: 20,
8: 40,
16: 60
}
cameras = {
12: 30,
24: 60
}
batteries = {
3000: 30,
4000: 40
}
# Generate combinations
combinations = list(itertools.product(processors.keys(), ram_memory.keys(), cameras.
keys(), batteries.keys()))
# Print all combinations
print("All possible combinations:")
for combination in combinations:
print(combination)
# Function to simulate performance evaluation
def evaluate_combination(combination):
processor, ram, camera, battery = combination
# Calculate costs
total_cost = (
processors[processor] +
ram_memory[ram] +
cameras[camera] +
batteries[battery]
)
# Calculate performance
performance = (
ram * 10 +
(camera / 12) * 20 +
(battery / 3000) * 15 +
(5 if processor == 'A' else 3 if processor == 'B' else 1)
)
# Calculate performance-to-cost ratio
ratio = performance / total_cost if total_cost > 0 else 0
return ratio
# Evaluate all combinations
results = {}
```

```
for combination in combinations:
score = evaluate_combination(combination)
results[combination] = score
# Sort results by score
sorted_results = sorted(results.items(), key=lambda x: x[1], reverse=True)
# Print the best combinations
print("\nTop combinations (performance-to-cost ratio):")
for combination, score in sorted_results[:5]: # Show top 5 combinations
print(f"Combination: {combination}, Ratio: {score:.2f}")
```

Another simple application example is the design of a restaurant's menu that meets customers' needs. To apply combinatorial design in this context, we could define the constraints (e.g., omnivorous option, vegetarian) starting from the available dishes and then generate the possible combinations of dishes by selecting the best ones. In this case, it is necessary to define the factors, including the diverse types of dishes, their levels (i.e., categories of dishes such as appetisers, first courses, and main courses), dietary options (omnivore and vegetarian), and prices.

Of course, a real situation is much more complex and other factors and levels should be considered.

Program 58 menu

```
from itertools import combinations
# Definition of available dishes
dishes = {
'starters': [
{'name': 'Bruschetta', 'type': 'vegetarian', 'price': 5},
{'name': 'Prosciutto and Melon', 'type': 'omnivore', 'price': 7},
{'name': 'Seafood Salad', 'type': 'omnivore', 'price': 8},
{'name': 'Tuna Carpaccio', 'type': 'omnivore', 'price': 20},
{'name': 'Hummus', 'type': 'vegetarian', 'price': 6}
],
'main_courses': [
{'name': 'Tomato Pasta', 'type': 'vegetarian', 'price': 9},
{'name': 'Mushroom Risotto', 'type': 'vegetarian', 'price': 11},
{'name': 'Lasagna Bolognese', 'type': 'omnivore', 'price': 12},
{'name': 'Spaghetti with Clams', 'type': 'omnivore', 'price': 13},
{'name': 'Bottarga Risotto', 'type': 'omnivore', 'price': 18},
{'name': 'Vegetable Agnolotti', 'type': 'vegetarian', 'price': 15}
],
'second_courses': [
{'name': 'Eggplant Parmigiana', 'type': 'vegetarian', 'price': 10},
{'name': 'Florentine Steak', 'type': 'omnivore', 'price': 18},
{'name': 'Baked Sea Bream', 'type': 'omnivore', 'price': 15},
{'name': 'Duck Breast', 'type': 'omnivore', 'price': 22},
{'name': 'Plant-Based Meat', 'type': 'vegetarian', 'price': 24}
]
}
def generate_menu(menu_type, max_budget):
valid_menus = []
for starter in dishes['starters']:
```

```
for main_course in dishes['main_courses']:
for second_course in dishes['second_courses']:
menu = [starter, main_course, second_course]
total_price = sum(dish['price'] for dish in menu)
if total_price <= max_budget:
if menu_type == 'omnivore' or all(dish['type'] in [menu_type, 'vegetarian'] for dish in
menu):
valid_menus.append((menu, total_price))
return valid_menus
def select_best_menus(valid_menus, num_menus=5):
return sorted(valid_menus, key=lambda x: x[1], reverse=True)[:num_menus]
def print_menu(menu):
for dish in menu:
print(f"{dish['name']} ({dish['type']}) - €{dish['price']}")
# User input
while True:
try:
max_budget = float(input("Enter the maximum menu cost: "))
break
except ValueError:
print("Please enter a valid number.")
while True:
print("\nChoose the menu type:")
print("1. Vegetarian")
print("2. Omnivore")
choice = input("Enter the number of your choice (1 or 2): ")
if choice == '1':
menu_type = 'vegetarian'
break
elif choice == '2':
menu_type = 'omnivore'
break
else:
print("Invalid choice. Please enter 1 or 2.")
# Generate and display menus
valid_menus = generate_menu(menu_type, max_budget)
best_menus = select_best_menus(valid_menus)
print(f"\nThe best {len(best_menus)} {menu_type} menus within a maximum budget of €{max_
budget:.2f}:")
for i, (menu, price) in enumerate(best_menus, 1):
print(f"\nMenu {i} (Total Price: €{price:.2f}):")
print_menu(menu)
if not best_menus:
print(f"No {menu_type} menus found within the budget of €{max_budget:.2f}.")
```

Another famous example is the combinatorics problem proposed by Thomas Kirkman (1806–1895) in 1850. The model we are trying to solve is a particular form of combinatorics known as "block design" (Schrijver, 2003; Dinitz, 2007), which can be solved using a block matrix. The problem states: " Fifteen young ladies in a school go out for walks three abreast for seven consecutive days: organize them daily so that no pair walks together more than once".

Formally, these are groups of subsets of dimension *k* of a set of dimension *n*, such that each subset of dimension *t* appears exactly in a subset of dimension *k*, they are called Steiner's system (Ray-Chaudhuri & Wilson, 1971; Keevash, 2018; Rosenberg et al., 2021) and are denoted by the notation S(t, k, n). In this case, the Kirkman problem can also be called S(2, 3, 15). There are exactly seven non-isomorphic solutions. Here is an example of a solution:

Giorno 1: [('A', 'B', 'C'), ('K', 'L', 'M'), ('H', 'L', 'N'), ('G', 'M', 'O'), ('F', 'N', 'I')]
Giorno 2: [('A', 'C', 'D'), ('L', 'M', 'N'), ('B', 'M', 'O'), ('H', 'N', 'I'), ('G', 'O', 'J')]
Giorno 3: [('A', 'D', 'E'), ('M', 'N', 'H'), ('C', 'N', 'I'), ('B', 'O', 'J'), ('H', 'I', 'K')]
Giorno 4: [('A', 'E', 'F'), ('N', 'H', 'I'), ('D', 'O', 'J'), ('C', 'I', 'K'), ('B', 'J', 'L')]
Giorno 5: [('A', 'F', 'G'), ('H', 'I', 'J'), ('E', 'I', 'K'), ('D', 'J', 'L'), ('C', 'K', 'M')]
Giorno 6: [('A', 'G', 'H'), ('I', 'J', 'K'), ('F', 'J', 'L'), ('E', 'K', 'M'), ('D', 'L', 'N')]
Giorno 7: [('A', 'H', 'I'), ('J', 'K', 'L'), ('G', 'K', 'M'), ('F', 'L', 'N'), ('E', 'M', 'O')]

This solution is derived from the following program:

Program 59 Kirkman's Schoolgirl Problem

```
from itertools import combinations
def kirkman_schoolgirl_problem():
girls = list('ABCDEFGHIJKLMNO') # 15 girls
days = 7 # Number of days
groups_per_day = 5 # Each day there are 5 groups
group_size = 3 # Each group has 3 girls
# Generate all possible combinations of 3 girls
all_triplets = list(combinations(girls, group_size))
# Function to check if an arrangement is valid
def check_validity(schedule, new_day):
used_pairs = set()
# Check all pairs already present in the schedule
for day in schedule:
for group in day:
for pair in combinations(group, 2):
used_pairs.add(frozenset(pair))
# Verify that the new pairs have not been used already
for group in new_day:
for pair in combinations(group, 2):
if frozenset(pair) in used_pairs:
return False
return True
# Recursively create the schedule
def generate_schedule(schedule, remaining_triplets):
# If we have filled all the days, return the schedule
if len(schedule) == days:
return schedule
# Try to fill the next day
for i in range(len(remaining_triplets)):
new_day = remaining_triplets[i:i + groups_per_day]
# Check if the new day is valid
```

```
if len(new_day) == groups_per_day and check_validity(schedule, new_day):
new_schedule = schedule + [new_day]
# Continue with the remaining schedule
result = generate_schedule(new_schedule, remaining_triplets[:i] + remaining_triplets[i +
groups_per_day:])
if result:
return result
return None
# Try to generate the schedule
final_schedule = generate_schedule([], all_triplets)
return final_schedule
# Execute the program and print the results
schedule = kirkman_schoolgirl_problem()
if schedule:
for day, triplets in enumerate(schedule, start=1):
print(f"Day {day}: {triplets}")
else:
print("Impossible to find a solution.")
```

The code employs a combination and backtracking approach, making it highly flexible. It can be adapted to:

- A different number of girls.
- Groups with different sizes.
- A different number of days.

However, once this problem is solved, some fundamental issues will remain. Are there any rules that allow us to determine whether a system defined by the three values (t, k, n) exists? Are there specific relationships between the three parameters for there to be at least one solution? The precise relationships between the parameters t, k, and n are complex and depend on the nature of the problem. In general, the more extensive t is, the more difficult it is to construct a Steiner system; the closer k is to n, the more difficult it is to construct a system.

Peter Keevash (1978-) has shown that, for sufficiently large values of n, a Steiner system S(t, k, n) exists if and only if two divisibility conditions are met:

1. (n-i) must be divisible by (k-i) for every i = 0, 1, ..., t-1
2. $\binom{n}{y}$ must be divisible by $\binom{k}{t}$.

Although seemingly trivial, this challenge has had profound implications in combinatorial optimization, a branch of mathematics focused on identifying the best solution from a set of alternatives (Papadimitriou & Steiglitz, 1998; Williamson & Shmoys, 2011).

Consider an example involving a Steiner system to provide a concrete context. Suppose an agronomist wants to assess the effect of seven different wheat varieties on a field, ensuring that each variety is grown on an equal number of plots and that every pair of varieties is compared across the same number of plots. Each variety corresponds to a point, and a "block" represents a group of plots on which three varieties are grown. This setup corresponds to the Steiner system S(2, 3, 7). A balanced and efficient experimental design is achieved by randomly assigning the varieties to plots within each group.

Program 60 agronomist

```
import random
def generate_blocks():
"""Generates the blocks of a Steiner system S(2, 3, 7).
Returns:
```

A list of lists, where each inner list represents a block.

```
"""
blocks = [[0, 1, 3], [1, 2, 4], [2, 3, 5], [3, 4, 6], [4, 5, 0], [5, 6, 1], [6, 0, 2]]
return blocks
def assign_varieties(varieties, blocks):
"""Randomly assigns varieties to the blocks.
Args:
varieties: A list of variety names.
blocks: A list of lists, representing the blocks.
Returns:
```

A list of lists, where each inner list contains the names of the varieties assigned to a block.

```
"""
plots = []
for block in blocks:
random.shuffle(block)
plots.append([varieties[index] for index in block])
return plots
def display_assignment(assignment):
"""Displays the assignment of varieties to the blocks.
Args:
assignment: A list of lists, representing the assignment of varieties.
"""
for i, block in enumerate(assignment):
print(f"Block {i+1}: {block}")
# Example usage
varieties = ['Senatore Cappelli', 'Timilia', 'Maris Wheeler', 'Vittorio', 'Rebecco',
'Triticum Compactum', 'Spelta']
blocks = generate_blocks()
assignment = assign_varieties(varieties, blocks)
display_assignment(assignment)
```

Returning to the field of what we call recreational mathematics, it is possible to examine a stimulating question proposed by Martin Gardner (1914-2010): is it possible to insert squares of sides 1, 2, 3,..., 24 into a single square of side 70 without overlapping them? Again, the problem under consideration can be traced back to a combinatorial optimisation problem and solved using an algorithm, such as the *Bottom-Left Algorithm* (BLA).

As can be seen from the graphic representation, the answer is negative.

Program 61 bottom-left graphical algorithm 1

```
import matplotlib.pyplot as plt
```

```
import matplotlib.patches as patches
class Rectangle:
def __init__(self, width, height, x=0, y=0):
self.width = width
self.height = height
self.x = x
self.y = y
def __repr__(self):
return f"Rectangle(width={self.width}, height={self.height}, x={self.x}, y={self.y})"
def bottom_left_algorithm(container_width, container_height, rectangles):
placed_rectangles = []
for rect in rectangles:
position = find_bottom_left_position(container_width, container_height, placed_rectangles,
rect)
if position is None:
print(f"Unable to place rectangle with width {rect.width} and height {rect.height}")
continue
rect.x, rect.y = position
placed_rectangles.append(rect)
return placed_rectangles
def find_bottom_left_position(container_width, container_height, placed_rectangles, rect):
best_x, best_y = None, None
min_y = container_height
for x in range(container_width - rect.width + 1):
y = 0
for placed in placed_rectangles:
if placed.x < x + rect.width and x < placed.x + placed.width:
y = max(y, placed.y + placed.height)
if y + rect.height <= container_height and y < min_y:
min_y = y
best_x, best_y = x, y
if best_x is None or best_y is None:
return None
return best_x, best_y
def draw_rectangles(placed_rectangles, container_width, container_height):
fig, ax = plt.subplots(1)
# Draw the container
ax.set_xlim(0, container_width)
ax.set_ylim(0, container_height)
ax.set_aspect('equal')
# Add the placed rectangles
for rect in placed_rectangles:
rect_patch = patches.Rectangle((rect.x, rect.y), rect.width, rect.height, edgecol-
or='black', facecolor='lightblue')
ax.add_patch(rect_patch)
# Add dimensions for each rectangle
ax.text(rect.x + rect.width/2, rect.y + rect.height/2, f'{rect.width}',
color='black', ha='center', va='center')
plt.gca().invert_yaxis() # Invert the y-axis to have the origin at the top left
```

```
plt.show()
def main():
container_width = 70
container_height = 70
# Create a list of squares with sides from 24 to 1
squares = [Rectangle(i, i) for i in range(24, 0, -1)]
# Apply the bottom-left algorithm
placed_squares = bottom_left_algorithm(container_width, container_height, squares)
# Draw the result
draw_rectangles(placed_squares, container_width, container_height)
if __name__ == "__main__":
main()
```

Figure 1. BLA

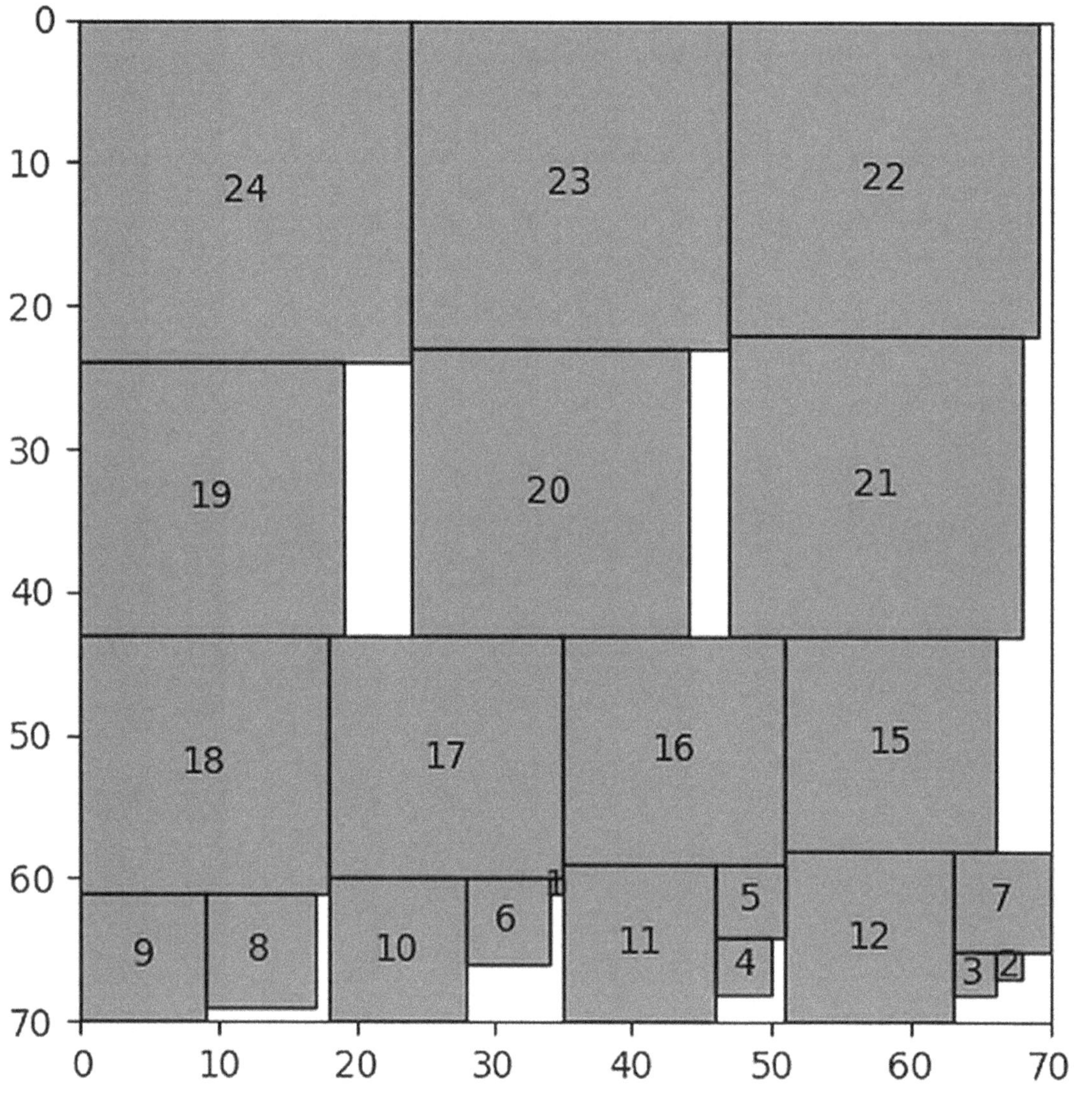

Since the problem of inserting squares is particularly interesting, the program can be modified to handle values as input.

Program 62 bottom-left graphical algorithm 2

```
import matplotlib.pyplot as plt
import matplotlib.patches as patches
class Rectangle:
def __init__(self, width, height, x=0, y=0):
self.width = width
self.height = height
```

```
self.x = x
self.y = y
def __repr__(self):
return f"Rectangle(width={self.width}, height={self.height}, x={self.x}, y={self.y})"
def bottom_left_algorithm(container_width, container_height, rectangles):
placed_rectangles = []
for rect in rectangles:
position = find_bottom_left_position(container_width, container_height, placed_rectangles,
rect)
if position is None:
print(f"Unable to place rectangle with width {rect.width} and height {rect.height}")
continue
rect.x, rect.y = position
placed_rectangles.append(rect)
return placed_rectangles
def find_bottom_left_position(container_width, container_height, placed_rectangles, rect):
best_x, best_y = None, None
min_y = container_height
for x in range(container_width - rect.width + 1):
y = 0
for placed in placed_rectangles:
if placed.x < x + rect.width and x < placed.x + placed.width:
y = max(y, placed.y + placed.height)
if y + rect.height <= container_height and y < min_y:
min_y = y
best_x, best_y = x, y
if best_x is None or best_y is None:
return None
return best_x, best_y
def draw_rectangles(placed_rectangles, container_width, container_height):
fig, ax = plt.subplots(1)
# Draw the container
ax.set_xlim(0, container_width)
ax.set_ylim(0, container_height)
ax.set_aspect('equal')
# Add the placed rectangles
for rect in placed_rectangles:
rect_patch = patches.Rectangle((rect.x, rect.y), rect.width, rect.height, edgecol-
or='black', facecolor='lightblue')
ax.add_patch(rect_patch)
# Add dimensions for each rectangle
ax.text(rect.x + rect.width/2, rect.y + rect.height/2, f'{rect.width}',
color='black', ha='center', va='center')
plt.gca().invert_yaxis() # Invert the y-axis to have the origin at the top left
plt.show()
def main():
# Ask the user to input the container dimensions and the maximum number of squares
container_width = int(input("Enter the side length of the container square: "))
container_height = container_width # Since it's a square, height = width
```

```
number_of_squares = int(input("Enter the maximum number of squares to insert (from 1 to 30): "))
if number_of_squares > 30 or number_of_squares < 1:
print("The number of squares must be between 1 and 30.")
return
# Create a list of squares with sides from number_of_squares to 1
squares = [Rectangle(i, i) for i in range(number_of_squares, 0, -1)]
# Apply the bottom-left algorithm
placed_squares = bottom_left_algorithm(container_width, container_height, squares)
# Draw the result
draw_rectangles(placed_squares, container_width, container_height)
if __name__ == "__main__":
main()
```

Code Theory can be applied in the restaurant context to assign a unique code to each menu item. Each code can represent a specific dish, and any combination of dishes that forms a complete meal can be considered a "word" in a coded alphabet. Each "word" corresponds to a unique menu option. By analysing the frequency of menu "words" used by customers, restaurants can identify the most popular dish combinations and gain insights into customer preferences, helping to optimize the menu by highlighting popular items and identifying areas for improvement.

Furthermore, Code Theory can be applied to reservations and table assignments to optimize seating arrangements and minimize waiting times. When a reservation is made, it is assigned a unique code, a combination of letters and numbers. This code contains information such as the reservation date, time, party size, type of reservation, and dietary preferences. The system can utilise this code to efficiently assign tables based on these factors, ensuring a smooth dining experience.

Additionally, the code can be used to verify the accuracy of the reservation, ensuring that the booking has been correctly received and processed. To further optimize data storage and processing, the restaurant could use a compression code to reduce the size of reservation data, making the system more efficient.

A simplified table schedule for managing ten tables might look like the following:

Program 63 restaurant codes

```
import random
class Table:
def __init__(self, number):
self.number = number
self.occupied = False
self.number_of_people = 0
class Reservation:
def __init__(self, number_of_people):
self.number_of_people = number_of_people
def generate_reservation_code():
return random.randint(10000, 99999)
def assign_reservation_to_table(reservation, tables):
for table in tables:
if not table.occupied:
table_number_of_people = table.number_of_people
if table_number_of_people >= reservation.number_of_people:
table.occupied = True
table.number_of_people = reservation.number_of_people
```

```
return table
return None
def print_table_status(tables):
for table in tables:
if table.occupied:
print(f"Table {table.number}: occupied by {table.number_of_people} people")
else:
print(f"Table {table.number}: free")
tables = [Table(i) for i in range(10)]
reservation_1 = Reservation(4)
reservation_2 = Reservation(2)
reservation_3 = Reservation(6)
table_1 = assign_reservation_to_table(reservation_1, tables)
table_2 = assign_reservation_to_table(reservation_2, tables)
table_3 = assign_reservation_to_table(reservation_3, tables)
for i in range(10):
if i % 2 == 0:
table = tables[i]
table.occupied = True
table.number_of_people = random.randint(1, 4)
print_table_status(tables)
```

The next step could refer to a problem that uses Geometric Combinatorics. The problem may arise from planning a layout for a dining room, where one needs to arrange a certain number of tables optimally within a limited space. In this context, combinatorial geometry (Dougherty, 2020) can be used to determine all possible table arrangements and assess which of them meets the specific requirements of the project, such as the minimum distance between tables to ensure guest privacy or access to emergency exits.

A well-known topic in geometric combinatorics is graph colouring, a special case of labelling where elements of a graph are assigned colours under certain constraints. In its simplest form, vertices are coloured so that adjacent ones have different colours, or edges are assigned distinct colours if they share a common vertex.

This problem has many practical applications. Think, for example, of frequency assignment in cellular networks: In a cellular network, two neighbouring radio stations cannot broadcast on the same frequency without causing interference. We can represent this situation with a graph where each vertex represents a cell, and there is an edge between two vertices if the corresponding cells are too close together. The goal is to colour the neighbouring vertices with different colours (different frequencies). If we use n colours, we can assign frequencies so there is no interference using n different frequencies. In reality, frequency assignments in cellular networks can be much more complex.

Let us use the NCTM High School Curriculum Project (2009) as an example: The Federal Communications Commission (FCC) must assign frequencies to seven new radio stations on the network. Such assignments are based on several considerations, including the possibility of creating interference by assigning the same frequency to stations that are too close to each other. In this simplified situation, we assume that transmissions from two stations located within 200 miles of each other will create interference if they transmit on the same frequency. In contrast, stations more than 200 miles apart can use the same frequency to transmit without causing interference with each other.

An example of the result, with the proposed data, can be viewed using a program such as:

Program 64 radio frequencies

```
import networkx as nx
import matplotlib.pyplot as plt
```

```
# Maximum distance within which two stations interfere with each other
MAX_DISTANCE = 200
# Radio station data (positions in x, y coordinates)
stations = {
'A': (50, 100),
'B': (100, 200),
'C': (100, 50),
'D': (200, 150),
'E': (300, 0),
'F': (350, 250),
'G': (300, 300)
}
# Function to calculate the Euclidean distance between two points
def euclidean_distance(p1, p2):
return ((p1[0] - p2[0]) ** 2 + (p1[1] - p2[1]) ** 2) ** 0.5
# Create the graph
G = nx.Graph()
# Add nodes
for station in stations:
G.add_node(station)
# Add edges between stations that are within MAX_DISTANCE miles of each other
for station1 in stations:
for station2 in stations:
if station1 != station2 and euclidean_distance(stations[station1], stations[station2]) <=
MAX_DISTANCE:
G.add_edge(station1, station2)
# Color the graph using the greedy algorithm
colors = nx.coloring.greedy_color(G, strategy="largest_first")
# Define a list of shades of gray
num_colors = len(set(colors.values()))
shades_of_gray = [str(i / (num_colors - 1)) for i in range(num_colors)]
# Map numerical colors to available colors
color_map = [shades_of_gray[colors[station]] for station in G.nodes]
# Visualize the graph with assigned colors
pos = nx.spring_layout(G) # Node positions for visualization
# Draw the graph with black borders for nodes
nx.draw(
G,
pos,
with_labels=True,
node_color=color_map,
edgecolors='black', # Node border color
font_weight='bold',
node_size=700,
font_size=12,
edge_color='gray'
)
plt.title("Radio Stations Graph with Shades of Gray and Black Borders")
plt.show()
```

Figure 2. A possible outcome of the assigned frequencies

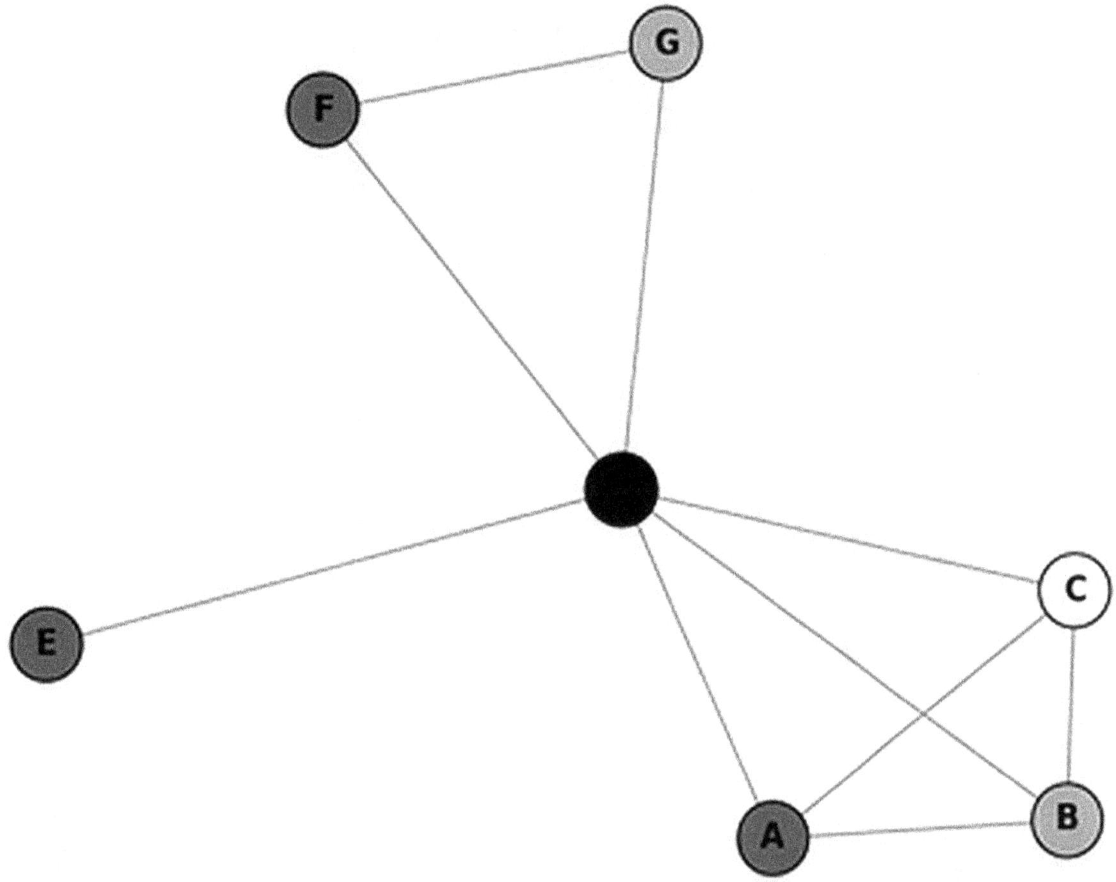

Another widely studied challenge is the surveillance of an art gallery (or museum), which can be modelled as a coverage or monitoring task in a geometric space. It can be defined as follows: How many video cameras located on the vertices of the plan of a room, having the shape of a simple polygon of n sides, are necessary and sufficient to monitor all the points of the room itself?

Figure 3. Art gallery map

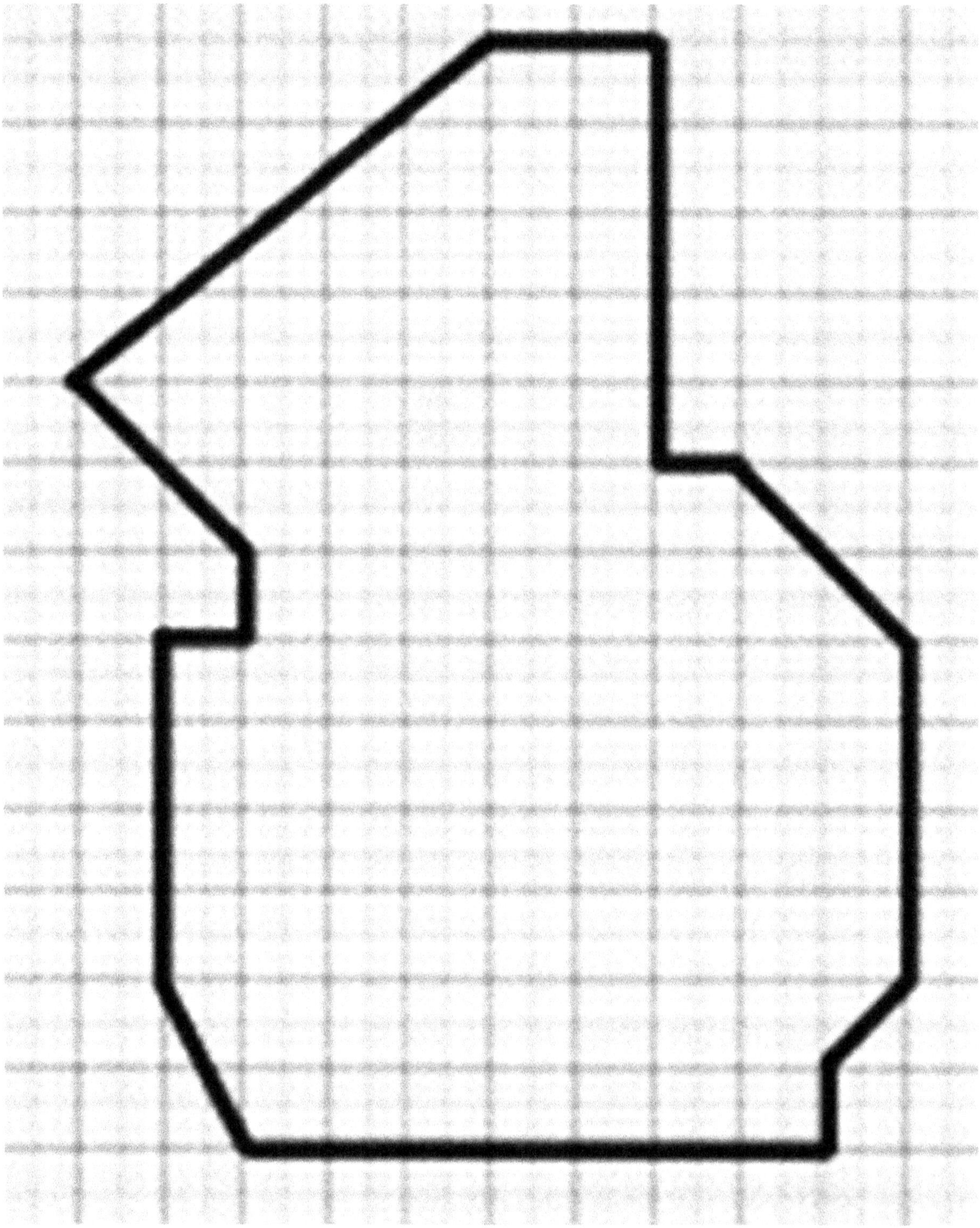

We divide the room into triangles and color each vertex so that we do not have contiguous vertices with the same color.

Figure 4. Camera placement

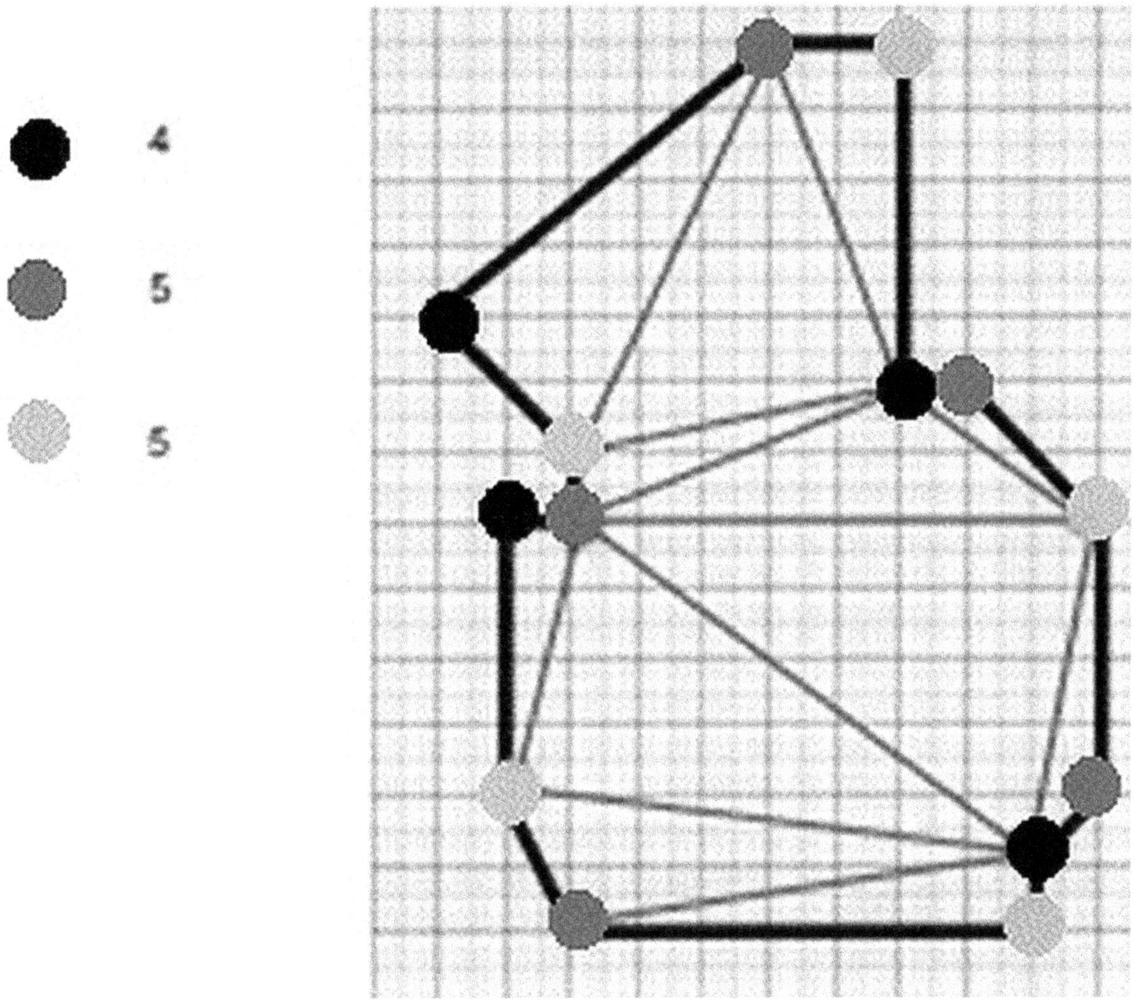

The cameras must be placed in the least coloured points and the number of them will be: cameras = int(vertices/3)

Algebraic Combinatorics is a particularly important sub-discipline of combinatorics.

Let us consider a simple problem. A classic example of algebraic combinatorics is the use of Stirling numbers of the second type to count partitions of sets of n elements into k non-empty subgroups. Suppose we want to determine how many ways we can divide the set (A, B, C, D) into two groups. We want to calculate S(4,2). These Stirling values can be found using the following recurrence relation:

$$S(n,k) = k \cdot S(n-1,k) + S(n-1,k-1)$$

We can calculate the Stirling numbers:

$S(3,2) = 2 \cdot S(2,2) + S(2,1) = 3$ since $S(2,2) = 1$ (two elements in separate groups) and $S(2,1) = 1$ (both in one group)

$S(3,1) = 1$ because all the elements are in a single group

In conclusion, we have $S(4,2) = 2 \cdot 3 + 1 = 7$

So, there are seven ways to partition a set of 4 items into two non-empty subgroups.

Of course, the problem becomes more complicated for higher values; for example, with n = 10 and k = 4, there are 34,105 modes..

Program 65 Stirling's partition

```
from itertools import combinations
def generate_partitions(n, k):
"""Generates all partitions of n elements into k non-empty groups."""
elements = list(range(1, n + 1))
results = []
def recursive_partitions(elements, k, current_partition):
# If we have exhausted the elements and required groups
if k == 0 and not elements:
results.append(current_partition)
return
if k <= 0 or not elements:
return
# Generate combinations for the first group
for i in range(1, len(elements) + 1):
for comb in combinations(elements, i):
new_elements = [x for x in elements if x not in comb]
recursive_partitions(new_elements, k - 1, current_partition + [list(comb)])
recursive_partitions(elements, k, [])
# Remove duplicates by sorting and converting to a set
unique_results = {tuple(sorted(map(tuple, part))) for part in results}
return [list(map(list, part)) for part in unique_results]
# Input
n = int(input("Enter the value of n: "))
k = int(input("Enter the value of k: "))
# Calculate the partitions
subgroups = generate_partitions(n, k)
# Print the partitions
print(f"Number of ways to partition a set of {n} elements into {k} subgroups: {len(sub-
groups)}")
for idx, partition in enumerate(subgroups, 1):
print(f"Partition {idx}: {partition}")
```

Algebraic combinatorics focuses on structures called graphs, which consist of objects known as vertices (or nodes) connected by arcs. A key concept in this theory is *graph colouring*, which is a specific type of graph labelling. This concept is fundamental, as it has numerous practical applications, such as resource allocation, network design, and scheduling.

A significant aspect of graph colouring is its connection to the *chromatic polynomial* (Dong & Koh, 2022; Sagan, 2022). This polynomial provides a method for counting the number of valid ways to colour a graph using a specified set of colours. Specifically, the chromatic polynomial, denoted as P(G,n), gives the total ways to assign *n* colours to a graph *G* such that no two adjacent vertices share the same colour.

For example, the graph shown in the image cannot be coloured using just two colours, but with three colours, somebody can colour it in 12 different ways.

Figure 5. Graph with four nodes

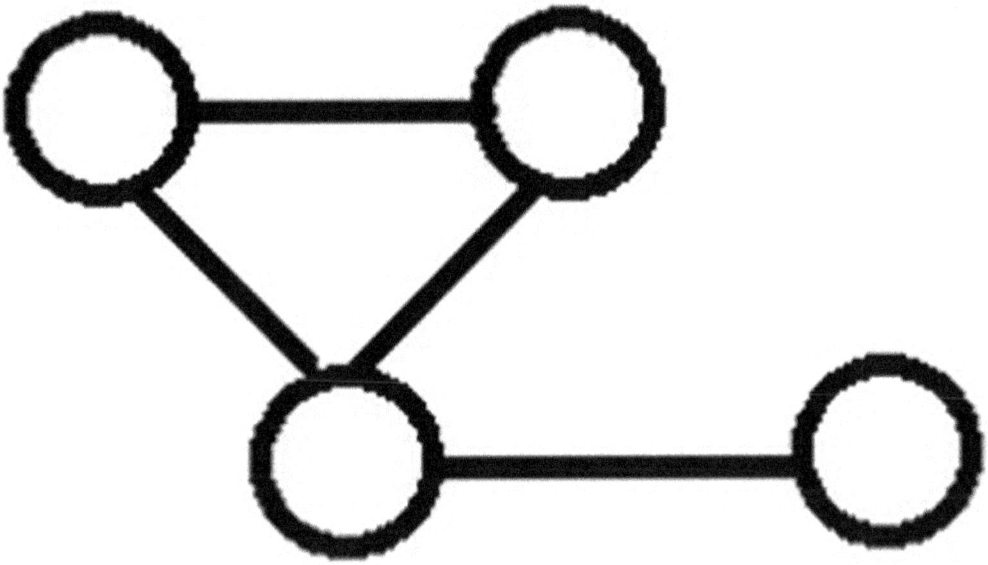

Returning to the problem of the restaurant, if it offers home delivery services, graph theory can be used to plan the service. Each delivery location and each rider's starting point can be represented as nodes in the graph, while the roads or streets between them can be represented as arcs. Delivery can be optimised using shortest route search algorithms to reduce delivery times and operating costs. Upon examining the problem, we observe that it is a classic situation related to graphs.

Given the complexity, which must consider numerous variables, it is possible to provide a trivial program that orders the deliveries of riders to four manufacturers based on the distance from the restaurant. In theory, there are 24 possible routes between the restaurant and the four houses. Each one represents a different route based on the order in which the houses are visited.

Program 66 rider route graph

```
import matplotlib.pyplot as plt
# Define the names of the houses
houses = ["House1", "House2", "House3", "House4"]
# Request travel times from input
travel_times = {}
for house in houses:
while True:
try:
time = float(input(f"Enter the travel time from Restaurant to {house} (in minutes): "))
if time < 0:
```

```
raise ValueError("Time must be a positive value.")
travel_times[house] = time
break
except ValueError as e:
print(f"Error: {e}. Please enter a valid numeric value.")
# Sort the houses based on travel times
sorted_houses = sorted(houses, key=lambda house: travel_times[house])
# Build the complete route from the restaurant to the houses based on the sorting
complete_route = ['Restaurant'] + sorted_houses + ['Restaurant']
print(f"Optimal route from the restaurant: {complete_route}")
# Extract the travel times based on the sorting of the houses
sorted_times = [travel_times[house] for house in sorted_houses]
# Create the plot with a white background and black graph
plt.figure(figsize=(10, 6))
plt.plot(sorted_times, 'ko-', markersize=8, linewidth=2, label='Travel time') # 'ko-' in-
dicates black line with black markers
plt.xticks(range(len(sorted_houses)), sorted_houses, color='black')
plt.yticks(color='black')
plt.xlabel("Houses", color='black')
plt.ylabel("Travel time (minutes)", color='black')
plt.title("Optimal Route Graph", color='black')
plt.gca().set_facecolor('white') # Set the background of the graph to white
plt.grid(True, color='gray', linestyle='--', linewidth=0.5) # Gray grid for contrast
plt.legend()
plt.show()
```

Another important example related to graph theory refers to Hasse diagrams (Davey & Priestley, 2002; Gratzer, 2011). These diagrams are a graphical representation of a *poset* (partially ordered set), a set of elements with a partial-order relationship. The main characteristic is that not every pair of elements must be comparable, as in a completely ordered set (e.g., natural numbers). In other words, a Hasse diagram illustrates the arrangement of components in a partial order without displaying all the derived orders. Only direct links between elements that are "immediately" connected, i.e., those with no other elements in between, are displayed.

Example: Suppose we have a set S = {1, 2, 3, 4} with the natural order (i.e., $a \leq b$ if and only if *a* divides *b*). The Hasse diagram of this poset will consist of nodes 1, 2, 3, 4 and the connections that show the immediate divisibility relationships. For example, one is connected to all other nodes, two to 4, and so on. Let us take a look at a real-world situation where diagrams find practical application in the context of project management. Imagine managing a project that involves multiple tasks to be completed. Some of these tasks must be completed before others can begin, and others must be completed in parallel. A Hasse diagram represents the dependency relationship between tasks. A program can quickly provide a representation, considering the required activities and the dependencies. Imagine working on a project that includes the following tasks:

A. Initial planning
B. Data collection
C. Data analysis
D. Design
E. Software development
F. Test
G. Writing Documentation
H. Final delivery

In addition, there are dependencies between tasks:

1. A must be completed before B and D.
2. B before C and E can begin.
3. C before F can begin.
4. D before E can begin.
5. E before F can begin.
6. F before H can begin.
7. G is parallel to F and must be completed before H can begin.

To represent these dependencies in a Hasse diagram, we draw a node for each task and connect the nodes with edges that indicate the predecessor relationship.

Program 67 Hasse diagram

```
import networkx as nx
import matplotlib.pyplot as plt
# Function to create the Hasse diagram
def create_hasse_diagram(activities, dependencies):
G = nx.DiGraph() # Directed graph to represent dependencies
# Add nodes (activities)
for activity in activities:
G.add_node(activity)
# Add edges based on dependencies
for predecessor, successor in dependencies:
G.add_edge(predecessor, successor)
return G
# Function to visualize the Hasse diagram
def visualize_hasse_diagram(G):
# Custom positions for the nodes
pos = {
'A': (0, 6), # Position A at the top
'B': (-1, 5), # Position B below A
'D': (1, 5), # Position D below A
'C': (-1, 4), # Position C below B
'E': (1, 4), # Position E below D
'F': (0, 3), # Position F below C and E
'G': (0, 2), # Position G below C (parallel to F)
'H': (0, 1) # Position H below F and G
}
# Draw the graph
plt.figure(figsize=(8, 6))
nx.draw(G, pos, with_labels=True, node_size=3000, node_color="grey", font_size=10, font_
weight="bold", arrowsize=20)
plt.title("Hasse Diagram of Project Activities", fontsize=14)
plt.show()
# Definition of activities
activities = ['A', 'B', 'C', 'D', 'E', 'F', 'G', 'H']
```

```
# Definition of dependencies: (previous activity, next activity)
dependencies = [
('A', 'B'), ('A', 'D'),
('B', 'C'), ('B', 'E'),
('C', 'F'),
('D', 'E'),
('E', 'F'),
('F', 'H'),
('G', 'H')
]
# Create the Hasse diagram
G = create_hasse_diagram(activities, dependencies)
# Visualize the diagram
visualize_hasse_diagram(G)
```

The resulting Hasse diagram will be, in this case:

Figure 6. Activity diagram

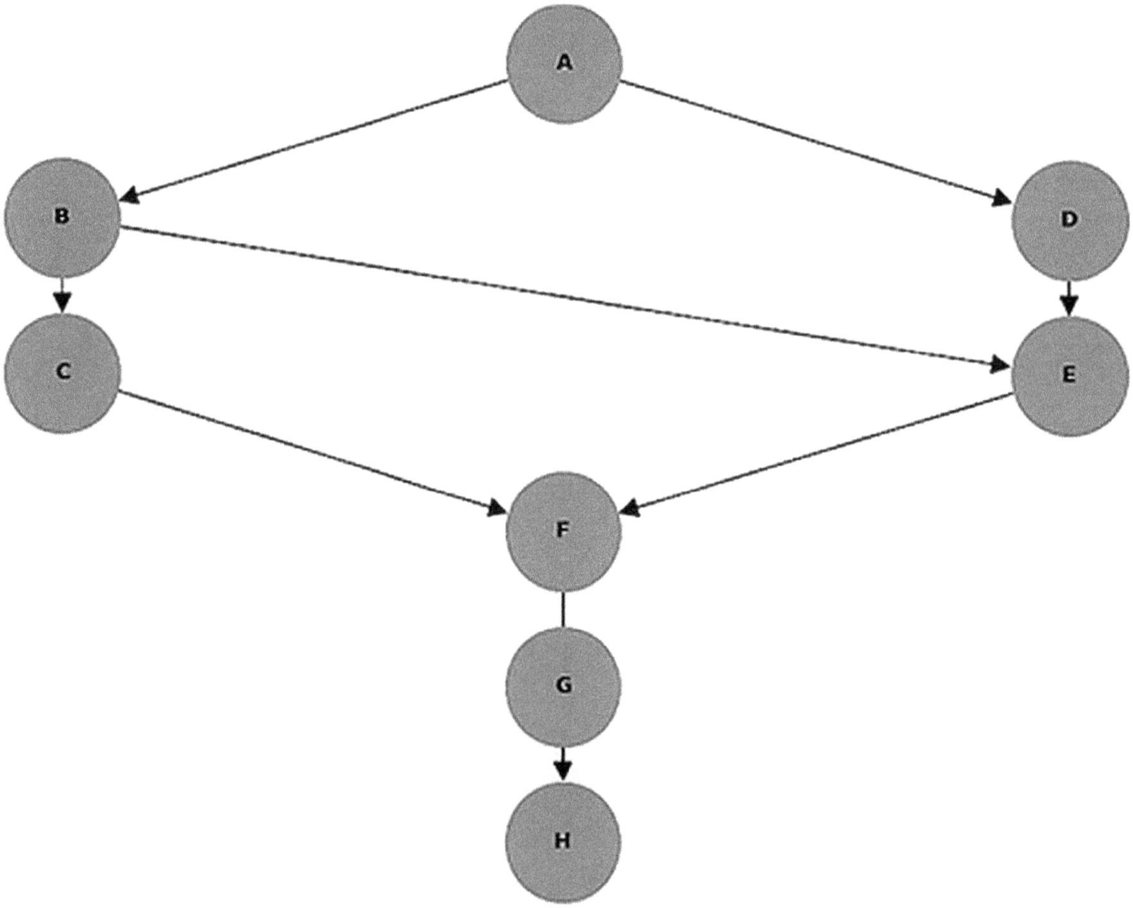

Hasse diagrams are fundamental visual tools for representing and analysing partial-order structures in combinatorics. A partial order is a binary relation on a reflexive, antisymmetric, and transitive set, but not necessarily total, i.e., not every pair of elements is comparable. Hasse diagrams, an effective way of visualizing these orders, are handy in various areas of combinatorics, where the relationships between elements of a set or structure need to be explored. A classic application of Hasse diagrams is the representation of partitions using integers. A partition of a natural number n is subdivided into sums of positive numbers, where we do not consider the order of the addends. In other words, two partitions that differ only in the order of the addends are considered equivalent. One can use Hasse diagrams to represent the order of inclusion between partitions of a number, where one partition is "less" than another if one is a subset of the other in terms of the sum of the addends. For example, the partitions of the 4 are: 4, 3 + 1, 2 + 2, 2 + 1 + 1, 1 + 1 + 1 + 1.

In a Hasse diagram, each node represents a partition, and there is a direct arc between two nodes if the lower partition can be transformed into the upper one by adding a single "1" or combining addends. This type of diagram provides a clear visual representation of all partitions of an integer, highlighting inclusion relationships and making it easier to explore combinatorial properties, such as counting the total partitions or identifying specific ones. In the Hasse diagram, the subgroups are arranged according to the inclusion relationship, with the arcs connecting each subgroup to those that contain it immediately. In this case, the diagram will display a pyramidal pattern, with the smallest subgroups at the base and the most significant ones at the apex. This type of diagram helps visualise a group's order structure and solve problems related to group theory, such as finding maximum or minimum subgroups or calculating the size of a subset compared to others.

Hasse diagrams are fundamental visual tools in combinatorics to represent the partial-order relationships between elements of various sets or structures. We found them in various applications, including representing partitions and displaying inclusions between subsets. Visualising order relations helps simplify and make complex combinatorial concepts easier to understand, making it easier to solve combinatorial problems and grasp the underlying mathematics.

REFERENCES

Colbourn, C. J., & Van Oorschot, P. C. (1989). Applications of combinatorial designs in computer science. *ACM Computing Surveys*, *21*(2), 223–250. DOI: 10.1145/66443.66446

Davey, B. A., & Priestley, H. A. (2002). *Introduction to lattices and order*. Cambridge university press. DOI: 10.1017/CBO9780511809088

Dinitz, J. H. (2007). *Handbook of combinatorial designs*. Chapman & Hall/CRC.

Dong, F., & Koh, K. M. (2022). Foundations of the chromatic polynomial. In *Handbook of the Tutte Polynomial and Related Topics* (pp. 213–251). Chapman and Hall/CRC. DOI: 10.1201/9780429161612-11

Dougherty, S. T. (2020). *Combinatorics and finite geometry* (Vol. 58). Springer. DOI: 10.1007/978-3-030-56395-0

Gratzer, G. A. (2011). *Lattice theory: foundation* (Vol. 2). Birkhäuser. DOI: 10.1007/978-3-0348-0018-1

Keevash, P. (2018). The existence of designs II. *arXiv preprint arXiv:1802.05900.*

Papadimitriou, C. H., & Steiglitz, K. (1998). *Combinatorial optimization: algorithms and complexity*. Courier Corporation.

Ray-Chaudhuri, D. K., & Wilson, R. M. (1971). Solution of Kirkman's schoolgirl problem. In *Proc. symp. pure Math* (Vol. 19, pp. 187-203). DOI: 10.1090/pspum/019/9959

Rosenberg, M., French, T., Reynolds, M., & While, L. (2021, June). A genetic algorithm approach for the Euclidean Steiner tree problem with soft obstacles. In *Proceedings of the Genetic and Evolutionary Computation Conference* (pp. 618-626). DOI: 10.1145/3449639.3459397

Sagan, B. E. (2022). The amazing chromatic polynomial. *The Mathematical Intelligencer*, *44*(3), 203–211. DOI: 10.1007/s00283-021-10136-z

Schrijver, A. (2003). Combinatorial optimization: polyhedra and efficiency: Vol. 24. *No. 2*. Springer.

Stinson, D. R. (2008). Combinatorial designs: Constructions and analysis. *ACM SIGACT News*, *39*(4), 17–21. DOI: 10.1145/1466390.1466393

Williamson, D. P., & Shmoys, D. B. (2011). *The design of approximation algorithms*. Cambridge university press. DOI: 10.1017/CBO9780511921735

Chapter 8
Graphs and Recursive Counting

ABSTRACT

The chapter introduces graphs as models for representing objects and relationships, defining concepts such as the Hamiltonian cycle and the Dirac theorem for their analysis. Next, the chapter explores recursive counting, a powerful method that solves complex problems by breaking them into simpler subproblems. This method is applied to the computation of permutations and combinations, highlighting how a recursive approach simplifies the solving of combinatorial problems. The chapter also discusses applications of graphs and recursive counting in various fields, such as network design, planning, and data analysis. Several examples show how graphs and recursive counting can be used to solve combinatorial problems. These examples include the analysis of social platforms, urban road systems, protein interactions, and the simulation of peer-to-peer (P2P) networks. Finally, a problem inspired by the Cretan labyrinth is examined, and several approaches to finding possible paths in a maze are discussed, such as the A algorithm, breadth-first search (BFS), and a recursive approach.*

GRAPHS

In Italy, the shortest line between two points is the arabesque

Ennio Flaiano

Let us imagine a group of people, and we want to represent the friendships between them. Each person is represented as a *node* (a circle), and every friendship is represented as an *arc* (a line) connecting the nodes. A combinatorial graph is an abstract graph representing a finite collection of objects and their relationships or connections. In a combinatorial graph, edges are defined to satisfy specific rules limiting their structure. Arc analysis is important in combinatorics and involves many concepts and theorems.

Examine the following problem: *2n* people sit at a circular table; everyone is friends with at least *n* other person. Can two friends always occupy two seats next to each other? To address this problem, we use graph theory (West, 2001; Chartrand & Zhang, 2013; Gross et al., 2018; Diestel, 2024). We consider each person as a node in a graph that we call *G*. An arc connects two nodes if the two corresponding people are friends. Then, we create a graph where the nodes represent individuals, and the edges represent the friendship relationships between them. The text states that "everyone is a friend of at least *n* people". In terms of graphs, this means that each node has a degree of at least *n* in *G*. The degree of a node is the number of arcs connected to it. So, each person (node) has at least n friends (connected arcs). If we can find a cycle that visits each node exactly once in our graph, known as a Hamiltonian cycle, then we can answer the question.

We can use Dirac's Theorem to prove that there is a Hamiltonian cycle in our graph: "A graph with *n* nodes has a Hamiltonian cycle if each node has a degree of at least n/2". We must verify if the condition of Dirac's Theorem is satisfied: is it true that $n \geq (2n)/2$? Yes, *n* equals $(2n)/2 = n$. Thus, Dirac's Theorem guarantees us that there is a Hamiltonian cycle in graph *G*. We can arrange people around the circular table following the order of the nodes in this cycle. Each person will sit next to two friends in this arrangement, solving the problem. In conclusion, using graph theory and Dirac's Theorem, we have shown that it is always possible to arrange *2n* people around a circular table so that two neighbours are friends if each is friends with at least *n* people.

DOI: 10.4018/979-8-3373-3089-1.ch008

Figure 1. Six friends around a circular table

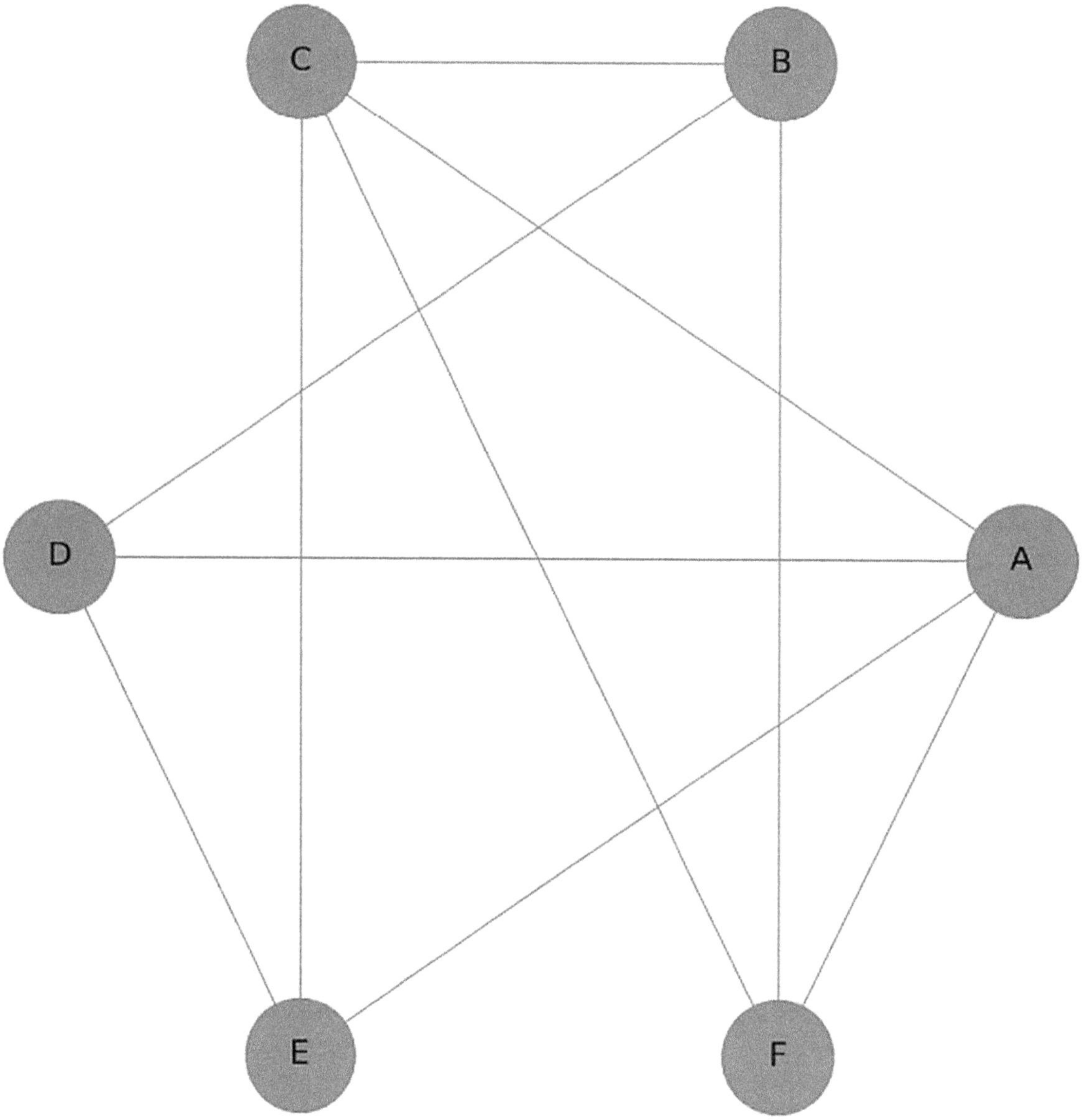

The graph is produced by a program that simulates the creation of friendships between a group of people and tests whether it is possible to arrange them around a circular table so that each person is seated next to two friends.

Friendships generated:

A: C, F, E, D
B: C, F, D
C: A, B, F, E
D: A, B, E
E: A, D, C

F: A, B, C

Valid arrangement found: ('A', 'C', 'E', 'D', 'B', 'F')

Program 68 friends

```
import networkx as nx
import matplotlib.pyplot as plt
import random
import itertools
# Data initialization
people = ['A', 'B', 'C', 'D', 'E', 'F']
num_people = len(people)
min_friends = 3
# Function to generate friendships
def generate_friendship(people, min_friends):
friendship = {person: set() for person in people}
num_people = len(people)
for person in people:
while len(friendship[person]) < min_friends:
possible_friends = set(people) - friendship[person] - {person}
if not possible_friends:
break
chosen_friend = random.choice(list(possible_friends))
friendship[person].add(chosen_friend)
friendship[chosen_friend].add(person)
return friendship
# Friendship generation
friendship = generate_friendship(people, min_friends)
# Print generated friendships
print("Generated friendships:")
for person, friends in friendship.items():
print(f"{person}: {', '.join(friends)}")
# Creating the friendship graph
G = nx.Graph()
# Adding nodes and edges
for person in people:
G.add_node(person)
for person, friends in friendship.items():
for friend in friends:
G.add_edge(person, friend)
# Circular positioning of nodes for visualization
pos = nx.circular_layout(G)
# Drawing the graph
plt.figure(figsize=(8, 8))
nx.draw(G, pos, with_labels=True, node_color='grey', node_size=3000, font_size=16, font_
color='black', edge_color='gray')
plt.title("Simulation of friendship relations")
```

```
plt.show()
# Check of the arrangement around the table
def check_arrangement(friendship, people):
n = len(people)
for i in range(n):
current = people[i]
left = people[(i - 1) % n]
right = people[(i + 1) % n]
if not (left in friendship[current] and right in friendship[current]):
return False
return True
# Try to find a valid arrangement
found = False
for perm in itertools.permutations(people):
if check_arrangement(friendship, perm):
print("Valid arrangement found:", perm)
found = True
break
if not found:
print("No valid arrangement found")
```

A broad class of problems refers to situations where the goal is to optimise tasks that must be performed sequentially. This optimisation deals with discrete "objects," such as graphs, by studying how to arrange, group, order, choose, and combine these elements most effectively. Among the classical problems involving combinatorial graphs are the search for the shortest path, the search for minimum coverage trees (i.e., the subset of edges of a graph that can connect all the vertices with the minimum total cost), and assigning colours so that adjacent vertices have different colours, among others.

There are many examples of applications: social platforms can be modelled as graphs, where the nodes are the users and the edges are friendship connections or other; the urban road system can be seen as a graph in which the nodes are the intersections, and the arcs are the roads; interactions between proteins involve biochemical networks in which proteins make up the nodes while the edges identify the interactions.

A special case is constituted by complete graphs in which each vertex has a direct connection to all the remaining vertices. In this case, the number of edges, given n vertices, is given (as we will see later) by the combinations of vertices taken in pairs. We could have the graphs visible in the figure:

Figure 2. Some graphs

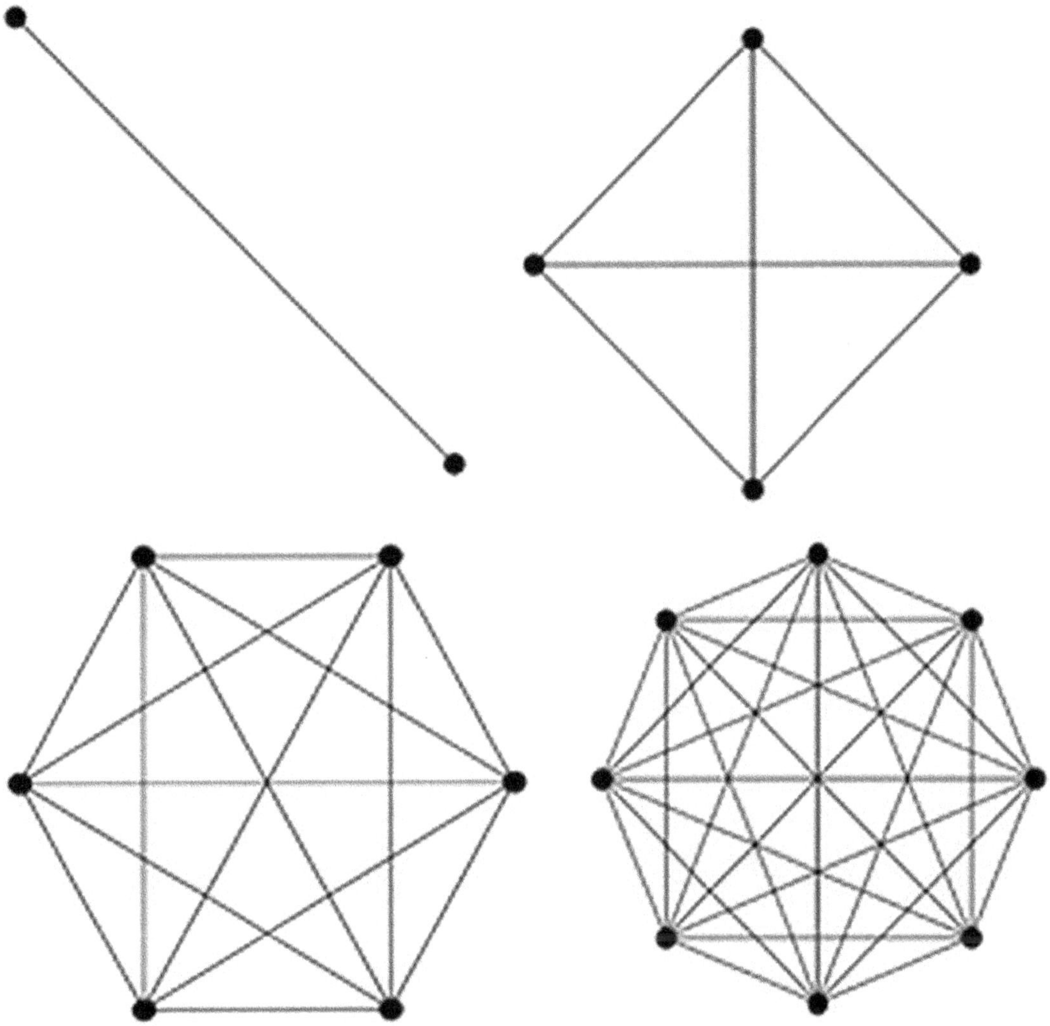

The arcs will be 1, 6, 15, and 28.

A significant example is the peer-to-peer (P2P) network (Manduva, 2024) simulation for communication between devices. In this case, there are several connections between combinatorics and here are some key points where combinatorics is relevant:

1. The P2P network can be represented as a graph, and analysing its properties requires combinatorial knowledge.
2. When one creates a network and devices are connected, we use combinatorics. The combinations $\binom{n}{2}$ gives the number of possible connections between n devices.
3. The function that simulates communication randomly chooses a sender and receiver between devices; this is a combinatorial problem of random selection without repetition.
4. Message routing algorithms in the network can be analysed and optimized using combinatorial techniques.

To reduce complexity, we can think of connecting only a part of the possible pairs, in the following program 50%.

Program 69 P2P network

```
import random
import time
import hashlib
import uuid
import networkx as nx
import matplotlib.pyplot as plt
# Attiva la modalità interattiva di matplotlib
plt.ion()
# Messaggi memorizzati
all_messages = []
class Message:
def __init__(self, sender, recipient, content):
self.id = str(uuid.uuid4())
self.sender = sender
self.recipient = recipient
self.content = content
self.timestamp = time.time()
self.hash = self._generate_hash()
def _generate_hash(self):
data = f"{self.id}{self.sender}{self.recipient}{self.content}{self.timestamp}"
return hashlib.sha256(data.encode()).hexdigest()[:10]
def __str__(self):
return f"[{self.hash}] {self.sender} -> {self.recipient}: {self.content}"
class Device:
def __init__(self, name):
self.name = name
self.connections = []
self.received_messages = set()
def add_connection(self, device):
if device not in self.connections:
self.connections.append(device)
device.add_connection(self)
def send_message(self, recipient, content, G):
message = Message(self.name, recipient.name, content)
if recipient in self.connections:
all_messages.append(str(message))
recipient.receive_message(message)
G[self.name][recipient.name]['weight'] += 1
else:
error = f"No direct connection between {self.name} and {recipient.name}"
all_messages.append(error)
def receive_message(self, message):
if message.id not in self.received_messages:
self.received_messages.add(message.id)
```

```
all_messages.append(f"Received: {message}")
Device.receive_message = receive_message # Aggiunto fuori dalla classe
def create_network():
return [Device(f"Node{num}") for num in range(1, 11)]
def connect_network(devices):
for i in range(len(devices)):
for j in range(i + 1, len(devices)):
if random.random() < 0.5:
devices[i].add_connection(devices[j])
return devices
# Inizializza figura matplotlib e assi
fig, (ax1, ax2) = plt.subplots(1, 2, figsize=(12, 6), gridspec_kw={'width_ratios': [2,
1]})
def draw_graph(G):
ax1.clear()
ax2.clear()
pos = nx.spring_layout(G, seed=42) # layout fisso per coerenza visiva
edges = G.edges()
weights = [G[u][v].get('weight', 1) for u, v in edges]
nx.draw(G, pos, ax=ax1, with_labels=True, node_size=1000, node_color='skyblue',
edge_color='gray', width=[w * 2 for w in weights], font_size=9)
ax1.set_title("P2P Network Graph")
ax2.axis("off")
ax2.text(0, 1, "Recent Messages", fontsize=10, fontweight='bold')
messages_to_show = all_messages[-15:]
for i, msg in enumerate(messages_to_show, start=1):
ax2.text(0, 1 - i * 0.06, msg, fontsize=8)
plt.tight_layout()
plt.draw()
plt.pause(0.5)
def generate_message_content():
contents = [
"Data sync request", "Ping", "Update status", "Share file", "Join request"
]
return random.choice(contents)
def simulate_communication(network, G):
sender = random.choice(network)
recipient = random.choice([d for d in network if d != sender])
content = generate_message_content()
sender.send_message(recipient, content, G)
draw_graph(G)
def main():
network = connect_network(create_network())
# Crea il grafo di rete
G = nx.Graph()
for device in network:
G.add_node(device.name)
for conn in device.connections:
G.add_edge(device.name, conn.name, weight=1)
```

```
draw_graph(G)
# Simula comunicazione
for _ in range(10):
simulate_communication(network, G)
time.sleep(3)
plt.ioff()
plt.show()
if __name__ == "__main__":
main()
```

To conclude this part, examining a program that simulates exam scheduling may be interesting. In this case, let us assume that a university needs to plan exams to avoid overlaps for students who follow standard courses. Each node represents an exam, and two nodes are connected if at least one student must take both exams.

The program automatically builds the conflict graph based on each student's exams by implementing a greedy algorithm that sorts exams by number of conflicts, assigns the first available time slot to each exam, and verifies that the schedule is valid by checking that no student has overlapping exams.

It is a GUI version that includes an input area for adding students and their exams, as well as a conflict graph display. Nodes represent exams, and arcs represent conflicts. The data summary displays a list of students and their exams, including conflicts for each exam, and shows the generated calendar.

The program code is sufficiently long to articulate the problem and accurately represent the graphical window.

Program 70 planning of exams

```
import tkinter as tk
from tkinter import ttk, messagebox
from math import cos, sin, pi
import random
class ExamScheduler:
def __init__(self):
self.exam_graph = {} # graph of exams
self.colors = {} # assigned time slots
self.students = {} # dictionary student -> list of exams
def add_exam(self, exam_name):
"""Adds a new exam to the graph."""
if exam_name not in self.exam_graph:
self.exam_graph[exam_name] = set()
def add_student_exams(self, student_id, exam_list):
"""Records a student's exams and updates the conflict graph."""
self.students[student_id] = exam_list
# Adds all exams to the graph if they do not exist
for exam in exam_list:
self.add_exam(exam)
# Creates connections between all exams of the student
for exam1 in exam_list:
for exam2 in exam_list:
if exam1 != exam2:
self.exam_graph[exam1].add(exam2)
self.exam_graph[exam2].add(exam1)
```

```
def find_available_color(self, exam):
"""Finds the first available time slot for an exam."""
used_colors = {self.colors.get(neighbor)
for neighbor in self.exam_graph[exam]
if neighbor in self.colors}
color = 1
while color in used_colors:
color += 1
return color
def color_graph(self):
"""Colors the graph using the greedy algorithm."""
self.colors = {}
# Sort exams by number of conflicts (node degree) in descending order
exams_by_conflicts = sorted(
self.exam_graph.keys(),
key=lambda x: len(self.exam_graph[x]),
reverse=True
)
for exam in exams_by_conflicts:
self.colors[exam] = self.find_available_color(exam)
def get_schedule(self):
"""Returns the exam schedule."""
if not self.colors:
self.color_graph()
schedule = {}
for exam, slot in self.colors.items():
if slot not in schedule:
schedule[slot] = []
schedule[slot].append(exam)
return schedule
def verify_schedule(self):
"""Verifies if the schedule is valid."""
if not self.colors:
return False
for student, exams in self.students.items():
used_slots = {}
for exam in exams:
slot = self.colors[exam]
if slot in used_slots:
return False
used_slots[slot] = exam
return True
class ExamSchedulerGUI:
def __init__(self, root):
self.root = root
self.root.title("Exam Scheduler")
self.root.geometry("1200x800")
self.scheduler = ExamScheduler()
self.create_gui()
```

```
self.load_sample_data()
def create_gui(self):
# Main frame divided into two columns
self.main_frame = ttk.PanedWindow(self.root, orient=tk.HORIZONTAL)
self.main_frame.pack(fill=tk.BOTH, expand=True, padx=5, pady=5)
# Left column for input and controls
self.left_frame = ttk.Frame(self.main_frame)
self.main_frame.add(self.left_frame)
# Right column for graph and schedule display
self.right_frame = ttk.Frame(self.main_frame)
self.main_frame.add(self.right_frame)
self.create_input_area()
self.create_data_view()
self.create_graph_canvas()
self.create_calendar_view()
def create_input_area(self):
input_frame = ttk.LabelFrame(self.left_frame, text="Data Entry", padding=5)
input_frame.pack(fill=tk.X, padx=5, pady=5)
# Student input
ttk.Label(input_frame, text="Student ID:").grid(row=0, column=0, padx=5, pady=5)
self.student_var = tk.StringVar()
ttk.Entry(input_frame, textvariable=self.student_var).grid(row=0, column=1, padx=5,
pady=5)
# Exams input
ttk.Label(input_frame, text="Exams (comma separated):").grid(row=1, column=0, padx=5,
pady=5)
self.exams_var = tk.StringVar()
ttk.Entry(input_frame, textvariable=self.exams_var).grid(row=1, column=1, padx=5, pady=5)
# Buttons
ttk.Button(input_frame, text="Add Student",
command=self.add_student).grid(row=2, column=0, padx=5, pady=5)
ttk.Button(input_frame, text="Generate Schedule",
command=self.generate_schedule).grid(row=2, column=1, padx=5, pady=5)
def create_data_view(self):
self.data_frame = ttk.LabelFrame(self.left_frame, text="Data Summary", padding=5)
self.data_frame.pack(fill=tk.BOTH, expand=True, padx=5, pady=5)
self.data_text = tk.Text(self.data_frame, height=10, width=40)
self.data_text.pack(fill=tk.BOTH, expand=True)
def create_graph_canvas(self):
graph_frame = ttk.LabelFrame(self.right_frame, text="Conflict Graph", padding=5)
graph_frame.pack(fill=tk.BOTH, expand=True, padx=5, pady=5)
self.graph_canvas = tk.Canvas(graph_frame, background='white')
self.graph_canvas.pack(fill=tk.BOTH, expand=True)
# Add handler for resizing
self.graph_canvas.bind('<Configure>', lambda e: self.update_display())
def create_calendar_view(self):
self.calendar_frame = ttk.LabelFrame(self.right_frame, text="Exam Schedule", padding=5)
self.calendar_frame.pack(fill=tk.BOTH, expand=True, padx=5, pady=5)
self.calendar_text = tk.Text(self.calendar_frame, height=10)
```

```
self.calendar_text.pack(fill=tk.BOTH, expand=True)
def draw_graph(self):
self.graph_canvas.delete('all')
# Calculate the actual dimensions of the canvas
canvas_width = self.graph_canvas.winfo_width()
canvas_height = self.graph_canvas.winfo_height()
# Wait until the canvas has valid dimensions
if canvas_width <= 1 or canvas_height <= 1:
self.root.after(100, self.draw_graph)
return
# Calculate the optimal radius
radius = min(canvas_width, canvas_height) * 0.35
# Central point of the canvas
center_x = canvas_width / 2
center_y = canvas_height / 2
nodes = list(self.scheduler.exam_graph.keys())
n = len(nodes)
if n == 0:
return
# Store node positions
positions = {}
for i, node in enumerate(nodes):
angle = 2 * pi * i / n
x = center_x + radius * cos(angle)
y = center_y + radius * sin(angle)
positions[node] = (x, y)
# Draw the edges
for node, neighbors in self.scheduler.exam_graph.items():
x1, y1 = positions[node]
for neighbor in neighbors:
if neighbor in positions:
x2, y2 = positions[neighbor]
self.graph_canvas.create_line(x1, y1, x2, y2, fill='gray')
# Draw the nodes and labels
node_size = min(40, radius * 0.2)
for node, (x, y) in positions.items():
# Create a circle for the node
self.graph_canvas.create_oval(
x - node_size / 2, y - node_size / 2,
x + node_size / 2, y + node_size / 2,
fill='lightblue', outline='darkblue', width=2
)
# Add the node label
self.graph_canvas.create_text(
x, y, text=node,
font=('Arial', int(node_size / 3)),
fill='black'
)
def add_student(self):
```

```
student = self.student_var.get().strip()
exams = [e.strip() for e in self.exams_var.get().split(',')]
if student and exams:
self.scheduler.add_student_exams(student, exams)
self.update_display()
self.student_var.set('')
self.exams_var.set('')
else:
messagebox.showwarning("Error", "Please enter student ID and at least one exam")
def generate_schedule(self):
schedule = self.scheduler.get_schedule()
if self.scheduler.verify_schedule():
self.show_schedule(schedule)
else:
messagebox.showerror("Error", "Unable to generate a valid schedule")
def update_display(self):
self.update_summary()
self.draw_graph()
def update_summary(self):
self.data_text.delete('1.0', tk.END)
self.data_text.insert(tk.END, "STUDENTS AND THEIR EXAMS:\n\n")
for student, exams in self.scheduler.students.items():
self.data_text.insert(tk.END, f"{student}: {', '.join(exams)}\n")
self.data_text.insert(tk.END, "\nCONFLICTS PER EXAM:\n\n")
for exam in sorted(self.scheduler.exam_graph.keys()):
conflicts = len(self.scheduler.exam_graph[exam])
self.data_text.insert(tk.END, f"{exam}: {conflicts} conflicts\n")
def show_schedule(self, schedule):
self.calendar_text.delete('1.0', tk.END)
self.calendar_text.insert(tk.END, "EXAM SCHEDULE:\n\n")
for slot, exams in sorted(schedule.items()):
self.calendar_text.insert(tk.END, f"Time Slot {slot}:\n")
for exam in exams:
self.calendar_text.insert(tk.END, f" - {exam}\n")
self.calendar_text.insert(tk.END, "\n")
def load_sample_data(self):
sample_data = [
("S1", ["Mathematics", "Physics", "Chemistry"]),
("S2", ["Physics", "Computer Science", "English"]),
("S3", ["Chemistry", "Biology", "English"]),
("S4", ["Mathematics", "Computer Science", "Biology"])
]
for student, exams in sample_data:
self.scheduler.add_student_exams(student, exams)
self.update_display()
def main():
root = tk.Tk()
app = ExamSchedulerGUI(root)
root.mainloop()
```

```
if __name__ == "__main__":
main()
```

Figure 3. Exam schedule

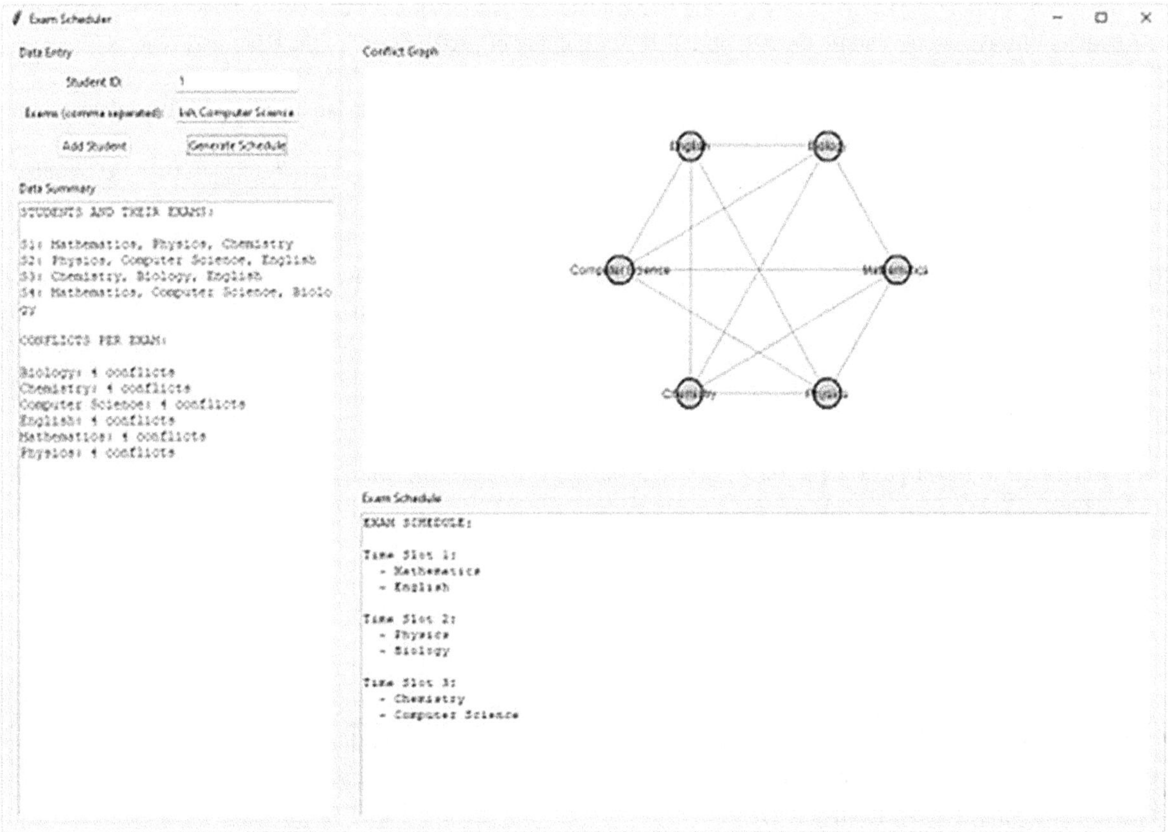

RECURSIVE COUNTING

Recursive counting is a powerful technique used to break down complex problems into smaller, more manageable steps. The key idea is that a solution can be expressed in terms of more minor instances of the same type. In other words, solving intricate tasks involves recursively addressing simpler cases of lower dimensions.

The fundamental principle of recursive counting is based on three elements:

1. Identify the basic cases: These are the simplest situations solved directly without resorting to other subproblems. Base cases are the starting point of recursion.
2. Find a recursive relationship: Establish a formula that connects the main problem with the subproblems, usually by reducing the size (e.g., considering $n-1$ instead of n).
3. Ensure that subproblems eventually lead to base cases: This ensures that the recursion ends correctly and does not continue indefinitely.

An interesting example of recursive counting is the concept of permutations. These refer to the different ways we can arrange *n* objects in a specific order. The recursive formula for calculating permutations, denoted by P(n), is based on the idea that we can fix one object and then determine the arrangement of the remaining n−1 elements. The recursive relation for permutations is:

1. Base case: P(0) = 1 (there is exactly one way to arrange zero elements: do nothing).
2. Recursive relation: P(n) = n · P(n−1), where to calculate the permutations of objects, we need to multiply *n* (the number of ways we can choose the first object) by the number of permutations of the other n−1 objects.

Another example related to combinatorial calculus in which recursive counting can be applied is the calculation of combinations of *k* elements from a set of *n*:

Base case: C(n,0) = 1 because there is only one way to choose zero elements from a set of, which is to choose nothing. Also, C(n,n) = 1 because there is only one way to choose all *n* elements from a set of *n*.

Recursive relation: C(n,k) = C(n−1,k−1) + C(n−1,k). This relationship suggests that we can consider two possibilities for calculating the number of combinations. Include the first element: We must choose k−1 elements from the remaining n−1, with C(n−1,k−1). Exclude the first element: We must choose *k* elements from the remaining n−1, with C(n−1,k).

If we want to calculate C(3,2), i.e., the number of ways to choose two elements from a set of 5, we use the recursive formula C(3,2) = C(2,1) + C(2,2).

Let us see the various steps:

C(2,1) = C(1,0) + C(1,1) = 1+ 1 = 2
C(2,2) = 1

Therefore:

C(3,2) = 2 + 1 = 3

There are three ways to choose two items from a set of 3.

Let us examine a classic problem that we will revisit: the Travelling Salesman Problem (TSP). This well-known puzzle can be described as follows: a travelling salesman must visit a set of cities, starting and ending at the city of residence, and aims to find the shortest possible route that passes through each city exactly once.

When only two cities are involved, there is no real choice to make. With three, there are just two possible routes. In general, for n cities, there are (n−1)! possible routes to consider. For instance, with 10 cities, the potential paths are 362,880.

Determining the total number of routes is straightforward; the real challenge lies in answering the key question (Bellman, 1962; Applegate et al., 2011): What is the optimal path?

The problem becomes exponentially more challenging when many cities require efficient algorithms (Van Beek, 2006; Cormen et al., 2009; Sedgewick & Wayne, 2011; Chaudhury, 2022, Kabir et al., 2024) that can explore and evaluate numerous alternative routes at increasingly complex levels. These algorithms use *backtracking*.

Backtracking (Zhang et al., 2024) explores different solutions and, when a dead end is reached, retraces the last steps to find new paths. It helps to systematically explore all possibilities while pruning the search space to avoid unnecessary calculations.

Example: Consider the following matrix of distances between cities:

Table 1. Distance matrix

	A	B	C
A	0	10	15
B	10	0	20
C	25	20	0

Let us walk through the process of exploring all possible routes between three cities (A, B, and C) using backtracking:

1. Starting from city A, we first select B as the next city:
   - Path: A → B
   - Possible next cities: C (A has already been visited)
   - Path: A → B → C
   - We have visited all cities, so we return to A: A → B → C → A
   - Total distance: 10 + 20 + 25 = 55
2. Now, backtrack and select C as the next city from A:
   - Path: A → C
   - Possible next cities: B (A has already been visited)
   - Path: A → C → B
   - We have visited all cities, so we return to A: A → C → B → A
   - Total distance: 15 + 20 + 10 = 45

Through backtracking, we generated two different routes: A → B → C (distance = 55) and A → C → B (distance = 45). This illustrates how backtracking can help evaluate multiple solutions, and the distance between cities influences the route length.

Backtrack is an exhaustive research strategy that considers all possible solutions to the problem but uses appropriate heuristics or pruning to avoid exploring branches of the decision tree that cannot lead to an optimal solution. Let us imagine having a maze: backtracking is like following all the possible corridors until one finds the exit, but with a specific rule: if a corridor leads to a dead end, one goes back and tries another route. Backtracking is especially useful for solving problems with many practical solutions to explore and can be solved by building a solution step by step.

Since this type of search can take a long time, a technique called *Branch and Bound* has been introduced. This technique allows the search to stop along a branch if it proves unable to improve the previous solution. It works by dividing the problem into smaller subproblems and then eliminating some based on limits on the optimal solution.

Branch and Bound is a powerful tool for solving combinatorial optimization. By effectively combining branching and boundaries, this algorithm offers a systematic approach to finding the optimal solution.

A new sequence recalculates the length, and if this value exceeds the previous one, the search is halted. Otherwise, the old admissible solution will be replaced. A concrete example could be the following: we have a set of tasks and a set of workers. Each task must be assigned to only one worker, and each worker can perform a maximum of one task at a time. Each pair (task, worker) has an associated cost, representing the time or effort required to complete the task. The goal is to assign tasks to workers in a way that minimizes the total cost.

The problem can be represented as a cost matrix, where rows represent tasks and columns represent workers. At each step, we select a task and a worker and create two subproblems: one in which the task is assigned to the worker and one in which the task is not assigned to the worker. It is necessary to calculate a lower bound on the total cost and keep track of the total cost of the best allocation found. If the lower bound of a subproblem is greater than the total cost of the best assignment found so far, then it can be discarded. Continue to branch and prune until all possible assignments have been explored or until one finds a solution that cannot be improved.

It is interesting to compare a recursive algorithm with an iterative one.

Suppose we have the following combinatorial situation: Put *2n* balls in a bag and number them from 1 to *2n*, then randomly select half of them. How many distinct sums are possible for n = 1, 2, 3, ...? (https://mathworld.wolfram.com/NumberPicking.html).

Let us consider two simple examples to understand how to get to the sequence.

Case n=1: There are two balls, 1 and 2; having to choose half of these balls, take only one randomly. The possible sums are 1 or 2; therefore, the distinct ones obtainable are 2.

Case n=2: There are four balls, 1, 2, 3, and 4; having to choose half of these balls, take two balls at random. The distinct sums that can be obtained are 5.

From an IT point of view, two algorithms—recursive or iterative—can address the problem..

Program 71 Distinct sums Recursive method

```
def distinct_sum_recursive(balls, n, start=0, selected=None, current_sum=0):
if selected is None:
selected = []
# Base case: if we have selected exactly n balls
if len(selected) == n:
# Add the sum of this combination to the set of distinct sums
return {current_sum}
# If we cannot select more balls, return an empty set
if start >= len(balls):
return set()
# 1. Select the current ball
selected.append(balls[start])
sums_with_ball = distinct_sum_recursive(
balls, n, start + 1, selected, current_sum + balls[start])
selected.pop()
# 2. Do not select the current ball
sums_without_ball = distinct_sum_recursive(
balls, n, start + 1, selected, current_sum)
# Combine the results of the two choices (selecting or not selecting)
return sums_with_ball | sums_without_ball
def num_distinct_sums_recursive(n):
# Create a list of balls from 1 to 2n
balls = list(range(1, 2 * n + 1))
# Call the recursive function
distinct_sums = distinct_sum_recursive(balls, n)
return len(distinct_sums)
# Execute the program for different values of n
for n in range(1, 11):
print(f"For n = {n}, the number of distinct sums is: {num_distinct_sums_recursive(n)}")
Program 72 Distinct sums Iterative method
from itertools import combinations
def num_distinct_sums(n):
# Create a list of numbers from 1 to 2n
balls = list(range(1, 2 * n + 1))
# Calculate all combinations of half the balls (i.e., n balls)
combs = combinations(balls, n)
```

```
# Calculate the sums for each combination
sums = set(sum(comb) for comb in combs)
# Return the number of distinct sums
return len(sums)
# Execute the program for different values of n
for n in range(1, 11):
print(f"For n = {n}, the number of distinct sums is: {num_distinct_sums(n)}")
```

Apart from the different lengths of the programs, it can be emphasized that the total combinatorial complexity is the same: $O(\binom{2n}{n} \cdot n)$, where $\binom{2n}{n}$ is the number of combinations of *n* balls of *2n*, and *n* is the complexity to calculate the sum of each combination. For large *n*, the iterative approach is generally faster than the recursive one. This is because the itertools library is written in C and optimized for performance while handling combinations recursively, which requires more memory and time to manage the recursion stack.

As we have seen when faced with a complex situation, a good practice is to start with the original problem and break it down into more straightforward and manageable cases, solving each separately. These more uncomplicated cases can, in turn, be broken down further, if necessary, until one reaches cases so simple that they can be solved directly. Once all the simplest cases are solved, they combine their solutions to get the solution to the original problem.

For example, suppose one wants to calculate the number of ways to arrange *n* objects in a sequence specifically. In that case, one can use a recursive approach to consider the options for the first object, then the options for the second object, and so on. The sum of these options would represent the total count of possible arrangements.

Let us look at the case where one wants to know how many ways one can climb a ladder in steps of 1 or 2 at a time. If one is at the bottom of the ladder, one can climb one or two steps at a time. The function calls itself to calculate the possible ways to take the following steps until one has reached the top.

Program 73 Step Count 1

```
def count_ways_to_climb_stairs(n, memo={}):
if n in memo:
return memo[n]
if n == 0:
return 1
elif n < 0:
return 0
else:
memo[n] = count_ways_to_climb_stairs(n - 1, memo) + count_ways_to_climb_stairs(n - 2,
memo)
return memo[n]
# Define the number of steps
n_steps = int(input("Enter the number of steps: "))
num_ways = count_ways_to_climb_stairs(n_steps)
print(f"The number of ways to climb a staircase with {n_steps} steps is {num_ways}.")
```

Instead of a recursive Program, one can use Dynamic Programming:

Program 74 Step Count 2

```
def count_ways_to_climb_stairs(n):
```

```
if n == 0:
return 1
elif n < 0:
return 0
# Array to store results up to n
dp = [0] * (n + 1)
dp[0] = 1 # There is 1 way to climb 0 steps (do nothing)
for i in range(1, n + 1):
dp[i] = dp[i - 1]
if i >= 2:
dp[i] += dp[i - 2]
return dp[n]
# Define the number of steps
n_steps = int(input("Enter the number of steps: "))
num_ways = count_ways_to_climb_stairs(n_steps)
print(f"The number of ways to climb a staircase with {n_steps} steps is {num_ways}.")
```

Another example is as follows: Suppose we need to calculate the number of ways to reach a certain sum, S, using a set of coins of different values. It is a classic challenge known as the "exchange rate problem" or "coin counting problem"." The goal is to find unique ways to combine coins to get the sum of *S*.

Suppose we have coins of 10, 20, and 50 cents, as well as 1 euro, and we want to determine how many ways we can combine them to reach a total sum of S.

Program 75 Coin Count 1

```
def countCoins(S, coins, current_combination=None):
if S == 0:
return 1
if S < 0 or not coins:
return 0
total_result = 0
# Option 1: Use the first coin
new_combination = current_combination.copy() if current_combination else []
new_combination.append(coins[0])
result1 = countCoins(S - coins[0], coins, new_combination)
total_result += result1
# Option 2: Do not use the first coin
result2 = countCoins(S, coins[1:], current_combination)
total_result += result2
return total_result
# Example
coins = [10, 20, 50, 100]
S = float(input("Enter the desired sum (in cents): "))
result = countCoins(S, coins)
print(f"Total number of ways to obtain {S} with the given coins: {result}")
```

The program employs a recursive approach to explore all possible combinations of coins that add up to the desired sum. The function divides the problem into more minor cases, considering whether to use or not use the first coin in the coin list and continues to invoke itself recursively until it reaches one of the basic conditions (a desired sum of 0 or a negative sum). Using recursion is a common approach for solving combination and permutation tasks, as it systematically explores all possible options. In the case of counting coin combinations, the recursive function keeps track of the coins used up to that point and prints all valid combinations when the desired sum becomes 0.

The program is slower than one that uses dynamic programming, such as the following. In dynamic programming (Powell, 2016), the problem can be divided into smaller sub-problems, which are repeated several times. Instead of solving these sub-problems every time they are encountered, they solve them once and store their solutions in a data structure (such as an array or table), allowing them to be reused.

Program 76 Coin Count 2

```
def countCoins(S, coins):
# Initialize an array to store intermediate results
results = [0] * (S + 1)
results[0] = 1 # There is 1 way to make 0
# Iterate over each available coin
for coin in coins:
# Update the results for all sums greater than or equal to the current coin
for i in range(coin, S + 1):
results[i] += results[i - coin]
return results[S]
# Example
coins = [10, 20, 50, 100]
S = int(input("Enter the desired sum (in cents): "))
result = countCoins(S, coins)
print(f"Total number of ways to obtain {S} with the given coins: {result}")
```

A simple problem with numerous applications is to count all the possible paths in a grid. Suppose we have a matrix n × n. We are in the upper left cell (0; 0) and want to reach another cell, with the constraint that we can only move to the right and down. How many possible routes are there?

Program 77 grid path

```
n = int(input("Grid dimensions: "))
F = []
for i in range(n):
F.append([0 for x in range(n)])
for j in range(n):
if i == 0 or j == 0:
F[i][j] = 1
else:
F[i][j] = F[i-1][j] + F[i][j-1]
print("The matrix of paths is:")
for i in range(n):
print(F[i])
```

The program, with n = 4, produces the following output in which, to reach the cell at the bottom right, one can follow 20 different paths:

[1, 1, 1, 1]
[1, 2, 3, 4]
[1, 3, 6, 10]
[1, 4, 10, 20]

A concrete application could take its cue from the Cretan labyrinth where the Minotaur was imprisoned. Let us examine the problem from two perspectives: searching for a single possible path or searching for all. There are several possible approaches. The first uses the A* algorithm to determine the shortest path to reach point B starting from A in a hypothetical labyrinth or implements an algorithm such as *Breadth-First Search* (BFS) to find paths, allowing the user to add/remove obstacles dynamically.

We are interested in exploring possible paths related to the second issue, which is the purpose of the Minotaur program.

Program 78 Labyrinth 1

```
import tkinter as tk
import heapq
import random
# Window settings for a 30x30 grid
window_width = 450 # 30 columns * 15 pixels
window_height = 450 # 30 rows * 15 pixels
cell_size = 15
columns = window_width // cell_size
rows = window_height // cell_size
# Colors
color_white = "#FFFFFF"
color_black = "#000000"
color_green = "#00FF00"
color_red = "#FF0000"
color_gray = "#A9A9A9"
color_blue = "#0000FF"
# Possible directions: up, down, left, right
directions = [(-1, 0), (1, 0), (0, -1), (0, 1)]
# Function to calculate Manhattan distance (heuristic)
def heuristic(a, b):
return abs(a[0] - b[0]) + abs(a[1] - b[1])
# Function to check if a cell is valid (inside the grid and not an obstacle)
def is_valid(grid, x, y):
return 0 <= x < len(grid) and 0 <= y < len(grid[0]) and grid[x][y] != 1
# A* algorithm to find the shortest path
def a_star(grid, start, goal):
open_list = []
heapq.heappush(open_list, (0 + heuristic(start, goal), 0, start)) # Initialize the queue
with the starting point
g_cost = {start: 0}
came_from = {}
```

```
while open_list:
_, current_g, current = heapq.heappop(open_list)
if current == goal:
path = []
while current in came_from:
path.append(current)
current = came_from[current]
path.append(start)
return path[::-1]
for direction in directions:
neighbor = (current[0] + direction[0], current[1] + direction[1])
if is_valid(grid, neighbor[0], neighbor[1]):
tentative_g = current_g + 1
if neighbor not in g_cost or tentative_g < g_cost[neighbor]:
g_cost[neighbor] = tentative_g
f = tentative_g + heuristic(neighbor, goal)
heapq.heappush(open_list, (f, tentative_g, neighbor))
came_from[neighbor] = current
return None # If no path exists
# Function to create a grid with random obstacles
def create_grid_with_obstacles(size_x, size_y, obstacle_percentage=30):
grid = [[0 for _ in range(size_y)] for _ in range(size_x)]
num_obstacles = (size_x * size_y * obstacle_percentage) // 100
obstacles_added = 0
while obstacles_added < num_obstacles:
x = random.randint(0, size_x - 1)
y = random.randint(0, size_y - 1)
if (x, y) != (0, 0) and (x, y) != (size_x - 1, size_y - 1) and grid[x][y] != 1:
grid[x][y] = 1
obstacles_added += 1
return grid
# Function to draw the grid
def draw_grid(canvas, grid, path=None):
canvas.delete("all") # Clears everything drawn on the canvas
for i in range(len(grid)):
for j in range(len(grid[i])):
x0 = j * cell_size
y0 = i * cell_size
x1 = x0 + cell_size
y1 = y0 + cell_size
if grid[i][j] == 1:
color = color_black
elif (i, j) == (0, 0):
color = color_green
elif (i, j) == (len(grid)-1, len(grid[0])-1):
color = color_red
elif path and (i, j) in path:
color = color_blue
else:
```

```
color = color_white
canvas.create_rectangle(x0, y0, x1, y1, fill=color, outline=color_gray)
# Function to run the program
def main():
# Create the window
root = tk.Tk()
root.title("A* Algorithm - Pathfinding")
canvas = tk.Canvas(root, width=window_width, height=window_height)
canvas.pack()
# Create a 30x30 grid with 25% of cells as obstacles
grid = create_grid_with_obstacles(30, 30, obstacle_percentage=30)
start = (0, 0)
goal = (29, 29)
path = None
path_found = False
def on_mouse_click(event):
nonlocal path, path_found
if not path_found: # Ensure the A* algorithm runs only once
path = a_star(grid, start, goal)
path_found = True
draw_grid(canvas, grid, path)
def on_spacebar(event):
nonlocal grid, path, path_found
grid = create_grid_with_obstacles(30, 30, obstacle_percentage=25)
path = None
path_found = False
draw_grid(canvas, grid)
# Add handlers for mouse click and spacebar
canvas.bind("<Button-1>", on_mouse_click)
root.bind("<space>", on_spacebar)
# Draw the initial grid
draw_grid(canvas, grid)
root.mainloop()
if __name__ == "__main__":
main()
```

Program 79 Labyrinth 2

```
import tkinter as tk
from collections import deque
import random # Add this line to import the random library
# Maze dimensions
width = 10
height = 10
cell_size = 40
# Shades of gray
very_light_gray = "#C8C8C8" # Very light gray
```

```
light_gray = "#969696" # Light gray
medium_gray = "#646464" # Medium gray
dark_gray = "#323232" # Dark gray
# Function to generate a random maze
def create_maze():
maze = [[0 for _ in range(width)] for _ in range(height)]
for i in range(height):
for j in range(width):
# Create random obstacles with a probability
if random.random() < 0.3 and (i, j) != (0, 0) and (i, j) != (width - 1, height - 1):
maze[i][j] = 1
return maze
# Function to draw the maze
def draw_maze(canvas, maze):
canvas.config(bg="white")
for row in range(height):
for col in range(width):
x1 = col * cell_size
y1 = row * cell_size
x2 = x1 + cell_size
y2 = y1 + cell_size
# If the cell is empty (0), color it white, otherwise gray
color = "white" if maze[row][col] == 0 else dark_gray
canvas.create_rectangle(x1, y1, x2, y2, fill=color, outline="black")
# Draw points A and B
canvas.create_text(20, 20, text="A", font=("Arial", 16, "bold"), fill="black")
canvas.create_text(20 + (width - 1) * cell_size, 20 + (height - 1) * cell_size, text="B",
font=("Arial", 16, "bold"), fill="black")
# Function to calculate the shortest path using BFS
def bfs(maze, start, goal):
queue = deque([(start, [start])]) # Queue that tracks position and path
visited = set([start])
while queue:
(x, y), path = queue.popleft()
if (x, y) == goal:
return path
for dx, dy in [(-1, 0), (1, 0), (0, -1), (0, 1)]:
nx, ny = x + dx, y + dy
if 0 <= nx < width and 0 <= ny < height and maze[ny][nx] == 0 and (nx, ny) not in visited:
visited.add((nx, ny))
queue.append(((nx, ny), path + [(nx, ny)]))
return [] # If there is no path
# Function to draw the found path
def draw_path(canvas, path):
gray_shades = [very_light_gray, light_gray, medium_gray, dark_gray]
for i, (x, y) in enumerate(path):
x1 = x * cell_size + cell_size // 2
y1 = y * cell_size + cell_size // 2
color = gray_shades[i % len(gray_shades)]
```

```
if i < len(path) - 1:
x2, y2 = path[i + 1]
x2 = x2 * cell_size + cell_size // 2
y2 = y2 * cell_size + cell_size // 2
canvas.create_line(x1, y1, x2, y2, fill=color, width=2)
# Main function
def main():
root = tk.Tk()
root.title("Interactive Maze")
canvas = tk.Canvas(root, width=width * cell_size, height=height * cell_size)
canvas.pack()
# Create the random maze
maze = create_maze()
# Draw the maze
draw_maze(canvas, maze)
# Calculate the path using BFS
start = (0, 0)
goal = (width - 1, height - 1)
path = bfs(maze, start, goal)
print(f"Path found: {path}")
# Draw the path
if path:
draw_path(canvas, path)
# Function to add/remove obstacles
def on_mouse_click(event):
col = event.x // cell_size
row = event.y // cell_size
if maze[row][col] == 0:
maze[row][col] = 1 # Add an obstacle
else:
maze[row][col] = 0 # Remove an obstacle
draw_maze(canvas, maze) # Redraw the maze
# Recalculate the path
path = bfs(maze, start, goal)
if path:
draw_path(canvas, path)
# Add the mouse click handler
canvas.bind("<Button-1>", on_mouse_click)
root.mainloop()
if __name__ == "__main__":
main()
```

Program 80 Minotaur

```
import tkinter as tk
# Maze definition
maze = [
```

```
[0, 1, 0, 0, 0, 1, 0, 0, 0, 1],
[0, 1, 0, 1, 0, 1, 0, 1, 0, 1],
[0, 0, 0, 1, 0, 0, 0, 1, 0, 0],
[1, 1, 0, 1, 1, 1, 0, 1, 1, 1],
[0, 0, 0, 0, 0, 1, 0, 0, 0, 0],
[0, 1, 1, 1, 0, 1, 1, 1, 1, 0],
[0, 0, 0, 1, 0, 0, 0, 0, 0, 0],
[1, 1, 0, 1, 1, 1, 1, 1, 1, 0],
[0, 0, 0, 0, 0, 0, 0, 0, 0, 0],
[0, 1, 1, 1, 1, 1, 1, 1, 1, 0]
]
# Maze dimensions
width = len(maze[0])
height = len(maze)
cell_size = 40
# Shades of gray
VERY_LIGHT_GRAY = "#C8C8C8" # Very light gray
LIGHT_GRAY = "#969696" # Light gray
MEDIUM_GRAY = "#646464" # Medium gray
DARK_GRAY = "#323232" # Dark gray
# Function to draw the maze
def draw_maze(canvas):
# Set the canvas background color to white
canvas.config(bg="white")
for row in range(height):
for col in range(width):
x1 = col * cell_size
y1 = row * cell_size
x2 = x1 + cell_size
y2 = y1 + cell_size
# If the cell is empty (0), color it white, otherwise gray
color = "white" if maze[row][col] == 0 else DARK_GRAY
canvas.create_rectangle(x1, y1, x2, y2, fill=color, outline="black")
# Draw points A and B
canvas.create_text(20, 20, text="A", font=("Arial", 16, "bold"), fill="black")
canvas.create_text(20 + (width - 1) * cell_size, 20 + (height - 1) * cell_size, text="B",
font=("Arial", 16, "bold"), fill="black")
# Function to calculate possible paths and trace them
def calculate_paths(maze, x, y, destination_x, destination_y, visited, path, paths):
if x == destination_x and y == destination_y:
paths.append(list(path))
return 1
if x < 0 or y < 0 or x >= width or y >= height or maze[y][x] == 1 or visited[y][x]:
return 0
visited[y][x] = True
path.append((x, y))
total_paths = (calculate_paths(maze, x + 1, y, destination_x, destination_y, visited,
path, paths) +
calculate_paths(maze, x - 1, y, destination_x, destination_y, visited, path, paths) +
```

```
calculate_paths(maze, x, y + 1, destination_x, destination_y, visited, path, paths) +
calculate_paths(maze, x, y - 1, destination_x, destination_y, visited, path, paths))
path.pop()
visited[y][x] = False
return total_paths
# Function to draw the paths
def draw_paths(canvas, paths, destination_x, destination_y):
gray_shades = [VERY_LIGHT_GRAY, LIGHT_GRAY, MEDIUM_GRAY, DARK_GRAY]
offset = 8 # Greater offset to avoid overlap
for i, path in enumerate(paths):
color = gray_shades[i % len(gray_shades)] # Choose a gray color for each path
for j in range(len(path) - 1):
x1, y1 = path[j]
x2, y2 = path[j + 1]
x1 = x1 * cell_size + cell_size // 2 + (i % 5 - 2) * offset
y1 = y1 * cell_size + cell_size // 2 + (i % 5 - 2) * offset
x2 = x2 * cell_size + cell_size // 2 + (i % 5 - 2) * offset
y2 = y2 * cell_size + cell_size // 2 + (i % 5 - 2) * offset
canvas.create_line(x1, y1, x2, y2, fill=color, width=2)
# Draw the final line to B
x1, y1 = path[-1]
x2, y2 = destination_x, destination_y
x1 = x1 * cell_size + cell_size // 2 + (i % 5 - 2) * offset
y1 = y1 * cell_size + cell_size // 2 + (i % 5 - 2) * offset
x2 = x2 * cell_size + cell_size // 2 + (i % 5 - 2) * offset
y2 = y2 * cell_size + cell_size // 2 + (i % 5 - 2) * offset
canvas.create_line(x1, y1, x2, y2, fill=color, width=2)
# Main function
def main():
root = tk.Tk()
root.title("Maze")
canvas = tk.Canvas(root, width=width * cell_size, height=height * cell_size)
canvas.pack()
# Draw the maze with a white background
draw_maze(canvas)
# Calculate paths and draw them
visited = [[False for _ in range(width)] for _ in range(height)]
paths = []
calculate_paths(maze, 0, 0, width - 1, height - 1, visited, [], paths)
print(f"Number of possible paths from A to B: {len(paths)}")
draw_paths(canvas, paths, width - 1, height - 1)
root.mainloop()
if __name__ == "__main__":
main()
```

Figure 4. Paths in the labyrinth

REFERENCES

Applegate, D. L., Bixby, R. E., Chvátal, V., & Cook, W. J. (2011). The traveling salesman problem: a computational study. In *The Traveling Salesman Problem*. Princeton university press.

Bellman, R. (1962). Dynamic programming treatment of the travelling salesman problem. [JACM]. *Journal of the Association for Computing Machinery*, *9*(1), 61–63. DOI: 10.1145/321105.321111

Chartrand, G., & Zhang, P. (2013). *A first course in graph theory*. Courier Corporation.

Chaudhury, B. R., Garg, J., McGlaughlin, P., & Mehta, R. (2022, July). Competitive equilibrium with chores: Combinatorial algorithm and hardness. *In Proceedings of the 23rd ACM Conference on Economics and Computation* (pp. 1106-1107). DOI: 10.1145/3490486.3538255

Cormen, T. H., Leiserson, C. E., Rivest, R. L., & Stein, C. (2009). *Introduction to Algorithms* (3rd ed.). MIT Press and McGraw-Hill.

Diestel, R. (2024). *Graph theory*. Springer.

Gross, J. L., Yellen, J., & Anderson, M. (2018). *Graph theory and its applications*. Chapman and Hall/CRC. DOI: 10.1201/9780429425134

Kabir, R., Watanobe, Y., Islam, M. R., & Naruse, K. (2024). Enhanced robot motion block of a-star algorithm for robotic path planning. *Sensors (Basel)*, *24*(5), 1422. DOI: 10.3390/s24051422 PMID: 38474956

Manduva, V. C. (2024). Review of P2P Computing System Cooperative Scheduling Mechanisms. *International Journal of Modern Computing*, *7*(1), 154–168.

Powell, W. B. (2016). Perspectives of approximate dynamic programming. *Annals of Operations Research*, *241*(1-2), 319–356. DOI: 10.1007/s10479-012-1077-6

Sedgewick, R., & Wayne, K. (2011). *Algorithms*. Addison-wesley professional.

Van Beek, P. (2006). Backtracking search algorithms. In *Foundations of artificial intelligence* (Vol. 2, pp. 85–134). Elsevier.

West, D. B. (2001). *Introduction to graph theory* (Vol. 2). Prentice hall.

Zhang, Y., Chi, J., Nguyen, H., Upasani, K., Bikel, D. M., Weston, J., & Smith, E. M. (2024). Backtracking improves generation safety. *arXiv preprint arXiv:2409.14586*.

Chapter 9
Combinatorics Games

ABSTRACT

This chapter examines combinatorial games, analyzing how decisions and strategies are based on the combination of different options. Combinatorial games involve creating and manipulating combinations of cards, tokens, or pieces on a board. These games focus on the organization and arrangement of available objects. Combinatorial games have a long history and are significant from both a research and educational perspective, introducing fundamental concepts such as combinations, permutations, and arrangements in a practical and accessible context. For example, in poker, players combine their cards to create the best possible hand, using combinatorics to determine probabilities. In chess, the Eight Queens Problem is a famous example of a combinatorial problem that requires placing eight queens on an 8x8 board so no queen can attack another. Other combinatorial games include Sudoku, Tower of Hanoi, and Rubik's Cube.

INTRODUCTION

Play is a free, circumscribed, uncertain, unproductive, regulated and fictive activity
Roger Caillois

Combinatorial games involve creating and manipulating combinations of elements, such as cards, pawns, or pieces on a chessboard. It is a class of abstract games involving strategies, decisions, and situations in which players must choose based on a finite set of options or configurations (Berlekamp et al., 2004; Demaine & Hearn, 2008; Hefetz et al., 2014; AA.VV., 2018; García Jiménez, 2018). These games focus on the combination and arrangement of available objects, resources, or moves, rather than the probability or randomness aspects found in many other games. These games have been played for centuries worldwide and hold considerable importance from both a research and educational perspective.

Combinatorial games introduce fundamental concepts, such as combinations, permutations, and arrangements, in a practical and accessible context. Consider card games; for example, poker requires players to combine their cards to form the best possible hand. The case history is vast, but it would be unfair to limit the analysis to "games" only; in reality, there are many applications in concrete situations. From financial investment strategies to the design of communication networks, combinatorial games play a crucial role in optimising decisions in complex situations (Niv & Patkin, 2020).

Listing all the possible uses would take too long; a few hints are enough to understand the extent of the phenomenon. In communication systems and computer networks, combinatorial games can be used to optimize data routing across a network. In the transportation and logistics industries, they can be used to optimize vehicle routes, reduce costs, and improve transportation efficiency. In statistics and experimental design, they are used to design experiments efficiently, selecting optimal combinations of variables to test. In computational biology, combinatorial games can be used to model interactions between molecules and proteins, contributing to the understanding of biological processes and the design of new drugs. In machine learning systems, combinatorial games model collaborative learning situations.

One of the main categorizations of combination games is the number of players involved. There are numerous works on games for two or more players, although many games in real life are essentially single player.

Let us examine some of the most popular combination games, beginning with those that involve more than one player and progressing to "automatic" games.

DOI: 10.4018/979-8-3373-3089-1.ch009

Astragals: The talus is a small bone in the foot joint. In some animal species, such as goats and rams, this bone has particularly regular proportions, and since ancient times, it has been used as a four-sided die. These ossicles were widely spread throughout the Mesopotamian region, while in Egypt, they spread around 1300 BC. The astragals were thrown, and based on the face facing upwards, it was possible to "read" a score: 1 (Monas), 3 (Trias), 4 (Tetras), 6 (Hexas). It is worth noting that the opposite sides added up to 7, just like traditional cubic dice.

Figure 1. Astragals

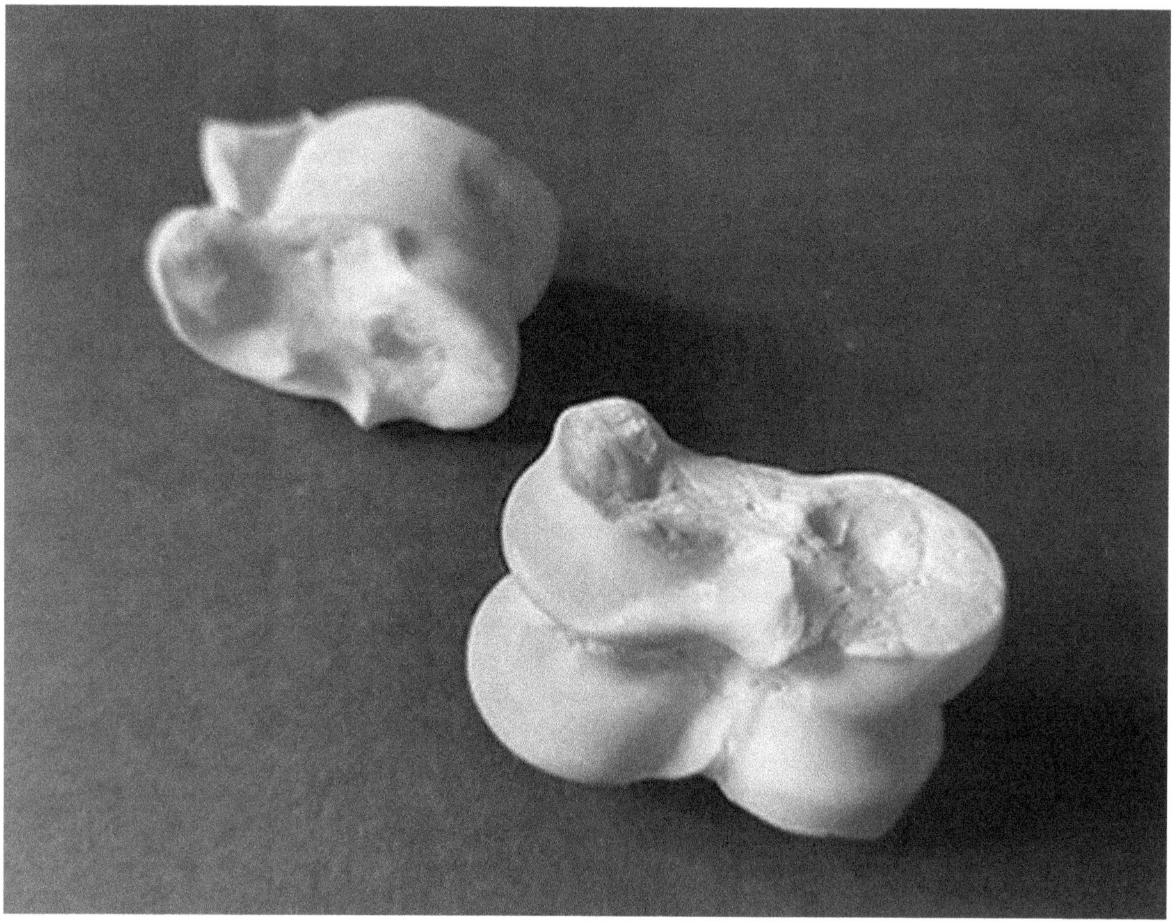

A name was assigned to each side and to each of the combinations that could be obtained by throwing four astragals. Since each talus can land on one of the four faces and there are four taluses, we can calculate the total number of possible combinations with the formula $4^4 = 256$. By throwing four astragals, there are several combinations of scores. The Romans assigned names and symbolic meanings to certain notable combinations:

1. Aphrodite's Strike: 1, 3, 4, 6; The most esteemed throw, representing harmony and beauty, was associated with the goddess of love.
2. Dog Strike: 1, 1, 1, 1; Considered the worst possible throw, symbolizing misfortune or disgrace.
3. Triple Threes: 3, 3, 3, 3; A rare and significant uniform roll.
4. Tetras: 4, 4, 4, 4; A uniform roll of fours, potentially lucky or neutral depending on context.
5. Six of Venus: 6, 6, 6, 6; A high and auspicious roll, sometimes associated with Venus or good fortune.

6. Lucky Rolls: 1, 1, 1, 3 or 4, 4, 4, 6; A particular combination with three equal faces and one different, often considered favourable.

Lu-Lu: The word Lu-Lu means "to shake". Young Hawaiians play Lu-Lu with volcanic stone discs about 2.5 cm in diameter. There are 1, 2, 3 or 4 points on one side, while the opposite face is empty. The basic version of the game consists of performing a single roll, in turn, of the four stones, after which only the points on the dice that fall face up are counted. The player must wave the dice in both hands and roll them. The first player to score 50 points wins.

Figure 2. Figure Lu-Lu

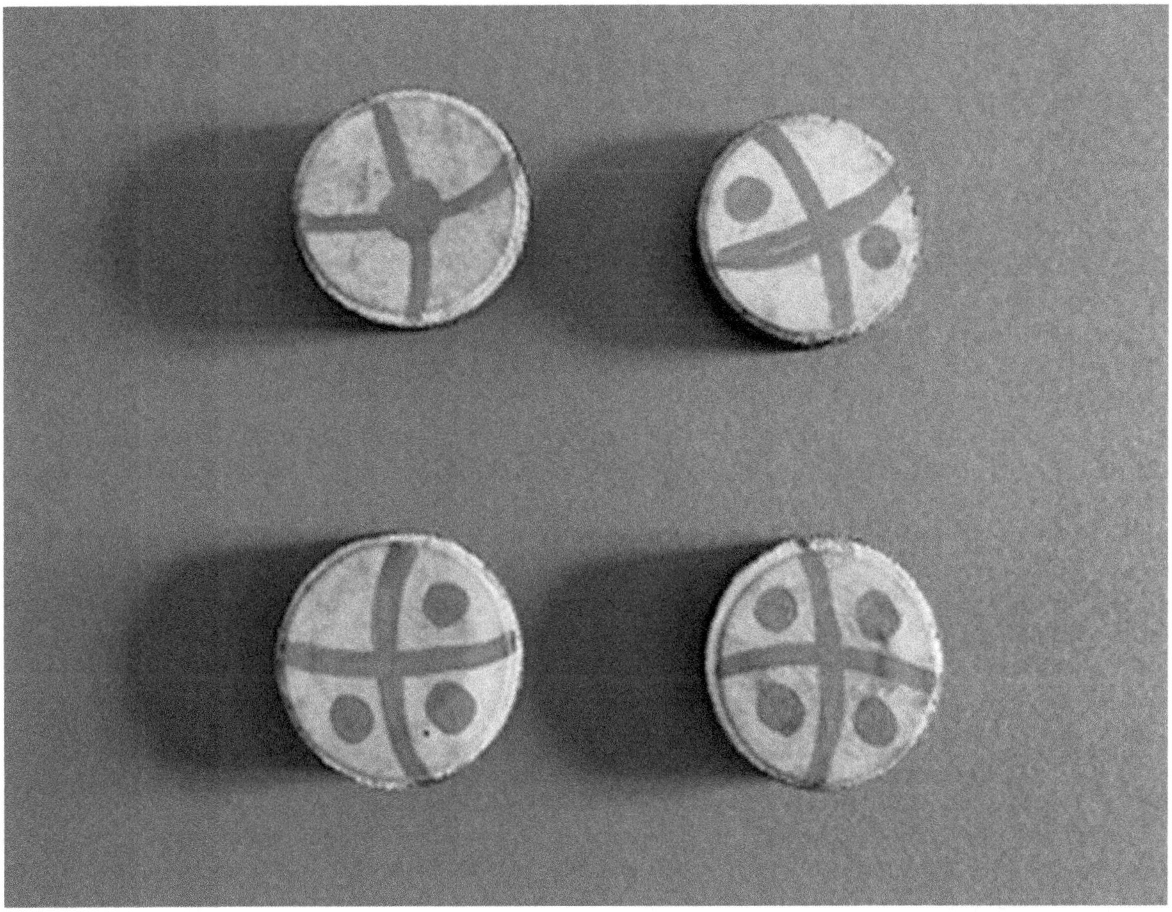

Some questions refer to the cases that may occur:

- How many and what are all the possible outcomes of a launch?
- How many and what scores are possible for a round?
- How many ways can one get each score?

Given the simplicity of the solution, it is possible to answer with a simple list.

There are sixteen possible results: 4-3-2-1; 0-4-3-2; 0-4-3-1; 0-4-2-1; 0-3-2-1; 0-0-3-4; 0-0-2-4; 0-0-2-3; 0-0-4-1; 0-0-3-1; 0-0-2-1; 0-0-0-4; 0-0-0-3; 0-0-0-2; 0-0-0-1; 0-0-0-0.

There are eleven possible scores, five given by the faces with possible dots upwards, six by the possible results: 1+2 = 3, 1+3 = 4, 1+4 = 5, 2+3 = 5, 2+4 = 6, 3+4 = 7, 1+2+3 = 6, 1+2+4 = 7, 1+3+4 = 8, 2+3+4 = 9, 1+2+3+4 = 10.

Scores can be obtained in sixteen ways: 0, 1, 2, 8, 9, 10 can occur in only one way (e.g., 10 = 4-3-2-1), while 3, 4, 5, 6, 7 can occur in two ways (e.g., 3 = 0-0-0-3 or 0-0-2-1).

Of course, there are combinatorial methods that allow us to determine how many ways there are to get each score without having to list all the combinations explicitly. A useful tool for problems like this is the generator function. We can represent the possible results for each disk using a generating function that considers the two possible outputs: a dotted face or a blank face. Since we have four disks, the total generating function G(x) is the product of the individual generating functions:

$$G(x) = 1 + x + x^2 + 2x^3 + 2x^4 + 2x^5 + 2x^6 + 2x^7 + x^8 + x^9 + x^{10}$$

The coefficients of the terms x^k will give the number of ways to get the *k* score.

Poker: It is one of the most famous card games. The game's objective is to achieve the best combination or persuade the other players to quit, thereby winning the full amount of bets. Poker is a game in which combinatorial calculus plays a significant role, as it is preparatory to the calculation of probabilities. It should be noted that there are many variations (Tang, 2018).

Draw Poker represents the classic form of the game. We need to calculate simple combinations to determine the number of card combinations that can occur in the game of Poker. Let us look at a few cases using all 52 cards in the deck. The possible combinations are $C_{52.5}$= n!/(n−k)k! = 2,598,960.

Royal Flush can only be done in four ways, one for each suit.

Straight flush in 36 ways: 4 suits • 9 possible sequences (A-2-3-4-5, 2-3-4-5-6, 3-4-5-6-7, 4-5-6-7-8, 5-6-7-8-9, 6-7-8-9-10, 7-8-9-10-J, 8-9-10-J-Q, 9-10-J-Q-K).

For Poker, the first card can be chosen in 52 ways, the second card in 3 ways, the third in 2 ways, and the fourth in one way. Since the order does not count, the possible four of a kind are (52 • 3 • 2 • 1) / 4! = 13. The fifth card can be chosen in 48 distinct ways; therefore, the possible combinations are 13 • 48 = 624.

The Full House consists of three of a kind that can be chosen in 13 ways multiplied by $C_{4.3}$ plus a pair that can be chosen in 12 ways by $C_{4.2}$, in total there are 3,744 combinations.

The Flush is obtained by multiplying the ways in which one can choose 5 cards for the four suits in a set of 13 cards and subtracting the possible royal flushes from the multiplication; this results in 5,108 cases.

To obtain the probabilities of the Simple Straight, multiply the 10 possible combinations to obtain a straight in the same suit by the four raised to the fifth (this represents all the possibilities with which you can choose the cards) and then subtract the number of royal flushes; So, you get 10,200.

For Three of a Kind, one can choose from 13 possible values for how one can choose three cards in a set of 4 cards, for how one can choose two cards (the remainder) in a set of 12 by 4^2: 13 • $C_{4.3}$ • $C_{12.2}$ • 4^2 = 54,912.

To calculate the ways to get a Two Pair, we follow these steps: we choose two distinct values from the 13 possible ones, so $C_{13,2}$. Once selected, for each, we must choose two cards from one of the 4 suits, in this case, there are $C_{4,2}$ ways. There are 44 cards left, and we can multiply everything to get 123,552.

We follow a similar process to calculate the number of ways to get a Pair, so we have 13 • $C_{4,2}$ • $C_{13,2}$ • 43 = 1,098,240.

The hands in which the player has no point (high card) remain to be calculated. In this case, it is sufficient to subtract the sum of all the other cases from the number of possible hands, obtaining 1,302,540.

Program 81 Draw poker

```
import math
# Function to calculate combinations
def nCr(n, r):
return math.comb(n, r)
```

```
# Total number of possible combinations of 5 cards from a 52-card deck
total_combinations = nCr(52, 5)
# Royal Flush: 4 possible combinations (one for each suit)
royal_flush = 4
# Straight Flush: 4 suits * 9 possible sequences
straight_flush = 4 * 9
# Four of a Kind: 13 possible values * 48 for the fifth card
four_of_a_kind = 13 * 48
# Full House: 13 possible values for the three of a kind * 4C3 (combinations of 4 cards chosen 3 at a time) *
# 12 possible values for the pair * 4C2 (combinations of 4 cards chosen 2 at a time)
full_house = 13 * nCr(4, 3) * 12 * nCr(4, 2)
# Flush: 4 suits * 13C5 (combinations of 13 cards chosen 5 at a time) - straight flush - royal flush
flush = 4 * nCr(13, 5) - straight_flush - royal_flush
# Straight: 10 possible sequences * 4^5 (possible combinations of suits) - straight flush - royal flush
straight = 10 * (4**5) - straight_flush - royal_flush
# Three of a Kind: 13 possible values for the three of a kind * 4C3 * 12C2 (combinations of 12 cards chosen 2 at a time) * 4^2 (suits for the other two cards)
three_of_a_kind = 13 * nCr(4, 3) * nCr(12, 2) * (4**2)
# Two Pair: 13C2 (combinations of 13 values chosen 2 at a time) * (4C2)^2 * 44 (remaining cards for the fifth card)
One = nCr(13, 2) * (nCr(4, 2)**2) * 44
# One Pair: 13 possible values for the pair * 4C2 (combinations of 4 cards chosen 2 at a time) * 12C3 * 4^3 (remaining suits for the other three cards)
one_pair = 13 * nCr(4, 2) * nCr(12, 3) * (4**3)
# High Card: subtract all other combinations from the total
high_card = total_combinations - (royal_flush + straight_flush + four_of_a_kind + full_house + flush + straight + three_of_a_kind + two_pair + one_pair)
# Print the results
print(f"Total combinations: {total_combinations}")
print(f"Royal Flush: {royal_flush}")
print(f"Straight Flush: {straight_flush}")
print(f"Four of a Kind: {four_of_a_kind}")
print(f"Full House: {full_house}")
print(f"Flush: {flush}")
print(f"Straight: {straight}")
print(f"Three of a Kind: {three_of_a_kind}")
print(f"Two Pair: {two_pair}")
print(f"One Pair: {one_pair}")
print(f"High Card: {high_card}")
# Calculate probabilities
print(f"Probability of Royal Flush: {royal_flush / total_combinations:.7f}")
print(f"Probability of Straight Flush: {straight_flush / total_combinations:.6f}")
print(f"Probability of Four of a Kind: {four_of_a_kind / total_combinations:.5f}")
print(f"Probability of Full House: {full_house / total_combinations:.4f}")
print(f"Probability of Flush: {flush / total_combinations:.4f}")
print(f"Probability of Straight: {straight / total_combinations:.4f}")
```

```
print(f"Probability of Three of a Kind: {three_of_a_kind / total_combinations:.4f}")
print(f"Probability of Two Pair: {two_pair / total_combinations:.4f}")
print(f"Probability of One Pair: {one_pair / total_combinations:.4f}")
print(f"Probability of High Card: {high_card / total_combinations:.4f}")
```

Texas Hold'em is a popular variant of poker played with a standard 52-card deck. For the first card dealt, there are 52 possibilities. For the second, only 51 remain. Therefore, the total ways to receive two cards is 52 × 51.

However, since the order in which the two hole cards are dealt does not matter—both contribute equally to the hand—we divide by 2 to avoid counting duplicates. Thus, the total of distinct two-card combinations is: 52 • 51/2 = 1,326. After all players have received their hole cards, the game proceeds to the "flop" phase. If there are four players, each receiving two cards, then 4 • 2 = 8 cards have been dealt, leaving 52 – 8 = 44 cards in the deck.

The number of possible three-card combinations that can be dealt on the flop is given by the combination formula: $C_{44,3}$ = 44! / (3! • (44-3)!) equals 14,190. The game continues by revealing one card on the "turn" and one on the "river."

There is also a version in which people use *poker craps*. It is played with five dice that bear Ace, King, Queen, Jack, Ten and Nine on their six sides. The game aims to make the best combination. Considering that from the first roll, all combinations can result (in total 6^5 = 7,776). The poker, i.e. five dice showing the same face, can only be done in 6 ways, the four of a kind comes in 6 • $C_{5,4}$ • 5 = 6 • 5 • 5 = 150 cases, the full house in 6 • $C_{5,3}$ • $C_{5,2}$/2 = 6 • 10 • 10/2 = 300 ways, the straight has 2 • 5! = 2 • 120 = 240 combinations, three of a kind 6 • $C_{5,3}$ •5 • 4 = 6 • 10 • 5 • 4 = 1,200, two pair 6 • $C_{5,2}$ • $C_{5,2}$ • 3 = 6 • 10 • 10 •3 = 1,800, pair 6 •$C_{5,2}$ • 5 • 4 • 3 = 6 • 10 • 5 • 4 • 3 = 3,600, There are still 480 cases in which we obtain no points.

Program 82 poker craps

```
import math
from itertools import combinations_with_replacement
# Function to calculate combinations
def nCr(n, r):
if r > n or r < 0:
return 0
return math.comb(n, r)
# Total number of possible combinations of 5 dice
total_combinations = 6**5
# Calculation of specific combinations
# Five of a Kind: 5 dice with the same face (6 possibilities)
five_of_a_kind = 6
# Four of a Kind: 4 dice with the same face and one die with a different face (6 faces *
C(5,4) * 5)
four_of_a_kind = 6 * nCr(5, 4) * 5
# Full House: 3 dice with the same face and 2 dice with a different same face (6 faces *
C(5,3) * C(5,2) / 2)
full_house = 6 * nCr(5, 3) * nCr(5, 2) // 2
# Straight: 5 dice with consecutive faces (2 possibilities: high and low straight)
straight = 2 * math.factorial(5)
# Three of a Kind: 3 dice with the same face, 1 die with a different face, and another die
with a different face (6 faces * C(5,3) * 5 * 4)
three_of_a_kind = 6 * nCr(5, 3) * 5 * 4
# Two Pair: 2 pairs of dice with the same face and 1 die with a different face (6 faces *
C(5,2) * C(5,2) * 3)
```

```
    two_pair = 6 * nCr(5, 2) * nCr(5, 2) * 3
    # One Pair: 1 pair of dice with the same face, 3 dice with different faces (6 faces *
C(5,2) * 5 * 4 * 3)
    one_pair = 6 * nCr(5, 2) * 5 * 4 * 3
    # High Dice: remaining combinations
    null = total_combinations - (five_of_a_kind + four_of_a_kind + full_house + straight +
three_of_a_kind + two_pair + one_pair)
    # Print results
    print(f"Total combinations: {total_combinations}")
    print(f"Five of a Kind: {five_of_a_kind}")
    print(f"Four of a Kind: {four_of_a_kind}")
    print(f"Full House: {full_house}")
    print(f"Straight: {straight}")
    print(f"Three of a Kind: {three_of_a_kind}")
    print(f"Two Pair: {two_pair}")
    print(f"One Pair: {one_pair}")
    print(f"High Dice: {null}")
    # Calculate probabilities
    print(f"Probability of Five of a Kind: {five_of_a_kind / total_combinations:.7f}")
    print(f"Probability of Four of a Kind: {four_of_a_kind / total_combinations:.6f}")
    print(f"Probability of Full House: {full_house / total_combinations:.4f}")
    print(f"Probability of Straight: {straight / total_combinations:.4f}")
    print(f"Probability of Three of a Kind: {three_of_a_kind / total_combinations:.4f}")
    print(f"Probability of Two Pair: {two_pair / total_combinations:.4f}")
    print(f"Probability of One Pair: {one_pair / total_combinations:.4f}")
    print(f"Probability of High Dice: {null / total_combinations:.4f}")
```

Chess: It is a widely used strategic game on a square board, called a "chessboard", divided into 64 squares, alternately light and dark in colour. 32 pieces on the board, 16 white and 16 black. Mathematics related to chess has many implications (Dvoryatkina & Simonovskaya, 2021; Toshniwal, 2024). Over time, some problems that take their cue from the rules of the game have become famous for their combinatorial content.

The *Game of the Eight Queens*: It is a well-known mathematical problem involving the position of eight queens on an 8×8 board so that no queen threatens each other, considering that the queens can move in all directions. The combinatorial nature of this challenge arises from the need to explore all possible queen arrangements and identify those that satisfy the given constraints (i.e., no queen can attack another). This requires searching through a set of valid solutions, a key feature of combinatorial analysis. Initially, one can place the first queen in any of the eight columns. The second queen can be placed in any of the seven remaining columns that the first queen does not threaten, and so on. The product, therefore, gives the total number of possible positions of the eight queens: $8 \cdot 7 \cdot 6 \cdot 5 \cdot 4 \cdot 3 \cdot 2 \cdot 1 = 40{,}320$. However, many of these positions are invalid because the queens threaten each other. In particular, there are 92 distinct arrangements for the game of the eight queens on an 8×8 board (Gauss managed to find 72 of these). If we do not consider rotations and reflections, the 92 becomes just 12. We use a "backtracking" algorithm, in which a partial attempt is made and reversed when a failure condition occurs. We place a queen in a column that does not generate conflict; if no such column is found, the program returns to the last valid configuration and tries a different one.

Michael Simkin of the Center of Mathematical Sciences and Applications at Harvard University approached the problem using an n x n chessboard. His research led to a formula that estimates the number of valid configurations: $(0.143 \cdot n)^N$. The factor 0.143 is an average value that does not yield an exact count, but it closely approximates it. Simkin's formula is designed to provide estimates for large values of n and is not meant to be accurate for smaller cases. Possible outcomes for the n-queen problem, avoiding duplicates and allowing custom board sizes, can be explored using the following program:

Program 83 chess

```
import tkinter as tk
from tkinter import ttk
class ChessSolver:
def __init__(self):
self.window = tk.Tk()
self.window.title(“Chessboard Solutions - Queens' Problem”)
self.window.geometry(“800x600”)
self.main_frame = ttk.Frame(self.window, padding=”10”)
self.main_frame.pack(fill=tk.BOTH, expand=True)
ttk.Label(self.main_frame, text=”Queens' Problem (92 Solutions)”, font=('Arial', 14)).
pack(pady=20)
ttk.Button(self.main_frame, text=”Solve”, command=self.solve_queens).pack(pady=5)
ttk.Button(self.main_frame, text=”Exit”, command=self.exit_program).pack(pady=20)
self.board_frame = None
self.current_solution = 0
self.solutions = []
def create_board_window(self):
if self.board_frame:
self.board_frame.destroy()
self.board_frame = ttk.Frame(self.window)
self.board_frame.pack(fill=tk.BOTH, expand=True)
self.canvas = tk.Canvas(self.board_frame, width=400, height=400)
self.canvas.pack(pady=20)
nav_frame = ttk.Frame(self.board_frame)
nav_frame.pack(pady=10)
ttk.Button(nav_frame, text=”Previous”, command=self.prev_solution).pack(side=tk.LEFT,
padx=5)
ttk.Button(nav_frame, text=”Next”, command=self.next_solution).pack(side=tk.LEFT, padx=5)
ttk.Button(nav_frame, text=”Menu”, command=self.show_menu).pack(side=tk.LEFT, padx=5)
self.solution_label = ttk.Label(self.board_frame, text=””)
self.solution_label.pack()
def draw_board(self, positions):
self.canvas.delete(“all”)
size = 50
for i in range(8):
for j in range(8):
color = “white” if (i + j) % 2 == 0 else “grey”
x1, y1 = j * size, i * size
self.canvas.create_rectangle(x1, y1, x1 + size, y1 + size, fill=color)
for pos in positions:
row, col = pos
x = col * size + size/2
y = row * size + size/2
self.canvas.create_text(x, y, text=”□”, font=('Arial', 24))
self.solution_label.config(
text=f”Soluzione {self.current_solution + 1} di {len(self.solutions)}”)
```

```
def show_menu(self):
if self.board_frame:
self.board_frame.destroy()
self.main_frame.pack()
def next_solution(self):
if self.solutions:
self.current_solution = (self.current_solution + 1) % len(self.solutions)
self.draw_board(self.solutions[self.current_solution])
def prev_solution(self):
if self.solutions:
self.current_solution = (self.current_solution - 1) % len(self.solutions)
self.draw_board(self.solutions[self.current_solution])
def is_safe_queen(self, board, row, col):
# Check this row on left side
for j in range(col):
if board[row][j] == 1:
return False
# Check upper diagonal on left side
for i, j in zip(range(row, -1, -1), range(col, -1, -1)):
if board[i][j] == 1:
return False
# Check lower diagonal on left side
for i, j in zip(range(row, 8, 1), range(col, -1, -1)):
if board[i][j] == 1:
return False
return True
def solve_queens_util(self, board, col):
if col >= 8:
positions = []
for i in range(8):
for j in range(8):
if board[i][j] == 1:
positions.append((i, j))
self.solutions.append(positions)
return
for i in range(8):
if self.is_safe_queen(board, i, col):
board[i][col] = 1
self.solve_queens_util(board, col + 1)
board[i][col] = 0
def solve_queens(self):
self.main_frame.pack_forget()
self.create_board_window()
self.solutions = []
self.current_solution = 0
board = [[0 for x in range(8)] for y in range(8)]
self.solve_queens_util(board, 0)
if self.solutions:
self.draw_board(self.solutions[0])
```

```
def exit_program(self):
self.window.destroy()
def run(self):
self.window.mainloop()
if __name__ == "__main__":
app = ChessSolver()
app.run()
```

Interestingly, there are how many distinct combinations there are based on the size of the board:

Table 1. Distinct combinations for the Queens' Game

n:	1	2	3	4	5	6	7	8	9	10	11	12	13	14
Distinct:	1	0	0	2	10	4	40	92	352	724	2,680	14,200	73,712	365,596

Another famous problem is the game of eight Rooks: one must arrange eight Rooks so they cannot attack each other, considering that the Rooks move horizontally or vertically on the board. As in the case of the eight queens, the possible solutions are 8! = 40,320. We have eight options on an empty board to place the first Rook on the first row. Since we cannot have more than two Rooks in the same column or row, we have seven options for placing the Rook in the second row. In the third row, we have six options, among others. Hence, the number of possible arrangements of eight identical Rooks is equal to $8 \cdot 7 \cdot 6 \cdot 5 \cdot 4 \cdot 3 \cdot 2 \cdot 1 = 8! = 40{,}320$.

It is different if we want to know how many ways the eight Rooks can be arranged without the constraint that they cannot be attacked. Using the formula for combinations (C64.8), we obtain 4,426,165,368 if the Rooks are indistinguishable, and $64 \cdot 63 \cdot 62 \cdot 61 \cdot 60 \cdot 59 \cdot 58 \cdot 57 = 178{,}462{,}987{,}637{,}760$ if they are distinguishable (e.g., they have different colours).

At this point, one can consider the Bishops. What is the maximum number of Bishops that can be placed on a board so that no one attacks another? There are seven white diagonals and eight black diagonals, but a Bishop placed in a corner square prevents the use of the opposite diagonal, and therefore, there are 14 diagonals available. Two squares remain available from the bottom right corner to the next black diagonal; the same happens in the following steps. Analysing the white diagonals, we conclude that only 14 Bishops can be inserted.

We can further extend the investigation by considering knights. In this case, we must consider that a knight on a white square can only reach black squares. Since there are 64 squares and half are white, the maximum number of knights is 64/2 = 32.

The *Knight Game*: This game, also known as the Knight's Tour or Knight's Problem, is another well-known mathematical problem involving the knight's movement on an 8 × 8 chessboard. The goal of the game is to find a path that allows the knight to cross each square of the board only once, considering that it moves in an L-shape (i.e., it can move two squares vertically or horizontally, and then one square in a direction perpendicular to the first). Combinatorial calculus can be used to determine the quantity of possible solutions to the horse game. Notably, there are more than 26 million possible paths for the knight on an 8 × 8 board, but only a few of these are valid. Here, a common technique is the "backtracking" algorithm, similar to the one used in the game of eight queens. This algorithm utilises combinatorial calculus to generate and evaluate all possible combinations of routes for the horse and then selects the correct route.

A final mention of chess and combinatorics can refer to a "new" game formula.

The name *Chess960* comes from the fact that there are 960 possible initial configurations for the pieces placed on the bottom rows, unlike the traditional board, which has a fixed initial position.

Bobby Fischer, the famous chess champion, introduced this game in 1996. Its purpose was to reduce the influence of opening memorization and place greater value on players' creativity and strategic skills.

The position of the pieces is decided randomly but with some constraints. The 960 possible positions are derived from the combination of two workpiece configuration rules:

1. Bishops must be placed on squares of different colours (one on a light square and the other on a dark square).
2. The King must be between the two Rooks.

Now, considering these constraints, the number of possible combinations for the arrangement of the pieces on the bottom line is calculated as follows:

- Bishops can be arranged in 4 • 4 = 16 ways.
- The Queen has 6 houses available
- Knights have 5 squares available and can be arranged in 5 • 4/ 2 = 10 ways.
- The King and the Rooks have only three squares left; given the second constraint, there is only 1 way to arrange them.

So, 16 • 6 • 10 • 1 = 960.

Risk: It is a strategic board game in which players compete to conquer territories and dominate the world. It involves elements of combat, territorial conquest, and strategic planning.

In the game, players deploy their troops across territories, attempt to expand their empire by conquering new territories, and attack opponents' troops to capture their territories. The strategy consists of balancing territorial expansion with the defence of one's borders and making tactical decisions on the distribution of troops and the attacks to be launched.

Combinatorial play in Risk involves the various combinations of choices that players can make. For example, they must decide how to distribute their troops among territories, which territory to attack, how many troops to deploy in an attack, and how to manage defences. Each decision significantly impacts the game's outcome, and numerous variables must be taken into account. The Risk game can be considered a combinatorial game as it involves a series of strategic choices, resource distribution, and combinations of actions that affect the game's outcome. Combinatorial calculus is applied to determine the possible combinations of dice outcomes and to evaluate the probabilities associated with each scenario.

To calculate these combinations, binomial coefficients are used. If we roll three dice and the opponent rolls two dice, the number of possible combinations of dice outcomes is given by the binomial coefficient: $C_{3,2} = 3! / (2! * (3-2)!) = 3$. It means there are three possible combinations for the number of wins and losses in combat rounds. The possible combinations are closely related to the calculation of probabilities; in this case, the probability of victory for the attacker is 0.3624.

A program could show the chances of an attacker winning based on how many tanks the attacker and defender have:

Program 84 Risk

```
import random
def risk(attack, defense):
while attack > 1 and defense > 0:
attack_dice = sorted([random.randint(1, 6) for _ in range(min(3, attack - 1))], re-
verse=True)
defense_dice = sorted([random.randint(1, 6) for _ in range(min(3, defense))], re-
verse=True)
for a, d in zip(attack_dice, defense_dice):
if a > d:
defense -= 1
else:
attack -= 1
return attack, defense
def simulate_risk(attack, defense, num_simulations=10000):
attack_wins = 0
for _ in range(num_simulations):
```

```
final_attack, final_defense = risk(attack, defense)
if final_defense == 0:
attack_wins += 1
return attack_wins / num_simulations
attack = int(input("Enter the number of attacking tanks: "))
defense = int(input("Enter the number of defending tanks: "))
print(f"Probability of attack victory: {simulate_risk(attack, defense)}")
```

Go: An abstract combinatorial strategy game in which two players, Black and White, take turns placing stones on the intersections of an empty grid. The primary objective is to claim as much territory as possible by enclosing areas of the board, while simultaneously reducing the opponent's territory.

A player may surround a group of the opponent's stones, thereby reducing its liberties—the adjacent empty points—to zero. When this occurs, the surrounded stones are captured and removed from the board.

Go exemplifies a complex combinatorial system with profound mathematical implications (Tromp & Farnebäck, 2007; Egri-Nagy & Törmänen, 2020). On a standard 19×19 board, there are 361 intersections available for play. The number of potential stone configurations grows exponentially, making the enumeration of all possible positions an extremely challenging computational task.

Each move introduces a vast number of possible continuations, creating a highly branched decision tree. Players must evaluate not only the immediate consequences of their actions but also anticipate and adapt to the opponent's future responses.

A critical aspect of strategic play in Go involves the recognition and manipulation of patterns and sequences of stones. These include living groups (which cannot be captured), dead groups (with no prospect of survival), and more ambiguous formations. Mastery of the game requires an understanding of how to construct, defend, or dismantle such patterns in order to influence the overall state of the game.

In sum, Go is an exceptionally sophisticated game that demands strategic foresight, combinatorial reasoning, and deep analytical thinking.

A program to play mini-Go 6×6:

Program 85 Go

```
import pygame
import numpy as np
import sys
# Pygame Initialization
pygame.init()
# Constants
window_size = 600
board_size = 6
cell_size = window_size // (board_size + 1) # Leave space for margins
BLACK = (0, 0, 0)
WHITE = (255, 255, 255)
BROWN = (165, 124, 82)
class Go:
def __init__(self):
self.board = np.zeros((board_size, board_size), dtype=int)
self.current_player = 1
self.previous_board = None
def place_stone(self, x, y):
if self.board[y, x] != 0:
```

```
return False
self.previous_board = self.board.copy()
self.board[y, x] = self.current_player
# Capture opponent's stones
for dx, dy in [(-1, 0), (1, 0), (0, -1), (0, 1)]:
self.capture_stones(x + dx, y + dy)
# Check for the suicide rule
if self.count_liberties(x, y) == 0:
self.board = self.previous_board
return False
# Check for the Ko rule
if np.array_equal(self.board, self.previous_board):
self.board = self.previous_board
return False
self.current_player = 3 - self.current_player
return True
def capture_stones(self, x, y):
if 0 <= x < board_size and 0 <= y < board_size and self.board[y, x] == 3 - self.current_player:
if self.count_liberties(x, y) == 0:
self.remove_group(x, y)
def count_liberties(self, x, y):
color = self.board[y, x]
visited = set()
return self._count_liberties(x, y, color, visited)
def _count_liberties(self, x, y, color, visited):
if (x, y) in visited:
return 0
if x < 0 or x >= board_size or y < 0 or y >= board_size:
return 0
if self.board[y, x] == 0:
return 1
if self.board[y, x] != color:
return 0
visited.add((x, y))
liberties = 0
for dx, dy in [(-1, 0), (1, 0), (0, -1), (0, 1)]:
liberties += self._count_liberties(x + dx, y + dy, color, visited)
return liberties
def remove_group(self, x, y):
color = self.board[y, x]
to_remove = set()
self._find_group(x, y, color, to_remove)
for x, y in to_remove:
self.board[y, x] = 0
def _find_group(self, x, y, color, group):
if (x, y) in group or x < 0 or x >= board_size or y < 0 or y >= board_size or self.board[y, x] != color:
return
group.add((x, y))
for dx, dy in [(-1, 0), (1, 0), (0, -1), (0, 1)]:
```

```
self._find_group(x + dx, y + dy, color, group)
def draw_board(screen, game):
screen.fill(BROWN)
# Draw grid
for i in range(board_size + 1):
pygame.draw.line(screen, BLACK, (cell_size, i * cell_size), (window_size - cell_size, i * cell_size), 1)
pygame.draw.line(screen, BLACK, (i * cell_size, cell_size), (i * cell_size, window_size - cell_size), 1)
# Draw stones
for y in range(board_size):
for x in range(board_size):
if game.board[y, x] == 1: # Black stone
pygame.draw.circle(
screen, BLACK,
(x * cell_size + cell_size, y * cell_size + cell_size),
cell_size // 4
)
elif game.board[y, x] == 2: # White stone
pygame.draw.circle(
screen, WHITE,
(x * cell_size + cell_size, y * cell_size + cell_size),
cell_size // 4
)
def play():
screen = pygame.display.set_mode((window_size, window_size))
pygame.display.set_caption("Go 6x6")
game = Go()
while True:
for event in pygame.event.get():
if event.type == pygame.QUIT:
pygame.quit()
sys.exit()
elif event.type == pygame.MOUSEBUTTONDOWN:
x, y = pygame.mouse.get_pos()
cell_x = (x - cell_size + cell_size // 2) // cell_size
cell_y = (y - cell_size + cell_size // 2) // cell_size
if 0 <= cell_x < board_size and 0 <= cell_y < board_size:
if game.place_stone(cell_x, cell_y):
print("Valid move")
else:
print("Invalid move. Try again.")
draw_board(screen, game)
pygame.display.flip()
if __name__ == "__main__":
play()
```

4-in-a-Row: A two-player board game consisting of a 6 x 7 vertical grid in which players alternate the placement of checkers of two different colours (to be inserted from the top of the grid. The goal is to line up four checkers of the same colour in a horizontal, vertical, or diagonal row before the opponent. Forza 4 can be considered a zero-sum combinatorial game with perfect information (Uiterwijk, 2019). Analysing the game from a combinatorial point of view, we can assign a value to each move based on its potential to create or block winning combinations. This allows us to develop strategies that identify the sequence of moves leading to victory with the highest probability using minimax methods.

Finally, it is possible to predict the winner of a game starting from a specific arrangement of the pieces. Minimax is a search algorithm often used in two-person games. This algorithm evaluates all possible moves up to a certain depth in the move tree, alternating between maximizing the current player's advantage and minimizing the opponent's advantage. However, given the game's complexity, Minimax may not be particularly efficient. Some modern approaches use neural networks trained on large game datasets to evaluate positions on the grid and determine the optimal move. In a 6 x 7 grid, there are 42 empty spaces for players to place their pawns in. Each space can be empty or occupied by one player, so there are three possibilities for each space, and the total number of possible combinations is 3^{42}. However, not all these combinations are valid in the game (a checker cannot "float" on top of an empty space). Through detailed computational analysis, experts have determined that there are 4,531,985,219,092 valid locations.

Program 86 Connect four

```
import pygame
import sys
import numpy as np
# Initialize Pygame
pygame.init()
# Constants
rows = 6
columns = 7
square_size = 100
width = columns * square_size
height = (rows + 1) * square_size
circle_size = int(square_size / 2 - 5)
circle_radius = int(circle_size / 2)
# Colors
blue = (0, 0, 255)
black = (0, 0, 0)
red = (255, 0, 0)
yellow = (255, 255, 0)
class Connect4:
def __init__(self):
self.board = np.zeros((rows, columns))
self.player = 1
def is_valid_move(self, column):
return self.board[rows - 1][column] == 0
def drop_disc(self, column):
for row in range(rows):
if self.board[row][column] == 0:
self.board[row][column] = self.player
return True
return False
def check_winner(self):
# Horizontal check
for c in range(columns - 3):
for r in range(rows):
```

```
if all(self.board[r][c + i] == self.player for i in range(4)):
return True
# Vertical check
for c in range(columns):
for r in range(rows - 3):
if all(self.board[r + i][c] == self.player for i in range(4)):
return True
# Positive diagonal check
for c in range(columns - 3):
for r in range(rows - 3):
if all(self.board[r + i][c + i] == self.player for i in range(4)):
return True
# Negative diagonal check
for c in range(columns - 3):
for r in range(3, rows):
if all(self.board[r - i][c + i] == self.player for i in range(4)):
return True
return False
def switch_player(self):
self.player = 3 - self.player
def draw_board(screen, game):
screen.fill(blue)
for c in range(columns):
for r in range(rows):
pygame.draw.rect(screen, blue, (c * square_size, (r + 1) * square_size, square_size,
square_size))
pygame.draw.circle(screen, black, (int(c * square_size + square_size / 2), int((r + 1) *
square_size + square_size / 2)), circle_radius)
for c in range(columns):
for r in range(rows):
if game.board[r][c] == 1:
pygame.draw.circle(screen, red, (int(c * square_size + square_size / 2), height - int(r *
square_size + square_size / 2)), circle_radius)
elif game.board[r][c] == 2:
pygame.draw.circle(screen, yellow, (int(c * square_size + square_size / 2), height - int(r
* square_size + square_size / 2)), circle_radius)
pygame.display.update()
def play():
screen = pygame.display.set_mode((width, height))
pygame.display.set_caption('Connect 4')
game = Connect4()
game_over = False
while not game_over:
for event in pygame.event.get():
if event.type == pygame.QUIT:
pygame.quit()
sys.exit()
if event.type == pygame.MOUSEMOTION:
pygame.draw.rect(screen, black, (0, 0, width, square_size))
```

```
posx = event.pos[0]
if game.player == 1:
pygame.draw.circle(screen, red, (posx, int(square_size / 2)), circle_radius)
else:
pygame.draw.circle(screen, yellow, (posx, int(square_size / 2)), circle_radius)
pygame.display.update()
if event.type == pygame.MOUSEBUTTONDOWN:
pygame.draw.rect(screen, black, (0, 0, width, square_size))
posx = event.pos[0]
column = int(posx // square_size)
if game.is_valid_move(column):
game.drop_disc(column)
if game.check_winner():
print(f"Player {game.player} wins!")
game_over = True
game.switch_player()
draw_board(screen, game)
else:
print("Invalid move. Try again.")
if __name__ == "__main__":
play()
```

Mancala: It is the name of an entire family of games of African origin; it is a traditional game in which two players take turns taking and distributing "suits" from a series of "holes" on the board. The goal is to collect as many seeds as possible. Each player can move the seeds from one hole to distribute them to other holes, following specific rules. The game ends when one of the players can no longer make moves, i.e., when he has no more suits on his line.

Figure 3. Mancala

Mancala is an example of a collection game whose combinatorial analysis involves configuration spaces and value functions that describe the number of seeds (or "suits") in each hole. A board configuration can be represented as a vector of integers, where each entry indicates the number of seeds in a corresponding hole. The optimal strategy depends on analysing the end positions (when all moves are made, and it is no longer possible to move) and optimizing the number of suits in the player's possession. Strategies in Mancala often involve planning sequences of moves to achieve advantageous configurations. This requires understanding the possible seed distributions and the consequences of future moves.

The game can be represented as a move tree, where each node represents a board configuration, and each arc represents a legal move. Combinatorial analysis of the move tree can help determine optimal strategies. Many variations of Mancala have symmetries and patterns that can be strategically exploited. For example, symmetrical configurations can help capture seeds or block the opponent. We can analyse the complexity of Mancala in terms of combinatorial *game theory* (Siegel, 2013; Larsson et al., 2016; Albert et al., 2019), considering that the board has two rows of 6 holes each, and each hole contains a variable number of stones.

Considering that each hole can contain a value from 0 to N (where N depends on the maximum of initial suits), the total possible combinations are immense. In the game, the Mancala holes (one for each player) play a unique and fundamental role, serving as "deposits" where players collect seeds. Unlike regular holes, where seeds are distributed and can be captured, Mancala holes accumulate the seeds earned during the game.

When a player distributes seeds from their holes (always starting from one in their row), they gain that suit if the last seed lands in their Mancala hole. This mechanism encourages strategic collection, as the goal is to gather as many seeds as possible in this position.

Calculating the exact variety of distinct configurations in a Mancala game is challenging due to the rules and interactions between the holes and deposits. However, an approximation based on seed distribution across available positions suggests that the possible arrangements reach into the billions. To estimate the different configurations, one can consider all possible distributions of the 48 seeds across the 14 positions (12 holes + 2 Mancala), a problem involving the placement of *n* identical objects into *k* distinct containers.

The formula for determining the ways to distribute objects in containers is given by $\binom{n+k-1}{k-1}$. In the case of Mancala, we have n = 48 (total seeds) and k = 14 (12 holes + 2 Mancala), so the number of distinct configurations is: $\binom{48+14-1}{14-1} = \binom{61}{13}$

Each move changes the state of the game, and we can analyse these moves in terms of the move tree: each node in the tree represents a state of the game, and each branch represents a possible move. With each move, the number of suits dealt changes and can affect future options. This tree is expandable at every stage of the game, but it quickly becomes complex. We can analyse the game of Mancala through game theory. The optimal strategy can be determined by search algorithms, such as the Minimax algorithm, which explores all possible states starting from an initial configuration to determine the best move, also considering the opponent's future moves.

Program 87 Mancala

```
import pygame
import random
# Initialize Pygame
pygame.init()
# Constants
WIDTH = 900
HEIGHT = 400
HOLE_RADIUS = 40
MARGIN = 150
HOLE_DISTANCE = 110
COLORS = {
'background': (184, 134, 11),
'hole': (255, 255, 255),
'seed': (0, 0, 0),
'mancala': (0, 255, 0),
'active_hole': (255, 215, 0),
}
# Define the game window
screen = pygame.display.set_mode((WIDTH, HEIGHT))
pygame.display.set_caption('Mancala Game')
# Function to draw the game board
def draw_board(state):
```

```
screen.fill(COLORS['background'])
# Draw the holes and mancalas
for i in range(6):
pygame.draw.circle(screen, COLORS['hole'], (MARGIN + i * HOLE_DISTANCE, HEIGHT - 100),
HOLE_RADIUS)
font = pygame.font.SysFont("Arial", 24)
text = font.render(str(state[i]), True, COLORS['seed'])
screen.blit(text, (MARGIN + i * HOLE_DISTANCE - text.get_width() // 2, HEIGHT - 100 -
text.get_height() // 2))
pygame.draw.circle(screen, COLORS['hole'], (MARGIN + i * HOLE_DISTANCE, 100), HOLE_RADIUS)
text = font.render(str(state[7 + i]), True, COLORS['seed'])
screen.blit(text, (MARGIN + i * HOLE_DISTANCE - text.get_width() // 2, 100 - text.get_
height() // 2))
# Draw the mancalas
pygame.draw.rect(screen, COLORS['mancala'], (MARGIN + 6 * HOLE_DISTANCE, HEIGHT - 200, 60,
100))
pygame.draw.rect(screen, COLORS['mancala'], (MARGIN - 120, 200, 60, 100))
# Display the score in the mancalas
font = pygame.font.SysFont("Arial", 24)
text = font.render(str(state[6]), True, COLORS['seed'])
screen.blit(text, (MARGIN + 6 * HOLE_DISTANCE + 30 - text.get_width() // 2, HEIGHT - 200 +
25))
text = font.render(str(state[13]), True, COLORS['seed'])
screen.blit(text, (MARGIN - 120 + 30 - text.get_width() // 2, 200 + 25))
# Function to execute the player's move
def execute_move(state, player, hole_index):
seeds = state[hole_index]
state[hole_index] = 0
index = hole_index
while seeds > 0:
index = (index + 1) % 14
if player == 1 and index == 13:
continue
if player == 2 and index == 6:
continue
state[index] += 1
seeds -= 1
# Capture seeds if the last seed lands in an empty hole on the player's side
if player == 1 and 0 <= index < 6 and state[index] == 1:
opposite_index = 12 - index
state[6] += state[opposite_index] + 1
state[index] = 0
state[opposite_index] = 0
elif player == 2 and 7 <= index < 13 and state[index] == 1:
opposite_index = 12 - index
state[13] += state[opposite_index] + 1
state[index] = 0
state[opposite_index] = 0
# Extra move if the last seed lands in the player's mancala
```

```
if player == 1 and index == 6:
return True
if player == 2 and index == 13:
return True
return False
# Function for the computer's move
def computer_move(state):
possible_moves = [i for i in range(7, 13) if state[i] > 0]
move = random.choice(possible_moves)
return move
# Function to check if the game is over and determine the winner
def check_game_over(state):
if sum(state[:6]) == 0 or sum(state[7:13]) == 0:
player_1_score = state[6]
player_2_score = state[13]
if player_1_score > player_2_score:
return "Player 1 wins!"
elif player_1_score < player_2_score:
return "Player 2 wins!"
else:
return "It's a tie!"
return None
# Function to manage the game
def game():
state = [4, 4, 4, 4, 4, 4, 0, 4, 4, 4, 4, 4, 4, 0]
turn = 1
while True:
screen.fill(COLORS['background'])
draw_board(state)
pygame.display.update()
result = check_game_over(state)
if result:
font = pygame.font.SysFont("Arial", 30)
text = font.render(result, True, (255, 0, 0))
screen.blit(text, (WIDTH // 2 - text.get_width() // 2, HEIGHT // 2 - text.get_height() //
2))
pygame.display.update()
pygame.time.delay(5000) # 5-second pause before closing
pygame.quit()
return
for event in pygame.event.get():
if event.type == pygame.QUIT:
pygame.quit()
return
if event.type == pygame.MOUSEBUTTONDOWN:
x, y = pygame.mouse.get_pos()
if HEIGHT - 150 < y < HEIGHT - 50:
hole_index = (x - MARGIN) // HOLE_DISTANCE
if 0 <= hole_index < 6 and state[hole_index] > 0 and turn == 1:
```

```
if execute_move(state, 1, hole_index):
continue
turn = 2
if 50 < y < 150:
hole_index = (x - MARGIN) // HOLE_DISTANCE
if 0 <= hole_index < 6 and state[7 + hole_index] > 0 and turn == 2:
if execute_move(state, 2, 7 + hole_index):
continue
turn = 1
if turn == 2:
move = computer_move(state)
if execute_move(state, 2, move):
continue
turn = 1
pygame.time.delay(200) # Pause between turns
# Start the game
game()
```

Gomoku: It is played on a 15 x 15 grid, where two players take turns placing a checker (black or white) to form a row of five consecutive checkers, either horizontally, vertically, or diagonally. Alternatively, they can put symbols such as crosses and circles; in this case, they need a squared sheet of paper and a pen. The rules are extremely simple, but this does not mean the game is trivial and devoid of mathematical considerations. Gomoku is a combinatorial game in which positions can be represented by a two-dimensional matrix of 0 (empty), 1 (first player's pawn), and -1 (second player's pawn). Each move can be evaluated based on how it affects the possible lines of five checkers.

The optimal strategy is based on creating winning lines and blocking opponents' lines. A winning strategy involves building chains of pawns where each move progressively reduces the opponent's options. Suppose a player has a chain of four consecutive checkers, the probability of winning increases significantly. Search and backtracking algorithms can determine the optimal move at each step. However, a perfect strategy relies on analysing local (checker proximity) and global (board-wide) configurations. For a 15 x 15 grid, there are 225 squares and three possible states for each square (empty, X, O). The number of possible configurations is 3^{225} (an astronomical amount). However, during a match, not all these configurations are achievable due to the alternating moves of the two players. Since Gomoku, however, has a large number of possible moves, the combinatorial analysis of the game positions results in a profound game tree. The game is theoretically solvable, but its computational complexity grows exponentially with the grid's size and the search's depth. Each move generates many possible configurations, and players explore the possible answers at each level of the tree.

To give an idea of the game's complexity, one can simulate a game against the computer. The computer uses a greedy algorithm with the following priorities:

1. Primary Defence: Search for and block sequences of three consecutive black checkers with free ends
2. Positional evaluation with a score based on:
   a. Number of consecutive checkers of the opponents
   b. Free spaces at the ends of sequences
   c. Increased weight for longer sequences
3. Choice of move:
   a. If it finds a sequence to lock -> lock
   b. Otherwise, choose the position with the highest score using a rating function
   c. If it does not detect any strategic moves, it chooses a random location near the centre

Program 88 Gomoku

```
import pygame
import sys
import random
class GomokuGUI:
def __init__(self):
pygame.init()
self.size = 15
self.cell_size = 40
self.margin = 40
self.window_size = self.size * self.cell_size + 2 * self.margin
self.screen = pygame.display.set_mode((self.window_size, self.window_size))
pygame.display.set_caption('Gomoku 15x15 - Player vs Computer')
self.board = [[' ' for _ in range(self.size)] for _ in range(self.size)]
self.current_player = 'X'
self.game_over = False
self.BLACK = (0, 0, 0)
self.WHITE = (255, 255, 255)
self.BROWN = (139, 69, 19)
def find_three_in_a_row(self):
directions = [(0, 1), (1, 0), (1, 1), (1, -1)]
for row in range(self.size):
for col in range(self.size):
if self.board[row][col] == 'X':
for dr, dc in directions:
count = 1
# Check forward
r, c = row + dr, col + dc
while 0 <= r < self.size and 0 <= c < self.size and self.board[r][c] == 'X':
count += 1
r += dr
c += dc
if count == 3:
# Check if one end is free
if (0 <= r < self.size and 0 <= c < self.size and
self.board[r][c] == ' '):
return r, c
# Check the other end
r = row - dr
c = col - dc
if (0 <= r < self.size and 0 <= c < self.size and
self.board[r][c] == ' '):
return r, c
return None
def computer_move(self):
# First check if there's a three-in-a-row to block
block_move = self.find_three_in_a_row()
if block_move:
row, col = block_move
self.board[row][col] = 'O'
```

```
if self.check_victory(row, col):
self.game_over = True
print('The computer has won!')
return True
best_score = -1
best_move = None
for row in range(self.size):
for col in range(self.size):
if self.board[row][col] != ' ':
for dr in [-1, 0, 1]:
for dc in [-1, 0, 1]:
r, c = row + dr, col + dc
if (0 <= r < self.size and 0 <= c < self.size and
self.board[r][c] == ' '):
attack_score = self.evaluate_position(r, c, 'O')
defense_score = self.evaluate_position(r, c, 'X')
score = attack_score + defense_score
if score > best_score:
best_score = score
best_move = (r, c)
if best_move is None:
center = self.size // 2
for _ in range(100):
row = random.randint(center-3, center+3)
col = random.randint(center-3, center+3)
if 0 <= row < self.size and 0 <= col < self.size and self.board[row][col] == ' ':
best_move = (row, col)
break
if best_move:
self.board[best_move[0]][best_move[1]] = 'O'
if self.check_victory(best_move[0], best_move[1]):
self.game_over = True
print('The computer has won!')
return True
return False
def evaluate_position(self, row, col, player):
score = 0
directions = [(0, 1), (1, 0), (1, 1), (1, -1)]
for dr, dc in directions:
sequence = []
for i in range(-4, 5):
r, c = row + dr * i, col + dc * i
if 0 <= r < self.size and 0 <= c < self.size:
sequence.append(self.board[r][c])
count = 0
open_ends = 0
for i in range(len(sequence)):
if sequence[i] == player:
count += 1
```

```
elif sequence[i] == ' ':
if count > 0:
open_ends += 1
score += count * (1 + open_ends)
count = 0
open_ends = 1
else:
if count > 0:
score += count * (1 + open_ends)
count = 0
open_ends = 0
if count > 0:
score += count * (1 + open_ends)
return score
def draw_board(self):
self.screen.fill(self.BROWN)
for row in range(self.size):
for col in range(self.size):
rect = pygame.Rect(
self.margin + col * self.cell_size,
self.margin + row * self.cell_size,
self.cell_size,
self.cell_size
)
pygame.draw.rect(self.screen, self.BLACK, rect, 1)
if self.board[row][col] != ' ':
center = (
self.margin + col * self.cell_size + self.cell_size // 2,
self.margin + row * self.cell_size + self.cell_size // 2
)
color = self.BLACK if self.board[row][col] == 'X' else self.WHITE
pygame.draw.circle(self.screen, color, center, self.cell_size // 2 - 4)
if self.board[row][col] == 'O':
pygame.draw.circle(self.screen, self.BLACK, center, self.cell_size // 2 - 4, 2)
def get_cell_from_mouse(self, pos):
x, y = pos
if self.margin <= x <= self.window_size - self.margin and \
self.margin <= y <= self.window_size - self.margin:
row = (y - self.margin) // self.cell_size
col = (x - self.margin) // self.cell_size
return row, col
return None
def check_victory(self, row, col):
directions = [(0, 1), (1, 0), (1, 1), (1, -1)]
player = self.board[row][col]
for dr, dc in directions:
count = 1
r, c = row + dr, col + dc
while 0 <= r < self.size and 0 <= c < self.size and \
```

```
self.board[r][c] == player:
count += 1
r += dr
c += dc
r, c = row - dr, col - dc
while 0 <= r < self.size and 0 <= c < self.size and \
self.board[r][c] == player:
count += 1
r -= dr
c -= dc
if count >= 5:
return True
return False
def run(self):
while True:
for event in pygame.event.get():
if event.type == pygame.QUIT:
pygame.quit()
sys.exit()
if event.type == pygame.MOUSEBUTTONDOWN and not self.game_over:
cell = self.get_cell_from_mouse(event.pos)
if cell:
row, col = cell
if self.board[row][col] == ' ':
self.board[row][col] = 'X'
if self.check_victory(row, col):
self.game_over = True
print('You won!')
else:
self.computer_move()
self.draw_board()
pygame.display.flip()
if __name__ == '__main__':
game = GomokuGUI()
game.run()
```

Nim: Two players take turns removing objects from different rows. In the film *Last Year at Marienbad* by Alain Resnais it was played by the protagonist by removing one or more matches from one of the four lines. The matches were 1, 3, 5, and 7, and in this variant, whoever removed the last element lost. The game is interesting from a combinatorial point of view because it involves strategies to maximize one's chances of winning and involves a certain amount of combinatorial calculation to determine the best move to make in a given situation (Ferguson, 2001).

Gambling strategies involve concepts such as "Nim-Sum". The Nim-Sum strategy uses binary sum without carryovers, also known as XOR sum. Here is how it works: convert the number of objects in each row into binary and perform the XOR (exclusive OR) without carryovers. A position is considered safe if the sum of all rows is zero. Players can adopt this strategy to force the opponent to make disadvantageous moves. This strategy requires understanding the possible moves and combinatorial consequences.

Let us consider the game using matches as objects.

Figure 4. Nim game

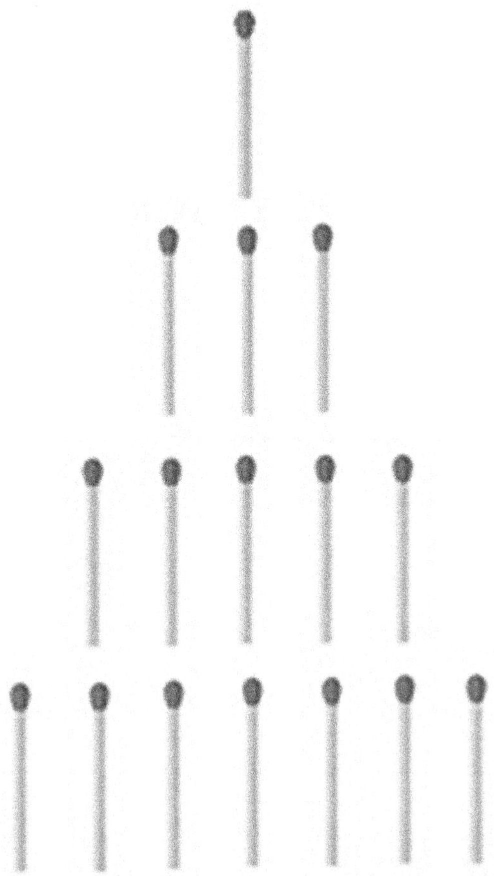

Example:

Row 1: 1 object (001 in binary)
Row 2: 3 Objects (011 in Binary)
Row 3: 5 Objects (101 in Binary)
Row 4: 7 Objects (111 in Binary)
Nim Sum: 0 (Safe Position)

This situation must be modified to make the position unsafe for the opponent. For example, reducing Line 2 from 3 to 0. In this way, the new situation is:

Row 1: 1 object (001 in binary)
Row 2: 0 objects (000 in binary)
Row 3: 5 Objects (101 in Binary)
Row 4: 7 Objects (111 in Binary)
New Nim Sum: 011 (Insecure Position)

For the record, the first Computer Game in history, the Ferranti NIMROD, was created (1951) precisely to play Nim. It was 3.6 meters wide, 2.7 meters deep and 1.5 meters high.

A program to play Nim using the graphics window that also shows the binary version of the number of matches can be:

Program 89 Nim

```
import tkinter as tk
import random
# Global declaration of the variable piles
piles = []
# Function to calculate the Nim-sum
def nim_sum(piles):
result = 0
for pile in piles:
result ^= pile # XOR between all the piles
return result
# Function for the computer's optimal move
def computer_move(piles):
# Calculate the Nim-sum
xor_result = nim_sum(piles)
# If the Nim-sum is 0, the computer is in a losing position, but it will make a move any-
way
if xor_result == 0:
# In a losing position, choose a random move
non_empty_piles = [i for i, pile in enumerate(piles) if pile > 0]
chosen_pile = random.choice(non_empty_piles)
remove = random.randint(1, piles[chosen_pile])
piles[chosen_pile] -= remove
return piles
else:
# Otherwise, make the winning move
for i, pile in enumerate(piles):
# Find the pile that, when reduced by the Nim-sum, results in 0
new_pile = pile ^ xor_result
if new_pile < pile:
piles[i] = new_pile
break
return piles
# Function to update the GUI
def update_gui():
global piles # Declare that we're using the global variable piles
for i, pile in enumerate(piles):
pile_labels[i].config(text=str(pile)) # Update the pile's text
binary_labels[i].config(text=format(pile, 'b')) # Show the binary representation
if check_victory(piles):
result_label.config(text="You won!")
return
# Computer's move
```

```
piles = computer_move(piles) # Make sure to update piles
for i, pile in enumerate(piles):
pile_labels[i].config(text=str(pile)) # Update the pile's text
binary_labels[i].config(text=format(pile, 'b')) # Show the binary representation
if check_victory(piles):
result_label.config(text="The computer won!")
return
# Function to check if someone has won
def check_victory(piles):
return all(pile == 0 for pile in piles)
# Function for the player's move
def player_move(pile_index, remove_count):
global piles # Declare that we're using the global variable piles
if piles[pile_index] >= remove_count:
piles[pile_index] -= remove_count
update_gui()
# Function for the player's move (click handling)
def on_click_pile(pile_index):
def on_confirm():
remove_count = int(remove_entry.get())
if remove_count > 0 and remove_count <= piles[pile_index]:
player_move(pile_index, remove_count)
confirm_window.destroy()
else:
error_label.config(text="Invalid number")
confirm_window = tk.Toplevel(root)
confirm_window.title(f"Remove from Pile {pile_index+1}")
tk.Label(confirm_window, text=f"Remove from Pile {pile_index+1} (1-{piles[pile_in-
dex]}):").grid(row=0, column=0)
remove_entry = tk.Entry(confirm_window)
remove_entry.grid(row=1, column=0)
tk.Button(confirm_window, text="Confirm", command=on_confirm).grid(row=2, column=0)
error_label = tk.Label(confirm_window, fg="red")
error_label.grid(row=3, column=0)
# Function to initialize the game window
def initialize_gui():
global piles, pile_labels, binary_labels, result_label, root
piles = [1, 3, 5, 7] # Initialize the piles with 1, 3, 5, 7 objects
pile_labels = []
binary_labels = []
# Create the main window
root = tk.Tk()
root.title("Nim Game")
# Result label
result_label = tk.Label(root, text="The game begins!", font=('Arial', 14))
result_label.grid(row=0, column=0, columnspan=4, pady=10)
# Create buttons for each pile
for i, pile in enumerate(piles):
label = tk.Label(root, text=str(pile), font=('Arial', 16), width=5, relief="solid")
```

```
label.grid(row=1, column=i, padx=10, pady=10)
pile_labels.append(label)
# Label for the binary representation below the pile
binary_label = tk.Label(root, text=format(pile, 'b'), font=('Arial', 12), width=5, re-
lief="solid")
binary_label.grid(row=2, column=i, padx=10, pady=5)
binary_labels.append(binary_label)
label.bind("<Button-1>", lambda event, index=i: on_click_pile(index))
root.mainloop()
# Start the game
if __name__ == "__main__":
initialize_gui()
```

Tsyan-shizi: This is an ancient Chinese game whose name means "collecting stones". It is closely related to the Nim. In this case there are two piles of stones, and the rule is to take one or more objects from one pile or the same number from both forcing the opponent to take the last object and therefore lose. By setting the restrictive condition of taking stones from a single pile, one can play against the computer that adopts the Nim-Sum strategy:

Program 90 Tsyan-shizi

```
import random
class TsyanShizi:
def __init__(self):
# Initialize the piles with a random number of stones between 5 and 15
self.piles = [random.randint(5, 15), random.randint(5, 15)]
self.player = 0 # 0 for the human player, 1 for the computer
def print_state(self):
"""Prints the current game state showing the piles and the number of stones."""
print("\nCurrent game state:")
for i, stones in enumerate(self.piles):
print(f"Pile {i + 1}: {'o' * stones}")
def valid_move(self, pile, number):
"""Checks if the move is valid for the chosen pile and number of stones."""
return 0 <= pile < len(self.piles) and 1 <= number <= self.piles[pile]
def make_move(self, pile, number):
"""Executes the move by subtracting the number of stones from the specified pile."""
self.piles[pile] -= number
def game_over(self):
"""Checks if the game is over (if all piles are empty)."""
return sum(self.piles) == 0
def computer_move(self):
"""Calculates the optimal move for the computer using Nim strategy."""
nim_sum = self.piles[0] ^ self.piles[1]
if nim_sum == 0:
# If XOR sum is 0, the computer is in a losing position
pile = random.randint(0, 1)
number = random.randint(1, self.piles[pile])
else:
```

```
# Find the winning move
for pile in range(2):
target = self.piles[pile] ^ nim_sum
if target < self.piles[pile]:
number = self.piles[pile] - target
return pile, number
return pile, number
def play_tsyan_shizi():
"""Manages the game loop between the player and the computer."""
game = TsyanShizi()
print("Welcome to Tsyan-Shizi!")
print("There are two piles of stones. You can remove as many stones as you want from a
single pile.")
print("The player who removes the last stone loses.")
while not game.game_over():
game.print_state()
if game.player == 0:
while True:
try:
pile = int(input("Choose a pile (1-2): ")) - 1
number = int(input("How many stones do you want to remove? "))
if game.valid_move(pile, number):
break
else:
print("Invalid move. Try again.")
except ValueError:
print("Please enter valid numbers.")
else:
pile, number = game.computer_move()
print(f"\nThe computer removes {number} stones from pile {pile + 1}")
game.make_move(pile, number)
if game.game_over():
game.print_state()
if game.player == 0:
print("The computer has won!")
else:
print("Congratulations! You have won!")
break
game.player = 1 - game.player # Alternate between 0 and 1
if __name__ == "__main__":
play_tsyan_shizi()
```

Magic squares: These are considered combinatorial games. A magic square is a grid of integers arranged in such a way that the sum of each row, column, and main diagonal of the grid gives the same value, called the magic constant. The game features various variations, including different grid sizes and numbers that can be used to fill in the boxes. The game requires logical thinking and deduction skills to solve the puzzle, and it does not require any advanced mathematical knowledge.

Magic squares have been known since ancient times and have been described in many different cultures. Shu is one of the oldest known magic squares, strongly associated with Chinese culture. Its name derives from an ancient Chinese legend that tells of its discovery. The legend associated with Shu is that the Chinese emperor Yu the Great discovered the square while meditating on the banks of the Luo River. He saw a turtle emerging from the river with this magic square on its back, and the digits 1-9 were arranged in the square. The fact that the sum of each row, column and main diagonal was always the same struck him deeply.

Program 91 Magic square

```
from itertools import permutations
def is_magic_square(square):
# Unpack the square into rows
a, b, c, d, e, f, g, h, i = square
# Check the sums of rows, columns, and diagonals
return (a + b + c == 15 and d + e + f == 15 and g + h + i == 15 and # rows
a + d + g == 15 and b + e + h == 15 and c + f + i == 15 and # columns
a + e + i == 15 and c + e + g == 15) # diagonals
# Generate all possible permutations of numbers from 1 to 9
all_permutations = permutations(range(1, 10))
# Count the number of magic squares
magic_square_count = 0
for perm in all_permutations:
if is_magic_square(perm):
magic_square_count += 1
# Total number of permutations of 9 elements
total_permutations = 9 * 8 * 7 * 6 * 5 * 4 * 3 * 2 * 1
# Calculate the probability
probability = magic_square_count / total_permutations
print(f"Total number of 3x3 magic squares: {magic_square_count}")
print(f"Total number of possible combinations: {total_permutations}")
print(f"Probability that a random permutation is a magic square: {probability:.6f}")
```

Today, many variations of magic square games exist, including digital formats and versions with different grid sizes and number sets. Combinatorial calculus plays a key role in analysing these variations, particularly in determining the number of possible magic squares that can be formed from a given set of integers.

A standard construction method is the substitution technique, where specific numbers are placed into selected positions within the grid, and the remaining values are calculated so that each row, column, and diagonal sums to the magic constant.

Combinatorial methods are also essential in estimating or computing the total number of distinct magic squares of a given size, which is valuable for constructing databases and studying their mathematical properties.

For instance, there is only one unique perfect 3×3 magic square (excluding rotations and reflections), whereas there are 880 distinct perfect 4×4 magic squares.

In the case of n = 5, the number of distinct perfect magic squares is 275,305,224, as determined in 1973 by the American mathematician Richard Schroeppel, with the aid of a computer.

Determining how many magic squares there are in order 6 is still an open problem today; however, it is estimated that the number of these magic squares is about 10^{19}. Drawing odd-order magic squares is not difficult while constructing even-order magic squares is more complex. It is sometimes referred to as the "Siamese magic square" as it is believed to have originated in Siam (present-day Thailand). Despite the name "Siamese", similar methods have been independently discovered in other cultures, such as India and China. The solution method was first described in the West by a French diplomat, Simon de la Loubère (1642–1729).

The algorithm is interesting: after creating an n x n matrix filled with zeros, the first number is placed in the centre cell of the first row and for each subsequent value, it tries to move diagonally up and to the right. If the target cell is already occupied, it moves one row down instead of diagonally.

Let us examine the following program:

Program 92 Siamese magic square

```
import numpy as np
def siamese_magic_square(n):
"""
```

Generates a magic square of size n using the Siamese method.

```
"""
if n % 2 == 0:
raise ValueError("The size must be odd.")
# Initialize an empty n x n square
square = np.zeros((n, n), dtype=int)
i, j = 0, n // 2 # Start in the middle of the first row
# Populate the square
for number in range(1, n * n + 1):
square[i, j] = number
ni, nj = (i - 1) % n, (j + 1) % n # Move diagonally up-right
if square[ni, nj] == 0:
i, j = ni, nj
else: # Move down if the spot is already filled
i = (i + 1) % n
return square
def print_magic_square(square):
"""
```

Prints the magic square in a nice format on the console.

```
"""
size = len(square)
for i in range(size):
for j in range(size):
print(f"{square[i, j]:2}", end=" ")
print() # Newline after each row
# Main program logic
if __name__ == "__main__":
try:
# Get user input
size = int(input("Enter the size of the magic square (must be odd): "))
if size % 2 == 0:
raise ValueError("The size must be odd.")
# Generate the magic square
magic_square = siamese_magic_square(size)
```

```
# Print the magic square in the console
print("Magic Square:")
print_magic_square(magic_square)
except ValueError as e:
print(f"Error: {e}")
```

Sudoku: It is a combinatorial game that involves completing a 9×9 grid of squares with numbers from 1 to 9, so that each row, column, and region of 3×3 squares contains all digits from 1 to 9 only once. Felgenhauer and Jarvis found that there are 6,670,903,752,021,072,936,960 Sudoku (Felgenhauer & Jarvis, 2005). In addition, Jarvis and Russell found that 5,472,730,538 Sudoku puzzles are essentially different (Russell & Jarvis, 2006). The game requires logic and deduction skills to solve the puzzle, and it does not require any advanced mathematical knowledge.

Sudoku has become popular worldwide over the past few decades, and many variations of the game are available, including large Sudoku puzzles, variants where some boxes are already filled in at the beginning, and even three-dimensional versions of the game. Sudoku-solving involves a mix of number placement, permutation analysis, and combinatorial strategy, so we can consider it an example of a combinatorial game that can be solved using some combinatorial techniques. The most common technique used to solve Sudoku puzzles is the "empty space counting" technique. This method estimates the number of possible ways numbers can fill empty cells. For example, if there are three blank spaces in a row, and we know that the integers from 1 to 9 have not yet been used, then there are $9 \cdot 8 \cdot 7 = 504$ possible combinations of numbers that can be placed in those three blank spaces. However, not all these combinations are valid, as the sum of each row, column, and block of 3×3 must equal 45 (since $1 + 2 + \ldots + 9 = 45$).

Another example used to solve Sudoku employs a "candidate elimination" technique, utilising the sum rule to examine the possible candidates in a row, column, or block and eliminate impossible solutions. Using combinatorial techniques can reduce the possible combinations to examine and solve Sudoku more quickly.

To simplify the analysis, we can work on a mini 4x4 Sudoku.

How many mini-Sudoku puzzles are there? They can be enumerated in many ways. If one considers the block in the upper left, the numbers must be 1, 2, 3 and 4, but they can be entered in any order. This gives $4! = 24$ ways to fill this block.

Figure 5. Mini sudoku

1	2		
3	4		
		*	
			*

The other items are determined by the choice of the two items marked with an asterisk (*). These two entries are arbitrary, except that they must be different; therefore, there are $4 \cdot 3 = 12$ ways to choose them. Now we can calculate the number of mini-Sudoku: $24 \cdot 12 = 288$ different mini-Sudoku.

This mini Sudoku (Arcos et al., 2010) can be programmed to show a random grid:

Program 93 mini Sudoku

```
import random
def create_grid():
return [[0 for _ in range(4)] for _ in range(4)]
def print_grid(grid):
for i, row in enumerate(grid):
if i % 2 == 0 and i != 0:
print("-" * 13)
for j, num in enumerate(row):
if j % 2 == 0 and j != 0:
print("|", end=" ")
if num == 0:
print(".", end=" ")
else:
print(num, end=" ")
print()
def is_valid(grid, num, pos):
row, col = pos
# Check the row
if num in grid[row]:
return False
# Check the column
if num in [grid[i][col] for i in range(4)]:
return False
# Check the 2x2 quadrant
quad_x = col // 2
quad_y = row // 2
for i in range(quad_y * 2, quad_y * 2 + 2):
for j in range(quad_x * 2, quad_x * 2 + 2):
if grid[i][j] == num and (i, j) != pos:
return False
return True
def find_empty(grid):
for i in range(4):
for j in range(4):
if grid[i][j] == 0:
return (i, j)
return None
def solve(grid):
empty = find_empty(grid)
if not empty:
return True
row, col = empty
for num in range(1, 5):
if is_valid(grid, num, (row, col)):
grid[row][col] = num
if solve(grid):
```

```
return True
grid[row][col] = 0
return False
def generate_sudoku(num_removed):
grid = create_grid()
solve(grid)
cells = [(i, j) for i in range(4) for j in range(4)]
random.shuffle(cells)
for i in range(num_removed):
row, col = cells[i]
grid[row][col] = 0
return grid
def play():
sudoku = generate_sudoku(8) # Generate a Sudoku with 8 empty cells
while True:
print_grid(sudoku)
try:
row = int(input("Enter the row (1-4): ")) - 1
col = int(input("Enter the column (1-4): ")) - 1
num = int(input("Enter the number (1-4): "))
except ValueError:
print("Invalid input. Please try again.")
continue
if 0 <= row < 4 and 0 <= col < 4 and 1 <= num <= 4:
if sudoku[row][col] == 0:
if is_valid(sudoku, num, (row, col)):
sudoku[row][col] = num
if not find_empty(sudoku):
print("Congratulations! You completed the Sudoku!")
break
else:
print("Invalid move. Try again.")
else:
print("This cell is already occupied. Choose another cell.")
else:
print("Invalid input. Please try again.")
if __name__ == "__main__":
play()
```

Polyominoes: The first traces of the polyomino game date back to the early twentieth century, with the publication of mathematical studies that explored the properties and combinations of geometric shapes composed of squares joined together. However, the popularisation of play as a recreational activity is largely attributed to the mathematician Solomon Golomb (1932-2016). The polyomino game is based on tiles formed by a variable number of squares joined together along the sides (Golomb, 1996). Depending on the number of squares that compose them, polyominoes receive specific names, the most famous being Domino (formed by two squares), Tetramino (formed by four squares), and Pentomino (formed by five squares). In the case of the Pentomino, there are 12 possible configurations, while for the Tetramino, there are only 5.

Program 94 Pentamino

```
import matplotlib.pyplot as plt
import matplotlib.patches as patches
# Define the pentominoes as lists of coordinates (x, y)
pentominoes = {
'F': [(0, 0), (1, 0), (1, 1), (2, 1), (1, 2)],
'I': [(0, 0), (0, 1), (0, 2), (0, 3), (0, 4)],
'L': [(0, 0), (0, 1), (0, 2), (0, 3), (1, 3)],
'N': [(0, 0), (1, 0), (1, 1), (1, 2), (2, 2)],
'P': [(0, 0), (1, 0), (0, 1), (1, 1), (1, 2)],
'T': [(0, 0), (1, 0), (2, 0), (1, 1), (1, 2)],
'U': [(0, 0), (0, 1), (1, 1), (2, 0), (2, 1)],
'V': [(0, 0), (0, 1), (0, 2), (1, 2), (2, 2)],
'W': [(0, 0), (1, 0), (1, 1), (2, 1), (2, 2)],
'X': [(1, 0), (0, 1), (1, 1), (2, 1), (1, 2)],
'Y': [(1, 0), (1, 1), (0, 2), (1, 2), (2, 2)],
'Z': [(0, 0), (1, 0), (1, 1), (2, 1), (2, 2)]
}
# Colors for each pentomino
colors = {
'F': 'red', 'L': 'blue', 'N': 'green', 'P': 'purple', 'T': 'orange',
'U': 'cyan', 'V': 'magenta', 'W': 'yellow', 'X': 'brown', 'Y': 'pink', 'Z': 'gray', 'I':
'black'
}
fig, ax = plt.subplots()
# Function to draw a pentomino
def draw_pentomino(ax, pentomino, color, offset_x, offset_y):
for (x, y) in pentomino:
rect = patches.Rectangle((x + offset_x, y + offset_y), 1, 1, edgecolor='black', facecol-
or=color)
ax.add_patch(rect)
# Draw all pentominoes
offset_x, offset_y = 0, 0
for label, pentomino in pentominoes.items():
draw_pentomino(ax, pentomino, colors[label], offset_x, offset_y)
offset_x += 4 # Shift the offset for the next pentomino
ax.set_aspect('equal')
ax.autoscale_view()
plt.axis('off')
plt.show()
```

A classic situation is the following: we want to arrange 12 pentominoes on an 8×8 board, leaving the four central squares empty; how many solutions are there? The problem, first solved by Dana Scott in 1958, was confirmed with Knuth's X-algorithm, implemented with the Dancing Links (DLX) technique. The answer is that there are 65 solutions.

Jigsaw: The term specifically refers to a puzzle in which one cuts an image into irregular pieces that must be assembled to reconstitute the original image. There are puzzles of many sizes. Commonly, puzzles are rectangular, with a ratio between the sides close to that of the A4 format. As for the number of pieces, at least as far as adults are concerned, they range from puzzles made up of a few hundred pieces to giants of several thousand pieces. The size of a puzzle depends not only on the quantity of pieces but also on their dimensions—puzzles with more pieces are often made physically smaller to keep the final image at a manageable size.

The connection to combinatorics appears in various aspects (Knuth, 2000). For example, determining the possible combinations of pieces to form the correct solution is a classic combinatorial problem. Calculating these combinations provides insight into the puzzle's complexity and informs solving strategies. In some cases, the objective goes beyond simple assembly and involves solving the puzzle optimally, such as minimizing the number of moves required.

Some algorithms use combinatorial principles to solve Jigsaw puzzles automatically. These algorithms analyse the edges of the pieces and their characteristics to find the correct way to assemble them. Each puzzle piece has edges that can be partially flat (for the outer edges) or irregular (for the inner pieces). For small puzzles, an exhaustive search can be used, considering all possible combinations of pieces, even if the number of possible assemblies grows dramatically. For a 2×2 puzzle, there are $4! = 24$ possible permutations; for a 3×3 puzzle, it goes up to $9!= 362,880$. A genetic algorithm can help us; in this case, 2000 iterations are enough to find the correct combination.

Program 95 Jigsaw 3x3

```
import random
# Define the 3x3 puzzle pieces as strings
puzzle_pieces = ['A', 'B', 'C', 'D', 'E', 'F', 'G', 'H', 'I']
target_solution = ('A', 'C', 'H', 'D', 'E', 'I', 'G', 'B', 'F') # Assume this is the cor-
rect configuration
# Ask the user to input the number of iterations
max_iterations = int(input("Number of iterations: "))
def generate_initial_population(puzzle_pieces, population_size=10):
population = []
for _ in range(population_size):
individual = random.sample(puzzle_pieces, len(puzzle_pieces))
population.append(individual)
return population
def evaluate_fitness(individual, target_solution):
fitness = sum(1 for i, piece in enumerate(individual) if piece == target_solution[i])
return fitness
def selection(population, fitness_scores):
total_fitness = sum(fitness_scores)
selection_probabilities = [score / total_fitness for score in fitness_scores]
selected = random.choices(population, weights=selection_probabilities, k=len(population))
return selected
def crossover(parent1, parent2):
crossover_point = random.randint(1, len(parent1) - 1)
pieces_parent1 = parent1[:crossover_point]
pieces_parent2 = parent2[:crossover_point]
child1 = pieces_parent1 + [piece for piece in parent2 if piece not in pieces_parent1]
child2 = pieces_parent2 + [piece for piece in parent1 if piece not in pieces_parent2]
return child1, child2
def mutation(individual, mutation_rate=0.01):
if random.random() < mutation_rate:
idx1, idx2 = random.sample(range(len(individual)), 2)
individual[idx1], individual[idx2] = individual[idx2], individual[idx1]
return individual
def solution_found(population, target_solution):
for individual in population:
```

```
if tuple(individual) == target_solution:
return True
return False
def best_solution(population, target_solution):
best_individual = max(population, key=lambda ind: evaluate_fitness(ind, target_solution))
return best_individual
def genetic_algorithm(puzzle_pieces, target_solution, max_iterations, population_size=10,
mutation_rate=0.01):
population = generate_initial_population(puzzle_pieces, population_size)
for iteration in range(max_iterations):
fitness_scores = [evaluate_fitness(individual, target_solution) for individual in popula-
tion]
if max(fitness_scores) == len(target_solution):
print(f"Solution found at iteration {iteration}")
break
selected_parents = selection(population, fitness_scores)
new_population = []
for i in range(0, len(selected_parents), 2):
parent1 = selected_parents[i]
parent2 = selected_parents[min(i + 1, len(selected_parents) - 1)]
child1, child2 = crossover(parent1, parent2)
new_population.extend([mutation(child1, mutation_rate), mutation(child2, mutation_rate)])
population = new_population
return best_solution(population, target_solution)
# Run the genetic algorithm
best_individual = genetic_algorithm(puzzle_pieces, target_solution, max_iterations)
print("Best solution found:", best_individual)
print("Correct solution:", target_solution)
```

Mastermind: This is a logic game where one has to guess a hidden combination of colours (or numbers) using clues without exceeding the limit of attempts allowed. Mastermind is another example of a combinatorial game where computation can be used to find a solution efficiently (Lamberti & Tovena, 2022). In this game, a player tries to guess the colour scheme chosen by the opposing player. The player has a series of attempts and receives feedback from the opposing player after each attempt, indicating how many colours are correct and how many are in the right place in the combination.

One can use Combinatorial calculus to determine the possible combinations formed using a given set of colours. If eight colours are available and the combination consists of 5, the number of possible combinations is $8^5 = 32{,}768$. A common strategy is to use the "minimax method" to minimise the attempts to guess the combination. This strategy involves choosing the next attempt to maximise the information obtained from the opposing player's feedback. One starts by choosing a random combination and comparing the result with the combination to be guessed. A colour in the correct position corresponds to a black dot, and in the wrong position, to a white dot. At this point, we identify which colours are correct and in the right place and which are the correct colours but in the wrong place. Based on this information, we reduce the possible combinations by choosing a move that minimises the remaining options.

A simulation using four numbers (also duplicates) on four places with ten attempts:

Program 96 Mastermind

```
import random
```

```
def generate_code():
return [random.randint(1, 4) for _ in range(4)]
def get_user_input():
while True:
try:
guess = list(map(int, input("Enter four numbers between 1 and 4 separated by spaces:
").split()))
if len(guess) != 4 or any(num < 1 or num > 4 for num in guess):
raise ValueError
return guess
except ValueError:
print("Invalid input. Make sure to enter four numbers between 1 and 4.")
def compare(code, guess):
black = sum(i == j for i, j in zip(code, guess))
white = sum(min(code.count(i), guess.count(i)) for i in set(guess)) - black
return black, white
def play():
code = generate_code()
attempts = 0
while attempts < 10:
guess = get_user_input()
black, white = compare(code, guess)
print(f"{black} black, {white} white")
if black == 4:
print("You won!")
return
attempts += 1
print(f"You lost! The code was {code}")
# Start the game
play()
```

Wordle: It is a game in which participants must guess a hidden five-letter word within a maximum of six attempts. Each attempt provides crucial information about the correct letters and their location. The game is played online, with players attempting to discover a new five-letter word each day. The game is currently only available in English. The guessing process is based on elimination and deduction techniques, and it has a combinatorial content. The challenge is significant, given that approximately 13,000 five-letter words exist in English. Since Wordle does not allow repeated letters in the same word, the initial possibilities are $C_{26.5} = 26•25•24•23•22/5! = 65{,}780$. It is important to note that many of these combinations will not make sense, as not all five-letter sequences form valid words in English. Players must form words on a grid of letters. Combinatorics calculates the number of possible words that can be formed.

The proposed program, connected to an online source of five-letter English words, can suggest ten initial words or load the list of words in Italian from the italiano.txt file.

Program 97 Wordle

```
import random
import requests
def download_english_words():
url_words = "https://norvig.com/ngrams/word.list" # URL for English words
```

```
try:
response = requests.get(url_words)
response.raise_for_status()
words = response.text.splitlines()
five_letter_words = [word.lower() for word in words if len(word) == 5 and len(set(word))
== 5]
return five_letter_words
except requests.exceptions.RequestException as e:
print(f"Error downloading words: {e}")
return []
def load_italian_words():
try:
with open('italiano.txt', 'r', encoding='utf-8') as file:
words = file.read().splitlines()
five_letter_words = [word.lower() for word in words if len(word) == 5 and len(set(word))
== 5]
return five_letter_words
except FileNotFoundError as e:
print(f"Error: {e}")
return []
def suggest_10_possible_words(five_letter_words):
if five_letter_words:
return random.sample(five_letter_words, k=10)
else:
print("Unable to retrieve words. Check the word source.")
return []
def menu():
print("Choose the language for the words:")
print("1. English (download from URL)")
print("2. Italian (load from local file)")
choice = input("Enter the number of your choice: ").strip()
if choice == '1':
five_letter_words = download_english_words()
elif choice == '2':
five_letter_words = load_italian_words()
if not five_letter_words:
print("Trying to download Italian words from URL as an alternative.")
five_letter_words = download_italian_words()
else:
print("Invalid choice.")
return
suggested_words = suggest_10_possible_words(five_letter_words)
if suggested_words:
print("10 possible starting words without repeated letters:")
for idx, word in enumerate(suggested_words, 1):
print(f"{idx}. {word}")
else:
print("No suitable words found.")
if __name__ == "__main__":
```

```
menu()
```

Boggle: It is a fun and intellectually stimulating word game in which players compete to find valid words within a limited amount of time using letters arranged in a 4 × 4 grid. Words are formed by tracing a sequence of adjacent letters—where adjacency includes horizontal, vertical, and diagonal connections—and each letter may be used only once per word.

Determining how many possible word combinations can be generated from a given grid is a nontrivial combinatorial problem. For example, in one specific 4 × 4 configuration, a computer program was able to identify 86 words of at least four letters in length.

Figure 6. An example of a grid

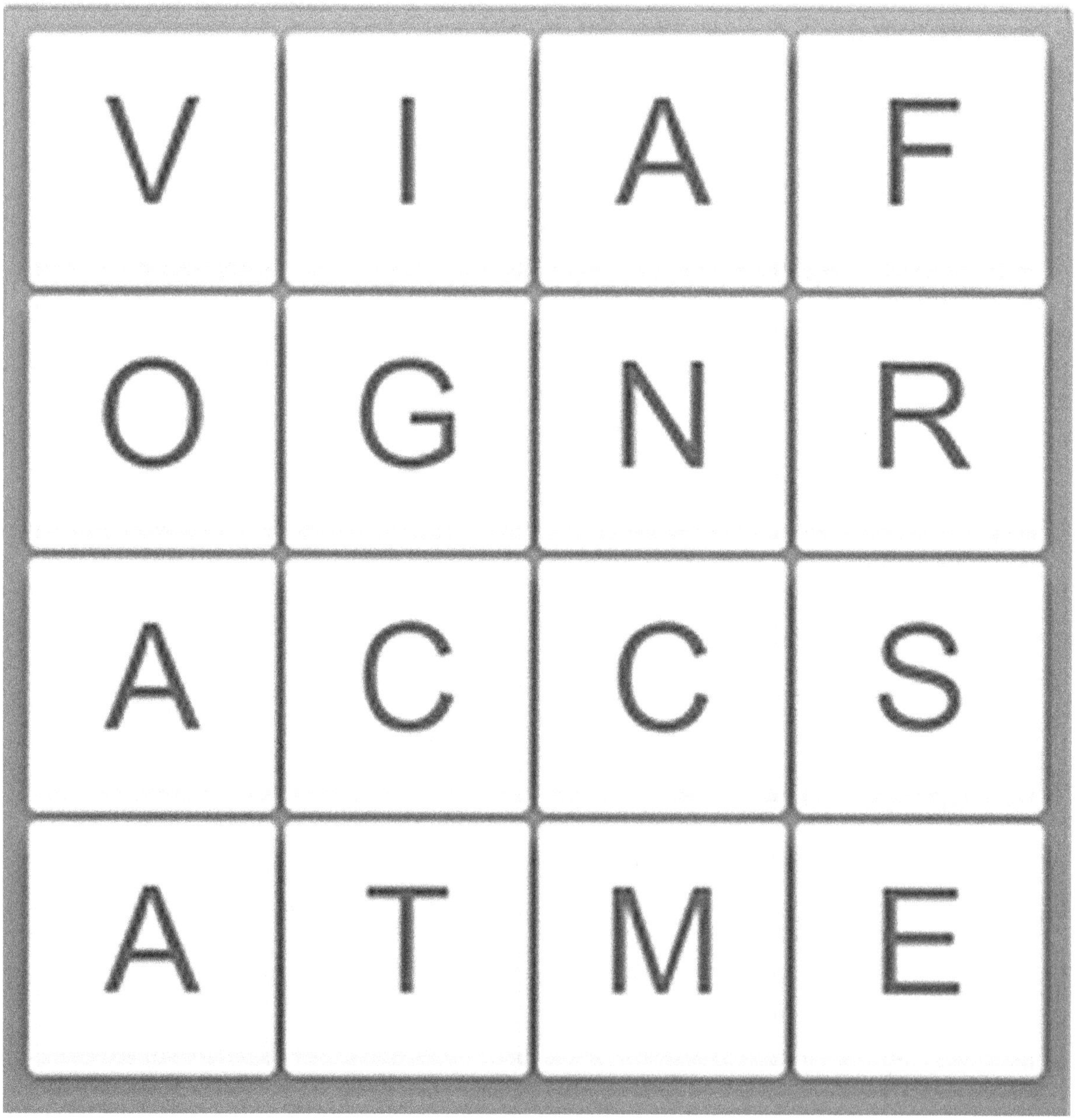

If one reduces the grid size to 2×2, one can count the number of possible paths of length 2 or 3. Of course, this does not mean that all possible paths provide sensible words. The count is evident if the grid is transformed into a graph. A path with two words can be done in $4 \cdot 3 = 12$ ways, and with three words, there are $4 \cdot 6 = 24$ possible combinations.

Figure 7. Boggle graph

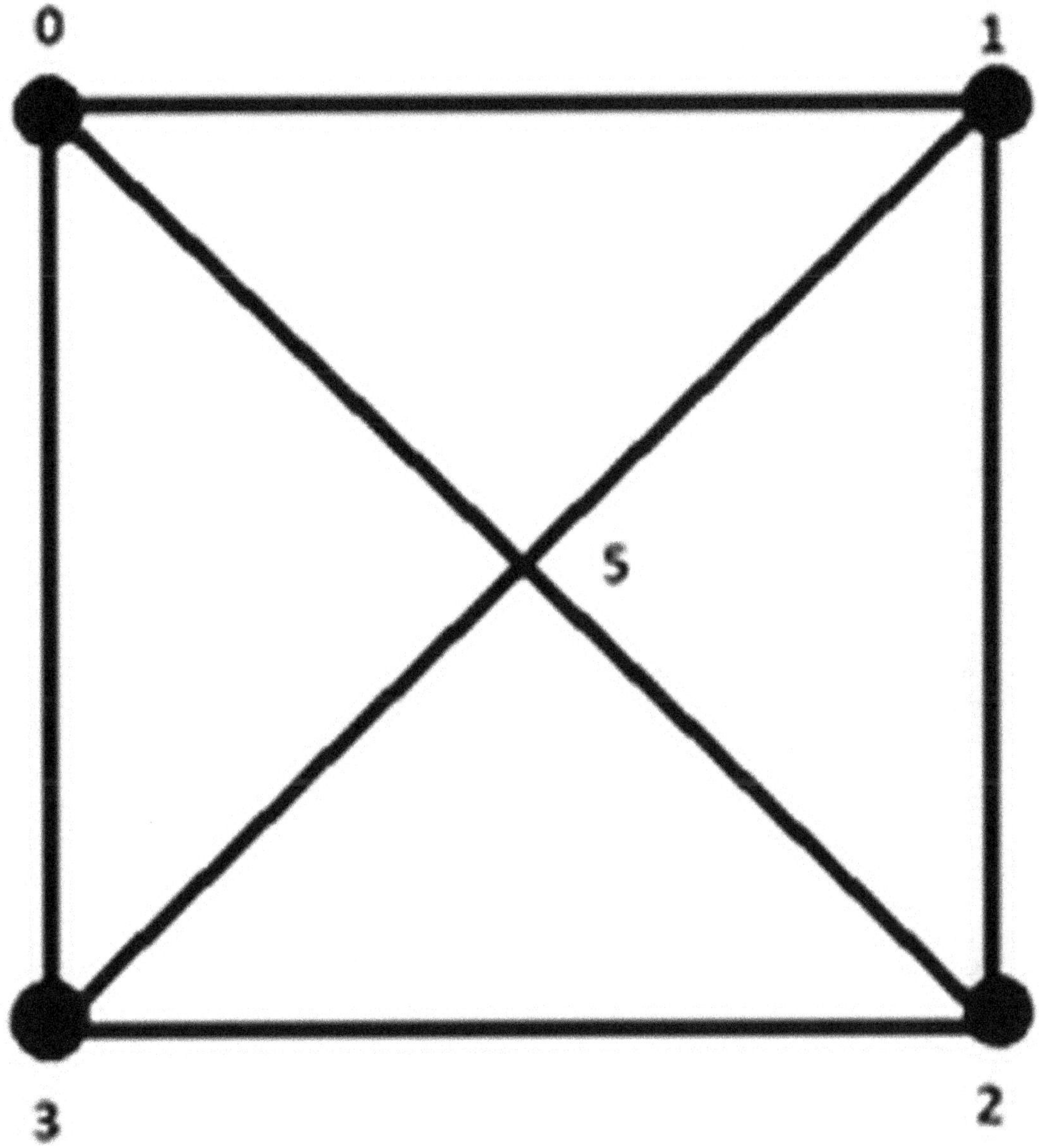

On a 3×3 grid, calculating paths of length 2 is still manageable. For instance: 4 corner nodes • 3 neighbours + 4 edge nodes • 5 neighbours + 1 center node • 8 neighbours = 40 paths.

For longer paths, however, computational assistance becomes essential.

Program 98 Boggle

```
import networkx as nx
import matplotlib.pyplot as plt
def generate_grid_edges(num_nodes):
edges = []
side = int(num_nodes**0.5)
for i in range(num_nodes):
for j in range(num_nodes):
if i != j:
row_i, col_i = divmod(i, side)
row_j, col_j = divmod(j, side)
if abs(row_i - row_j) <= 1 and abs(col_i - col_j) <= 1:
edges.append((i, j))
return edges
def create_graph(num_nodes):
G = nx.Graph()
G.add_nodes_from(range(num_nodes))
edges = generate_grid_edges(num_nodes)
G.add_edges_from(edges)
return G
def count_paths(graph, length):
return sum(count_paths_from_node(graph, node, length) for node in graph.nodes)
def count_paths_from_node(graph, current, length, path=None):
if path is None:
path = [current]
if len(path) == length:
return 1
return sum(
count_paths_from_node(graph, neighbor, length, path + [neighbor])
for neighbor in graph.neighbors(current)
if neighbor not in path
)
# User input
num_nodes = int(input("Enter number of nodes (e.g., 4 for 2x2, 9 for 3x3, 16 for 4x4): "))
length = int(input("Enter path length: "))
# Build and analyze graph
G = create_graph(num_nodes)
print("Number of possible paths:", count_paths(G, length))
```

The results are interesting, for example:

9 nodes, length 4 = 496 paths
16 nodes, length 8 = 183472 paths

Tower of Hanoi: In a Hindu temple, monks constantly move 64 gold discs from one column to another on three diamond columns, one at a time, so the smaller discs are always positioned on top of the larger ones. When the monks complete the work, the world will end. In reality, it is a legend invented by the company that first marketed the puzzle, invented in 1883 by the French mathematician Édouard Lucas. Combinatorial calculus can determine the minimum number of moves needed to solve the puzzle based on the number of discs involved. The "divide and conquer" strategy involves breaking down the problem into smaller sub-problems and solving each one individually. In this way, the puzzle solution can be found efficiently, even for many discs.

Usually, the tower of Hanoi is sold with discs arranged on rod A and two rods B (support) and C (the one on which the discs must end). It is advisable to start with a disc and then progressively increase the number of discs to understand the algorithm that leads to the solution with the fewest moves. Indicating with *n*, as already mentioned, the pieces, with *m* the moves and with *p* the initial one (this data helps us in the resolution procedure), we can construct the following table:

Table 2. How to fix Tower of Hanoi with n disks

n	m	p
1	1	A→C
2	3	A→B
3	7	A→C
4	15	A→B
5	31	A→C
6	63	A→B
7	127	A→C
8	255	A→B

As one can see, the solution strategy changes according to the number of disks; the algorithm can be summarized as follows: remembering not to move the same disk twice in a row and that a disk with a larger diameter cannot overlap one with a smaller diameter:

- We indicate the disks with a number (1 is the largest)
- If n is odd, the first move is from peg A to peg C.
- If n is even, the first move is from peg A to peg B.
- No odd disk can be placed directly on an odd disk.
- No even disk can be placed directly on an even disk.
- If there are two possible moves, move the disk to the non-empty tower.

The minimum number of moves needed is given, recursively, by $m(n) = m(n-1) + 1 + m(n-1) = 2 \cdot m(n-1) + 1$. Note that the moves are equal to the powers of two minus one unit. So, the minimum number of moves needed to solve the Hanoi puzzle with *n* discs is $2^n - 1$. This result can be demonstrated using combinatorics. At each pass, one can move a disc from the starting peg to one of the two remaining pegs, meaning there are two possibilities for each disc. Therefore, the total of possible moves is 2^n. However, the goal is to move the complete set of discs to the finished peg, which requires making one move less than the total of possible moves. So, the minimum number of moves required to solve the Tower of Hanoi with *n* discs is just $2^n - 1$. The monks of legend would have to make 18,446,744,073,709,551,615 moves ($2^{64} - 1$) to complete the job; Assuming the monks make one move per second, the world will end in 5,845,580,504 centuries.

To confirm that the minimum number of moves to solve the Tower of Hanoi with *n* disks is $2^n - 1$, one can use the principle of induction; in this case, following these steps:

1. Basic step: We can start from n = 1. To solve this case, move (with one move) the only disk available to us in the final post. For n = 1, we have $2^1 - 1 = 1$.
2. Inductive step: assume it takes at least $2^n - 1$ moves to solve the problem with n disks and prove (substituting n+1) that it takes $2^{n+1} - 1$ moves to solve the puzzle with n+1 disks. To solve the problem with n+1 disks, we must first move n disks, leaving only the largest disk in place. Once this is done, move the larger disk to a new post and finish by answering the first n smaller disks above it. In total, there is a $2^n - 1$ moves to shift the n discs except for the large

one, one to move the large one to a new post, and another 2^n - 1 moves to move all the discs above the large disc. $2^n - 1 + 1 + 2^n - 1 = 2 \bullet (2^n) - 1 = 2^{n+1} - 1$.

To consider the necessary moves, one can use the program:

Program 99 Hanoi

```
def tower_of_hanoi(n, from_peg, to_peg, aux_peg):
global count
if n > 0:
# Move a tower of n - 1 disks from from_peg to aux_peg, using to_peg.
tower_of_hanoi(n - 1, from_peg, aux_peg, to_peg)
print("Move", count + 1, ": Move the disk from peg", from_peg, "to peg", to_peg)
count += 1
# Move the tower of n - 1 disks from aux_peg to to_peg using from_peg.
tower_of_hanoi(n - 1, aux_peg, to_peg, from_peg)
count = 0
n = int(input("Enter the number of disks: "))
tower_of_hanoi(n, 'A', 'B', 'C')
print("Total number of moves:", count)
```

To visualize the necessary moves, however, it is more interesting to use the graphic functions:

Program 100 graphic Hanoi

```
from turtle import *
import math
move_count = 0 # Counter for the number of moves
class Disk(Turtle):
def __init__(self, n):
Turtle.__init__(self, shape="square", visible=False)
self.pu()
self.shapesize(1.5, n * 1.5, 2) # Square --> Rectangle
self.fillcolor(get_color(n))
self.st()
class Tower(list):
def __init__(self, x):
self.x = x
def add(self, disk):
disk.setx(self.x)
disk.sety(-150 + 34 * len(self))
self.append(disk)
def remove(self):
global move_count
move_count += 1
disk = list.pop(self)
disk.sety(150)
```

```
return disk
def get_color(n):
red = (math.sin(n) / 2 + 0.5)
green = (math.sin(n + 2) / 2 + 0.5)
blue = (math.sin(n + 4) / 2 + 0.5)
return (red, green, blue)
def hanoi(n, from_, with_, to_):
if n > 0:
hanoi(n - 1, from_, to_, with_)
to_.add(from_.remove())
hanoi(n - 1, with_, from_, to_)
def play():
global move_count
onkey(None, "space")
clear()
try:
num_disks = int(textinput("Number of disks", "Enter the number of disks: "))
if num_disks <= 0:
raise ValueError("The number of disks must be positive.")
for i in range(num_disks, 0, -1):
t1.add(Disk(i))
hanoi(num_disks, t1, t2, t3)
write(f"Moves made: {move_count}",
align="center", font=("Courier", 16, "bold"))
except ValueError as e:
print(f"Error: {e}")
except Terminator:
pass
def main():
global t1, t2, t3
ht()
penup()
goto(0, -225)
t1 = Tower(-250)
t2 = Tower(0)
t3 = Tower(250)
write("Press space to start",
align="center", font=("Courier", 16, "bold"))
onkey(play, "space")
listen()
return "Program running"
if __name__ == "__main__":
msg = main()
print(msg)
mainloop()
```

Rubik's Cube: It is considered a three-dimensional combinatorial game. The Rubik's Cube (Diego-Mantecon et al., 2019; Corli et al., 2021)is a puzzle game that involves solving a grid of coloured cubes, where each side of the cube has nine coloured squares. The game's goal is to solve the cube so that each side of the cube contains nine squares of the same colour. The game requires logical thinking and spatial analysis skills to solve the puzzle and does not require any advanced mathematical knowledge. The Hungarian designer Ernő Rubik invented the Rubik's Cube in 1974 and became popular worldwide in the 1980s. Many variations of the game are available, including simpler versions and more complex ones with more cubes. Rubik's Cube is a classic example of a combinatorial game that can be solved using combinatorial techniques. Solving the Rubik's Cube requires knowledge of some permutation and combination techniques. In the original Rubik's Cube there are 8! ways to combine the eight vertices.

Seven of these can be oriented independently, and the orientation of the eighth will depend on this, giving $3^7 = 2,187$ chances. There are 12!/2 ways to arrange the 12 pieces in two colours located in the centre of the physical borders. Eleven edges can be turned independently, and the rotation of the twelfth edge will depend on the previous ones, giving $2^{11} = 2,048$ chances. In conclusion, the Rubik's Cube has 43,252,003,274,489,856,000 possible configurations, which makes the solution a significant challenge.

However, combinatorial techniques can significantly reduce the number of configurations that the Rubik's Cube needs to examine and solve efficiently. One of the standard combinatorial techniques used to solve the Rubik's Cube is the "group theory" technique, which involves analysing the properties of the symmetry group of the Rubik's Cube. It is composed of several subgroups, each representing a set of symmetrical transformations that can be applied without altering its shape. For example, one of the symmetry group subgroups is the rotation group, which represents all possible rotations that can be applied to the Rubik's Cube. This subgroup consists of 18 elements, i.e. the six possible rotations for each face. Using group theory, one can identify common permutations and permutation cycles in the Rubik's Cube that can be used to solve it efficiently. The "Fridrich" method uses a series of common permutations and orientation algorithms to solve the Rubik's Cube in less than 20 moves on average. The following program visually represents the cube in 2D and allows one to rotate it using the keyboard.

F: Rotate the front face
R: Rotate the right face
U: Rotate the upper face
M: Stir the cube
Program 101 Rubik' Cube

```
import pygame
import sys
import random
# Pygame Initialization
pygame.init()
# Constants
width, height = 600, 800
cube_size = 50
margin = 5
text_margin = 10
# Colors
WHITE = (255, 255, 255)
RED = (255, 0, 0)
BLUE = (0, 0, 255)
ORANGE = (255, 165, 0)
GREEN = (0, 255, 0)
YELLOW = (255, 255, 0)
BLACK = (0, 0, 0)
# Color Mapping
COLOR_MAP = {
```

```
'w': WHITE,
'r': RED,
'b': BLUE,
'o': ORANGE,
'g': GREEN,
'y': YELLOW
}
class Cube2x2:
def __init__(self):
self.faces = {
'f': [['r', 'r'], ['r', 'r']],
'b': [['o', 'o'], ['o', 'o']],
'u': [['w', 'w'], ['w', 'w']],
'd': [['y', 'y'], ['y', 'y']],
'l': [['g', 'g'], ['g', 'g']],
'r': [['b', 'b'], ['b', 'b']]
}
def rotate_face(self, face):
self.faces[face] = [list(row) for row in zip(*self.faces[face][::-1])]
def rotate_f(self):
self.rotate_face('f')
temp = self.faces['u'][1]
self.faces['u'][1] = self.faces['l'][1][::-1]
self.faces['l'][1] = self.faces['d'][1]
self.faces['d'][1] = self.faces['r'][1][::-1]
self.faces['r'][1] = temp
def rotate_r(self):
self.rotate_face('r')
temp = [row[1] for row in self.faces['u']]
for i in range(2):
self.faces['u'][i][1] = self.faces['f'][i][1]
self.faces['f'][i][1] = self.faces['d'][i][1]
self.faces['d'][i][1] = self.faces['b'][i][1]
self.faces['b'][i][1] = temp[i]
def rotate_u(self):
self.rotate_face('u')
temp = self.faces['f'][0]
self.faces['f'][0] = self.faces['r'][0]
self.faces['r'][0] = self.faces['b'][0]
self.faces['b'][0] = self.faces['l'][0]
self.faces['l'][0] = temp
def shuffle(self, moves=20):
rotations = [self.rotate_f, self.rotate_r, self.rotate_u]
for _ in range(moves):
random.choice(rotations)()
def draw_cube(screen, cube):
# Space for the cube
screen.fill(BLACK)
# Face positions
```

```
positions = {
'f': (225, 400),
'r': (325, 400),
'u': (225, 300),
'l': (125, 400),
'd': (225, 500),
'b': (425, 400)
}
for face, pos in positions.items():
for i in range(2):
for j in range(2):
x = pos[0] + j * (cube_size + margin)
y = pos[1] + i * (cube_size + margin)
color = COLOR_MAP[cube.faces[face][i][j]]
pygame.draw.rect(screen, color, (x, y, cube_size, cube_size))
def draw_text(screen):
font = pygame.font.Font(None, 36)
text = [
"Press F to rotate the Front face",
"Press R to rotate the Right face",
"Press U to rotate the Upper face",
"Press M to shuffle the cube"
]
y_offset = 10
for line in text:
text_surface = font.render(line, True, WHITE)
screen.blit(text_surface, (text_margin, y_offset))
y_offset += 40
# Screen Initialization
screen = pygame.display.set_mode((width, height))
pygame.display.set_caption("Rubik's Cube")
# Create the cube
cube = Cube2x2()
# Main Loop
running = True
while running:
for event in pygame.event.get():
if event.type == pygame.QUIT:
running = False
elif event.type == pygame.KEYDOWN:
if event.key == pygame.K_f:
cube.rotate_f()
elif event.key == pygame.K_r:
cube.rotate_r()
elif event.key == pygame.K_u:
cube.rotate_u()
elif event.key == pygame.K_m:
cube.shuffle()
draw_cube(screen, cube)
```

```
draw_text(screen)
pygame.display.flip()
pygame.quit()
sys.exit()
```

Fifteen puzzle: The Game is a classic puzzle game invented by Samuel Loyd (1841-1911) in 1878. The game consists of a 4 x 4 square multiplication table on which 15 square tiles are placed, numbered progressively starting from 1. Incidentally, there are 15 possible permutations, totalling 1,307,674,368,000 cases. The tiles can slide horizontally or vertically, but movement is obviously limited by a single empty space (denoted by 0). The game aims to rearrange the tiles in an ascending order after having "shuffled" them randomly.

Let us ask ourselves: Is it always possible to do this? To answer, it is necessary to analyse the scheme of the game from a mathematical point of view. Each move, understood as a permutation of two tiles, changes the system's parity (or disparity). The solvability depends on the configuration of the tiles and the location of the empty space. The puzzle is solvable if the number of inversions is even and the white space is in an odd row (counted from bottom to top), or the number of inversions is odd, and the white space is in an even row.

Figure 9. Initial setup

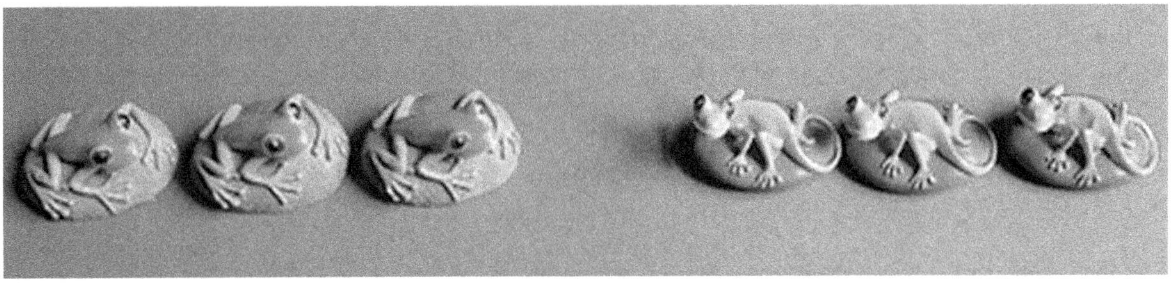

Now, consider the configuration expressed as a list according to the standard reading scheme from left to right and from top to bottom: 1,2,3,4,5,6,7,8,9,12,14,15,13,10,11,0. In this case, the configuration is solvable as there are 10 inversions (12 10),(12 11),(14,13),(14 10),(14 11),(15 13),(15 10),(15 11),(13 10),(13 11) and the empty space is in the first line from the bottom.

A program can easily generate random configurations and determine whether they are solvable or not.

Program 102 Fifteen puzzle

```
import random
def count_inversions(tiles):
inversions = 0
for i in range(len(tiles)):
for j in range(i + 1, len(tiles)):
if tiles[i] != 0 and tiles[j] != 0 and tiles[i] > tiles[j]:
inversions += 1
return inversions
def find_blank_row_from_bottom(grid):
for i in range(4):
if 0 in grid[i]:
```

```
return 4 - i # Count from the bottom: bottom row is 1
def is_solvable(grid):
flat = sum(grid, []) # Flatten the 2D grid into a list
inversions = count_inversions(flat)
blank_row = find_blank_row_from_bottom(grid)
return (inversions + blank_row) % 2 == 0
def generate_random_grid():
numbers = list(range(16))
random.shuffle(numbers)
return [numbers[i:i+4] for i in range(0, 16, 4)]
# Run the test
grid = generate_random_grid()
for row in grid:
print(row)
if is_solvable(grid):
print("□ The configuration is solvable.")
else:
print("□ The configuration is NOT solvable.")
```

Frogs and Lizards: The game, better known as Frogs and Toads, is considered a combinatorial game that involves discrete decisions and sequences of moves that can be analysed through the combination of possible moves and states of the game. Three frogs are on the first squares of a strip of seven squares. Three lizards occupy the last three squares on the right. In the middle there is a free box. Symbolically, we can write: RRR_LLL. The game aims to transfer the lizards to the squares initially occupied by the frogs and vice versa to get to LLL_RRR.

Players can apply the following moves:

1. Slide: A frog or lizard may move to an adjacent empty square.
2. Jump: A frog or lizard may jump over one adjacent piece (frog or lizard) into the next empty square.

There is no element of chance—no randomness or hidden information. The game is fully deterministic and has perfect information: all pieces and potential moves are visible at all times. The challenge is to discover a sequence of legal moves that transforms the initial configuration into the final goal state.

This structure makes the game a prime example of a combinatorial problem that can be solved through systematic exploration of possible moves and paths. It exemplifies a key feature of combinatorial games: the search for an optimal strategy or a definitive solution, often through backtracking, state space graphs, or recursive algorithms.

Figure 10. Frogs and lizards

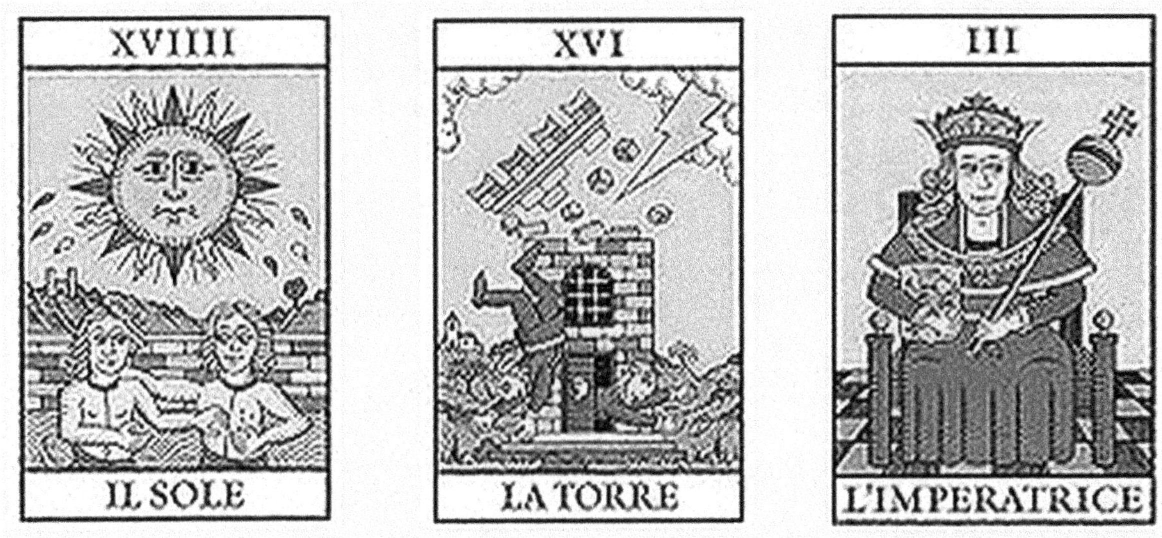

Some tips for solving the game:

1. Jump-jump is a configuration to avoid whichever of the two jumps one chooses; on the next move, one will be blocked.
2. If one must choose between a jump and a step, always opt for the jump; otherwise, one will be left stuck after the step.
3. If one must choose between two steps, choose the one the that does not take the skip-jump configuration

Tarots: This is a deck of cards used mainly for divination and meditation, but also for card games. A traditional tarot deck consists of 78 cards, divided into two main groups: the Major Arcana and the Minor Arcana.

The first group consists of 22 cards, each with a specific name and symbol (in Italian): “Il Bagatto”, “Il Carro”, “Il Diavolo”, “Il Giudizio”, “Il Matto”, “Il Mondo”, “Il Papa”, “Il Sole”, “La Forza”, “La Giustizia”, “La Luna”, “La Morte”, “La Papessa”, “la Ruota della Fortuna”, “La Stella”, “La Torre”, “L'Appeso”, “L'Eremita”, “L'Imperatore”, “L'Imperatrice”, “L'Innamorato”, “Temperanza”.

“The Castle of Crossed Destinies” by Italo Calvino is a work that explores narration using tarot cards. The book is an example of combinatorial literature, where stories are constructed by combining tarot cards in different ways. Each character uses tarot cards to tell their story, and the number of cards drawn varies depending on the story they tell. The cards are arranged in sequence, and the meaning of each card depends on its position relative to the others. Tarot cards represent the possibilities and directions that the characters' lives can take.

A program can quickly show different combinations of three cards.

Program 103 Tarots

```
import tkinter as tk
from tkinter import messagebox
from PIL import Image, ImageTk
import random
```

```
# List of names of Marseille tarot cards
tarot_cards = [
"Il Bagatto", "Il Carro", "Il Diavolo", "Il Giudizio", "Il Matto", "Il Mondo",
"Il Papa", "Il Sole", "La Forza", "La giustizia", "La Luna", "La Morte",
"La Papessa", "la Ruota della Fortuna", "La Stella", "La Torre", "L'Appeso", "L'Eremita",
"L'Imperatore", "L'Imperatrice", "L'Innamorato", "Temperanza"
]
# Function to load an image
def load_image(card_name):
try:
img = Image.open(f"tarocchi/{card_name}.jpg")
img = img.resize((100, 150), Image.LANCZOS)
return ImageTk.PhotoImage(img)
except FileNotFoundError:
messagebox.showerror("Error", f"Image for {card_name} not found.")
return None
# Function to display a combination of tarot cards
def show_combination():
combination = random.sample(tarot_cards, 3)
images = []
for card_name in combination:
img = load_image(card_name)
if img:
images.append(img)
for i, img in enumerate(images):
label = tk.Label(root, image=img)
label.image = img
label.grid(row=0, column=i, padx=10, pady=10)
# Create the main window
root = tk.Tk()
root.title("The Tavern of Crossed Destinies")
# Button to show a combination of tarot cards
btn_show = tk.Button(root, text="Show Combination", command=show_combination)
btn_show.grid(row=1, column=0, columnspan=3, pady=10)
# Start the main loop
root.mainloop()
```

An example of extraction:

Figure 11. Combination of three major Arcana

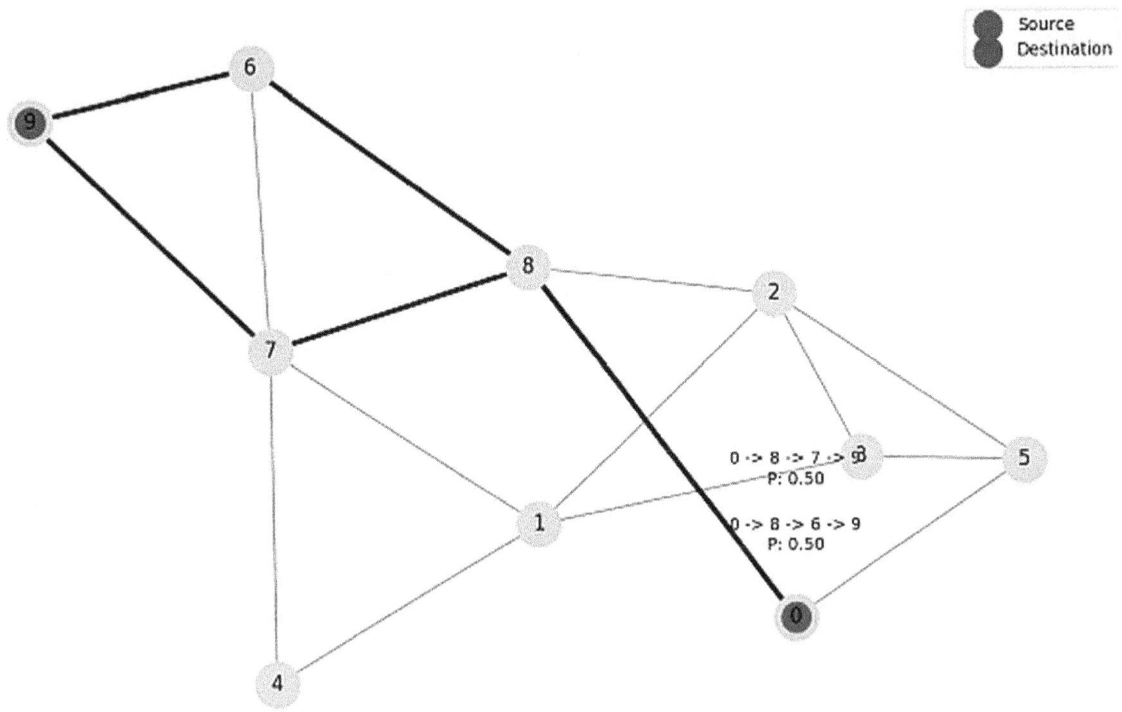

Tangram: It is a puzzle that involves creating a specific shape using seven pieces of different shapes. The pieces are two large isosceles, right triangles, a medium and two small isosceles, right triangles, a square and a parallelogram. Tangram is a combinatorial game that creates geometric figures using seven pieces of wood of different shapes, called tans. Combinatorial calculus can be used to determine the number of possible figures that can be created using the seven pieces of the Tangram. Specifically, there are 2,339,684 possible unique configurations of the Tangram, making the creation of new figures a considerable challenge. In addition, combinatorial calculus can identify the mathematical properties of the Tangram figures. For example, Euclidean geometry and combinatorial calculus can be used to show that each figure of the Tangram can be divided into seven congruent triangles. In addition, combinatorial calculus can create new Tangram challenges. Permutation and combination techniques can create new configurations of Tangram pieces that must be used to create a specific figure. In this way, players can use combinatorial calculus to create and solve new Tangram challenges. However, the total of figures that can be created using the seven tans of the Tangram is a complex combinatorial problem.

Calculating how many figures one can create is complex because Tangram is a game based on creativity and the free assembly of Tangram pieces to create an infinite variety of shapes. The possible figures that can be created with the Tangram are not limited to simple geometric shapes but also include complex figures, silhouettes, animals, objects, and even alphabet characters, among many other possibilities.

It can be interesting to "play" with the Tangram pieces in a graphic window by moving them with the mouse (click the right mouse to rotate):

Program 104 Tangram

```
import matplotlib.pyplot as plt
```

```
import matplotlib.patches as patches
import numpy as np
# Definition of Tangram pieces
pieces = {
'large_triangle1': [(0, 0), (2, 0), (1, 1)],
'large_triangle2': [(0, 0), (2, 0), (1, 1)],
'medium_triangle': [(0, 0), (1, 0), (1, 1)],
'small_triangle1': [(0, 0), (0.7, 0), (0.7, 0.7)],
'small_triangle2': [(0, 0), (0.7, 0), (0.7, 0.7)],
'square': [(0, 0), (0.7, 0), (0.7, 0.7), (0, 0.7)],
'parallelogram': [(0, 0), (0.7, 0), (1.4, 0.7), (0.7, 0.7)]
}
# Colors for the pieces
colors = ['red', 'green', 'blue', 'yellow', 'purple', 'orange', 'cyan']
# Initial positions of the pieces
positions = {
'large_triangle1': (0, 4),
'large_triangle2': (4, 4),
'medium_triangle': (2, 2),
'small_triangle1': (0, 2),
'small_triangle2': (3, 0),
'square': (1, 1),
'parallelogram': (2, 3)
}
# Initial rotation angles
angles = {name: 0 for name in pieces}
# Rotation points
rotation_points = {name: (x + 0.5, y + 0.5) for name, (x, y) in positions.items()}
# Function to draw the Tangram pieces
def draw_tangram(ax):
ax.clear()
ax.set_xlim(-1, 5)
ax.set_ylim(-1, 5)
ax.set_aspect('equal')
for name, vertices in pieces.items():
x, y = positions[name]
angle = angles[name]
vertices = np.array(vertices)
vertices = vertices - vertices.mean(axis=0) # Center the vertices
vertices = np.dot(vertices, np.array([[np.cos(angle), -np.sin(angle)], [np.sin(angle), 
np.cos(angle)]])) # Rotate the vertices
vertices = vertices + np.array([x, y]) # Translate the vertices
polygon = patches.Polygon(vertices, closed=True, edgecolor='black', facecolor=col-
ors[list(pieces.keys()).index(name)])
ax.add_patch(polygon)
# Draw the rotation point
rotation_point = patches.Circle(rotation_points[name], radius=0.1, color='black')
ax.add_patch(rotation_point)
# Global variables for dragging
```

```
selected_piece = None
offset = (0, 0)
# Function to handle mouse click
def on_click(event):
global selected_piece, offset
if event.inaxes is not None:
for name, (x, y) in positions.items():
if np.hypot(event.xdata - x, event.ydata - y) < 0.5: # Tolerance distance for the click
selected_piece = name
offset = (event.xdata - x, event.ydata - y)
break
# Function to handle mouse release
def on_release(event):
global selected_piece
selected_piece = None
# Function to handle mouse movement
def on_motion(event):
global selected_piece, offset
if selected_piece is not None and event.inaxes is not None:
x, y = event.xdata - offset[0], event.ydata - offset[1]
positions[selected_piece] = (x, y)
rotation_points[selected_piece] = (x + 0.5, y + 0.5)
draw_tangram(ax)
fig.canvas.draw()
# Function to handle rotation (Right-click)
def on_rotate(event):
if event.inaxes is not None and event.button == 3: # Check if it's a right-click (button
3)
for name, (x, y) in rotation_points.items():
if np.hypot(event.xdata - x, event.ydata - y) < 0.3: # Increased tolerance for rotation
angles[name] += np.radians(5) # Rotate by 5 degrees
draw_tangram(ax)
fig.canvas.draw()
# Create figure and axis
fig, ax = plt.subplots()
ax.set_xlim(-1, 5)
ax.set_ylim(-1, 5)
ax.set_aspect('equal')
# Draw Tangram pieces
draw_tangram(ax)
# Connect mouse events
fig.canvas.mpl_connect('button_press_event', on_click)
fig.canvas.mpl_connect('button_release_event', on_release)
fig.canvas.mpl_connect('motion_notify_event', on_motion)
fig.canvas.mpl_connect('button_press_event', on_rotate)
# Show the figure
plt.show()
```

Stomachion: It is a combinatorial game that dates back to ancient Greece. Archimedes invented it, and it has been described as a geometric puzzle that involves filling a complex shape consisting of 14 pieces within a square. The Stomachion comprises 14 polygons: 11 triangles, two quadrilaterals and a pentagon. Each piece has a unique shape and can be rotated and translated within the square. The game's goal is to fill the square with all 14 pieces, leaving no gaps and overlapping pieces. The game requires logical thinking and spatial analysis skills to solve the puzzle and does not require any advanced mathematical knowledge. Stomachion was one of Archimedes' favourite games and was only rediscovered in the twentieth century when fragments of parchment containing the game's instructions were found.

Today, many versions of the game are available, including digital versions and variants with different numbers of pieces. Stomachion is also a combinatorial game that can be solved using some combinatorial techniques. One of the standard techniques used to resolve Stomachion is the configuration counting technique. This technique uses the product rule to calculate the possible placement combinations of the pieces on the square of the Stomachion. A group of researchers has established that the number of obtainable configurations is 17,152. In 2003, Bill Cutler showed that there are exactly 536 diverse ways to form a perfect square considering the 32 possible rotations and reflections. Combinatorial techniques can consider the symmetries of the puzzle. Two workpiece arrangements that differ only by rotation or reflection are considered equivalent, so one needs to reduce the combinations by counting unique configurations. Group theory may analyse the symmetries, simplifying the calculation of unique configurations. This approach is common in combinatorial problems involving symmetry.

The program is conceptually similar to the previous one.

Program 105 Stomachion

```
import matplotlib.pyplot as plt
import matplotlib.patches as patches
import numpy as np
# Definition of Stomachion pieces
pieces = {
'piece1': [(0, 0), (0, 12), (2, 10)],
'piece2': [(0, 0), (2, 10), (4, 8)],
'piece3': [(0, 0), (3, 0), (2, 4)],
'piece4': [(3, 0), (2, 4), (3, 6)],
'piece5': [(0, 12), (6, 12), (4, 8)],
'piece6': [(4, 8), (6, 12), (6, 6)],
'piece7': [(6, 6), (6, 12), (9, 6), (8, 4)],
'piece8': [(8, 4), (9, 6), (12, 0)],
'piece9': [(6, 0), (8, 4), (6, 6)],
'piece10': [(9, 6), (12, 6), (12, 0)],
'piece11': [(9, 6), (12, 6), (12, 8)],
'piece12': [(9, 6), (6, 12), (12, 12), (12, 8)],
'piece13': [(3, 0), (3, 6), (4, 8), (6, 6), (6, 0)],
'piece14': [(6, 0), (8, 4), (12, 0)]
}
# Colors for the pieces
colors = ['red', 'green', 'blue', 'yellow', 'purple', 'orange', 'cyan', 'magenta',
'brown', 'pink', 'gray', 'olive', 'lime', 'teal']
# Initial positions of the pieces
positions = {
'piece1': (15, 15),
'piece2': (-10, -10),
```

```
'piece3': (-15, -5),
'piece4': (-15, 5),
'piece5': (-15, 15),
'piece6': (-5, -15),
'piece7': (-5, -5),
'piece8': (-5, 5),
'piece9': (-5, 15),
'piece10': (5, -15),
'piece11': (5, -5),
'piece12': (5, 10),
'piece13': (15, -15),
'piece14': (15, 0)
}
# Initial rotation angles
angles = {name: 0 for name in pieces}
# Rotation points
rotation_points = {name: (x + 0.5, y + 0.5) for name, (x, y) in positions.items()}
# Function to draw the pieces of the Stomachion
def draw_stomachion(ax):
ax.clear()
ax.set_xlim(-20, 20)
ax.set_ylim(-20, 20)
ax.set_aspect('equal')
for name, vertices in pieces.items():
x, y = positions[name]
angle = angles[name]
vertices = np.array(vertices)
vertices = vertices - vertices.mean(axis=0) # Center the vertices
vertices = np.dot(vertices, np.array([[np.cos(angle), -np.sin(angle)], [np.sin(angle),
np.cos(angle)]])) # Rotate the vertices
vertices = vertices + np.array([x, y]) # Translate the vertices
polygon = patches.Polygon(vertices, closed=True, edgecolor='black', facecolor=col-
ors[list(pieces.keys()).index(name)])
ax.add_patch(polygon)
# Draw the rotation point
rotation_point = patches.Circle(rotation_points[name], radius=0.5, color='black')
ax.add_patch(rotation_point)
# Global variables for dragging
selected_piece = None
offset = (0, 0)
# Function to handle mouse click
def on_click(event):
global selected_piece, offset
if event.inaxes is not None:
if event.button == 1: # Left mouse button
for name, vertices in pieces.items():
x, y = positions[name]
angle = angles[name]
vertices = np.array(vertices)
```

```
vertices = vertices - vertices.mean(axis=0) # Center the vertices
vertices = np.dot(vertices, np.array([[np.cos(angle), -np.sin(angle)], [np.sin(angle),
np.cos(angle)]])) # Rotate the vertices
vertices = vertices + np.array([x, y]) # Translate the vertices
polygon = patches.Polygon(vertices, closed=True)
if polygon.contains_point((event.xdata, event.ydata)):
selected_piece = name
offset = (event.xdata - x, event.ydata - y)
break
elif event.button == 3: # Right mouse button for rotation
for name, (x, y) in rotation_points.items():
if np.hypot(event.xdata - x, event.ydata - y) < 0.5: # Tolerance for rotation point
angles[name] += np.radians(5) # Rotate by 5 degrees
draw_stomachion(ax)
fig.canvas.draw()
break
# Function to handle mouse release
def on_release(event):
global selected_piece
selected_piece = None
# Function to handle mouse motion
def on_motion(event):
global selected_piece, offset
if selected_piece is not None and event.inaxes is not None:
x, y = event.xdata - offset[0], event.ydata - offset[1]
# Prevent dragging outside of the canvas area
x = max(min(x, 18), -18) # Prevent going beyond the x limits
y = max(min(y, 18), -18) # Prevent going beyond the y limits
positions[selected_piece] = (x, y)
rotation_points[selected_piece] = (x + 0.5, y + 0.5)
draw_stomachion(ax)
fig.canvas.draw()
# Create the figure and axis
fig, ax = plt.subplots()
ax.set_xlim(-20, 20)
ax.set_ylim(-20, 20)
ax.set_aspect('equal')
# Draw the Stomachion pieces
draw_stomachion(ax)
# Connect mouse events
fig.canvas.mpl_connect('button_press_event', on_click)
fig.canvas.mpl_connect('button_release_event', on_release)
fig.canvas.mpl_connect('motion_notify_event', on_motion)
# Show the figure
plt.show()
```

REFERENCES

AA.VV. (2018). Math-GAMES - Guidebook. *Erasmus.*

Albert, M., Nowakowski, R., & Wolfe, D. (2019). *Lessons in play: an introduction to combinatorial game theory.* AK Peters/CRC Press. DOI: 10.1201/9781482243048

Arcos, C., Brookfield, G., & Krebs, M. (2010). Mini-sudokus and groups. *Mathematics Magazine, 83*(2), 111–122. DOI: 10.4169/002557010X482871

Berlekamp, E. R., Conway, J. H., & Guy, R. K. (2004). *Winning ways for your mathematical plays* (Vol. 4). AK Peters/CRC Press. DOI: 10.1201/9780429487309

Corli, S., Moro, L., Galli, D. E., & Prati, E. (2021). Solving rubik's cube via quantum mechanics and deep reinforcement learning. *Journal of Physics. A, Mathematical and Theoretical, 54*(42), 425302. DOI: 10.1088/1751-8121/ac2596

Demaine, E. D., & Hearn, R. A. (2008, June). Constraint logic: A uniform framework for modeling computation as games. In *2008 23rd Annual IEEE Conference on Computational Complexity* (pp. 149-162). IEEE. DOI: 10.1109/CCC.2008.35

Diego-Mantecon, J. M., Arcera, O., Blanco, T. F., & Lavicza, Z. (2019). An engineering technology problem-solving approach for modifying student mathematics-related beliefs: Building a robot to solve a Rubik's cube. *The International Journal for Technology in Mathematics Education, 26*(2), 55–64. DOI: 10.1564/tme_v26.2.02

. Dvoryatkina, S. N., & Simonovskaya, G. A. (2021). Using Chess for Identifying and Correcting "Problem Areas" in the School Math Course. *TEM Journal, 10*(1).

Egri-Nagy, A., & Törmänen, A. (2020). The game is not over yet—Go in the post-alphago era. *Philosophies, 5*(4), 37. DOI: 10.3390/philosophies5040037

Felgenhauer, B., Jarvis, F. (2005). There Are 6670903752021072936960 Sudoku Grids.

Ferguson, T. S. (2001). Another form of matrix Nim. *the electronic journal of combinatorics*, R9-R9.

García Jiménez, J. C. (2018). Juegos combinatorios parciales.

Golomb, S. W. (1996). *Polyominoes: puzzles, patterns, problems, and packings* (Vol. 16). Princeton University Press.

Hefetz, D., Krivelevich, M., Stojaković, M., & Szabó, T. (2014). *Positional games* (Vol. 44). Birkhäuser. DOI: 10.1007/978-3-0348-0825-5

Knuth, D. E. (2000). Dancing links. *arXiv preprint cs/0011047.*

Lamberti, L., & Tovena, F. (2022). The Mastermind Game: A Combinatorial Approach. In *Handbook of Research on International Approaches and Practices for Gamifying Mathematics* (pp. 145-169). IGI Global. DOI: 10.4018/978-1-7998-9660-9.ch008

Larsson, U., Nowakowski, R. J., & Santos, C. P. (2016). Absolute combinatorial game theory. *arXiv preprint arXiv:1606.01975.*

Niv, A. N., & Patkin, D. P. (2020). Using Combinatorial Games to Explore Equivalences and Differences. *Learning and Teaching Mathematics, 28*, 8–11.

Russell, E., & Jarvis, F. (2006). Mathematics of sudoku II. *Mathematical Spectrum, 39*(2), 54–58.

Siegel, A. N. (2013). *Combinatorial game theory* (Vol. 146). American Mathematical Soc. DOI: 10.1090/gsm/146

Tang, A. (2018). Analysis of Chances of Winning in a Poker Game: Combinatorial Probability and the Law of Large Numbers.

Toshniwal, V. (2024). Combinatorics and Group Theory in Chess and Music.

Tromp, J., & Farnebäck, G. (2007). Combinatorics of Go. In *Computers and Games* (pp. 84–99). Springer. DOI: 10.1007/978-3-540-75538-8_8

Uiterwijk, J. W. (2019, August). Solving Strong and Weak 4-in-a-Row. In *2019 ieee conference on games (cog)* (pp. 1-8). IEEE.

Chapter 10
Random Combinations

ABSTRACT

The chapter analyses the Random combinations. Random combinations in mathematics offer a creative approach by using probabilistic methods to generate new selections of elements from a given set. Unlike classical combinations, which follow strict rules and produce all possible selections, random combinations use probabilistic techniques to create subsets randomly. Applications of random combinations include statistical sampling, marketing tests, games of chance, and financial analysis. These methods can stimulate creativity and innovation in fields such as creative painting, music, storytelling games, and brainstorming. For example, an artist might randomly mix colors and geometric shapes to create new compositions, or a musician might generate new melodies by randomly combining instruments. Additionally, random combinations are used in scientific contexts, such as generating random graphs, simulating complex events, and optimizing hyperparameters in machine learning.

INTRODUCTION

Nothing happens by chance, but everything according to reason and necessity

Leucippus

When we think of mathematics, we often imagine strict rules and precise calculations. However, there is a more creative side: random combinations allow us to mix elements to open up to new discoveries. Sometimes, the most unexpected paths lead to the most interesting discoveries. Random combinations, therefore, are a way to get off the beaten track and let us be guided by the unexpected.

In combinatorics, random combinations are an unstructured approach to generating new selections of elements from a given set or multiple sets (Deng, 2022). Unlike classical combinations, which follow stringent mathematical rules and produce all possible selections, random combinations use probabilistic methods to create subsets randomly.

Table 1. Differences between combinations

Characteristic	Classic Combinations	Random Combinations
Selection Method	Systematic and mathematical	Random and probabilistic
Rules	Follows fixed rules	Case-based
Reproducibility	Identical results with the same parameters	Variable results even with the same parameters
Example of Use	Calculation of possibilities	Statistical sampling

Classic Combinations follow fixed rules, often defined by deterministic algorithms, ensuring the reproducibility of the results. Random Combinations are ideal in statistical or simulation contexts, such as random sampling for data analysis or experiments. Random combinations offer flexibility and adaptability, making them particularly useful in

DOI: 10.4018/979-8-3373-3089-1.ch010

various fields, including statistics, creativity, computer security, and complex simulations. These advantages make them preferable to classical combinations where unpredictability and variability are required.

Random combinations represent a significant frontier in combinatorics. Before examining some particular cases, it may be significant to report some concrete examples:

- *Statistical sampling*: A researcher must select a random sample of n people from a population of n individuals. In this case, a random number generator draws individuals randomly, ensuring that each member of the population has an equal chance of being chosen.
- *Testing in marketing*: A company wants to test two versions of an advertisement (A and B) on a random sample of users. Users are randomly assigned to one of two groups (A or B). This prevents results from being influenced by external factors, ensuring an equal distribution among groups.
- *Games of chance*: One could take a set of numbers and generate random combinations. This method is used in the Italian Superenalotto or the English National Lottery to propose six random numbers.
- *Financial analysis*: An analyst can use random simulations to generate thousands of scenarios, utilising probabilistic distributions for the economic variables that affect the markets.

A particularly interesting example is the following: one could combine a set of words and images to generate new stories or poems. This process can be likened to using a "mental blender" to discover new combinations of available elements. One starts with a list (concepts, objects, ideas) and proceeds with a selection without being bound to a specific order. Next, a random shuffle of the selected elements is conducted. This process can lead to new associations, inspirations, or creative solutions.

List items such as words, colours, images, or concepts and then randomly select a couple to explore new associations or ideas. This process of random selection and mixing can stimulate creativity and innovation.

Some examples can be given to help understand this approach.

- *Brainstorming*: This creative group thinking technique generates new and innovative ideas in many situations. Example: In an advertising company, a team wants to generate new ideas for a product. A list of attributes could include the words "Sustainability", "Innovation", "Ecology", "Future", "Efficiency", and "Digital". A random combination could lead to three elements, such as "Ecology", "Innovation", and "Sustainability", which would lead to a campaign that highlights an environmentally friendly product that can contribute to a more sustainable future through innovation.
- *Creative painting*: An artist takes a series of colours and geometric shapes, randomly mixes them, and arranges them on a canvas to see what composition emerges. Example: Having "Yellow," "Red," "Blue," "Circle," "Square," or "Triangle" available, you could obtain, by selecting two elements, a blue square, a red triangle, or a completely green square (blue + yellow).
- *Creative music*: Random combinatorics can generate new melodies, chords, and arrangements. Example: Consider a set of instruments such as "Piano", "Guitar," "Saxophone", and "Drums." We can use random combinatorics to generate an arrangement by randomly combining three instruments from this set. For example, we can generate a new arrangement with (Saxophone, Piano, Guitar) or (Guitar, Saxophone, Drums).
- *Creative play*: There are creative storytelling games on the market in which players roll nine special dice with images and then have to create a story based on the visuals displayed using all or even just a part of the dice. Example: Once the dice have been rolled, each player will be the Storyteller, who must create a story starting with the classic "Once upon a time..."
- *Creative writing*: An author wants to draft an original story. He writes a list of keywords, then randomly selects two or more items from the list and tries to develop a story based on those combinations. Example: starting from five words such as "Moon", "Solitude", "Adventure", "Forest", and "Storm", one can extract all combinations of two words. Supposing to extract "Luna" and "Storm", one could create a story in which a traveller, under the rays of the moon, finds himself hit by a terrible storm. With the words "Solitude" and "Adventure", the story could be of a traveller who, walking under the weight of solitude, suddenly found himself involved in a magical adventure.

The idea of language as a combinatorial machine has inspired many writers.

Raymond Queneau wrote, among other things, a text with a combinatorial architecture. The book consists of ten sheets, each divided into fourteen horizontal bands. By rotating the strips like pages, the reader can choose one of the ten versions of the poem proposed by Queneau (Queneau, 1961). All ten versions have the same scansion and the same rhymes, and as already mentioned, each sonnet is regular in its form. From the combinatorial point of view, there are 10^{14}, that is, one hundred thousand billion possible poems.

The phenomenon can be understood using a simulation that, in a small way, uses the first four verses of some poems by William Blake (1757-1827).

Program 106 Creative Writing

```
import random
# Each poem is a list of lines.
poems = [
["Tyger Tyger, burning bright,", "In the forests of the night;", "What immortal hand or eye,", "Could frame thy fearful symmetry?"],
["Little Lamb who made thee?", "Dost thou know who made thee,", "Gave thee life, and bid thee feed", "By the stream and o'er the mead;"],
["To see a World in a Grain of Sand,", "And a Heaven in a Wild Flower,", "Hold Infinity in the palm of your hand,", "And Eternity in an hour."],
["I was angry with my friend:", "I told my wrath, my wrath did end.", "I was angry with my foe:", "I told it not, my wrath did grow."],
["I wander thro' each charter'd street,", "Near where the charter'd Thames does flow,", "And mark in every face I meet", "Marks of weakness, marks of woe."],
]
def generate_poem(poems):
poem_length = len(poems[0]) # Assumes all poems have the same number of lines
mixed_poem = [random.choice([poem[i] for poem in poems]) for i in range(poem_length)]
return "\n".join(mixed_poem)
def generate_multiple_poems(poems, num_poems):
return "\n\n".join(generate_poem(poems) for _ in range(num_poems))
# Ask the user how many poems to generate
num_poems = int(input("How many poems do you want to generate? "))
print("\n" + generate_multiple_poems(poems, num_poems))
```

Italo Calvino tells of a group of wayfarers who wish to tell their adventures (Calvino, 1994). However, having lost the word, they use tarot cards. By placing the various cards on the table in sequence, different narratives for different arrangements are obtained. Each character chooses a theme and places a card on the table to tell their story. The meaning of each individual card depends on its place in the succession of cards that precede and follow it.

The cut-up method consists of cutting pages from a text and rearranging them in random combinations. William Burroughs used the cut-up method in some novels (Burroughs, 1960). The work consisted of taking cues from the works of others in selecting the stylistically interesting material.

In conclusion, random combinatorics is a variant of classical combinatorics. The primary difference between the two disciplines is that classical combinatorics focuses on combinations of elements that can be generated deterministically. In contrast, random combinatorics focuses on combinations of elements that can be randomly generated.

A program as a generative example could be the following:

Program 107 Random combinations

```
import random
def generate_combinations():
elements = input("Enter the elements separated by a space: ").split()
number = int(input("Enter the number of elements per combination: "))
num_combinations = int(input("Enter the number of combinations to generate: "))
combinations = []
for _ in range(num combinations):
combinations.append(random.sample(elements, number))
return combinations
# Example usage
print(generate_combinations())
```

Another example of literature based on combinatorial reasoning is in the novel *The Nine Billion Names of God* (Clarke, 1967) by Arthur C. Clarke (1917-2008). The concept of name combinations comes from an incredible machine designed by Tibetan monks to generate possible combinations of names for God based on specific mathematical rules. These names are based on symbols representing the different possible configurations of characters or letters, resulting in an enormous number of combinations.

The monks believe that if they can write down all the possible names of God, they will have fulfilled their divine mission, and the world will come to an end, fulfilling the divine will. In the book, the specific constraints that lead to the exact number of nine billion combinations are not explained in detail. Clarke does not go into in-depth technical details about exactly how names are generated or the precise rules that limit the combinations. Instead, the author focuses on the story's more philosophical and religious view. The quantity of combinations is chosen for its symbolic meaning; the nine billion names are not presented as the result of a complex mathematical calculation made explicit but as the completion of a spiritual mission. The machine is programmed to perform this task.

The previous examples have provided an idea of the role of random combinations, but they find application mainly in several fundamental scientific fields. For example, random combinations are fundamental in the study of random graphs (Cameron, 1997; Bollobás, 1981; Frieze & Karoński, 2016) in which the edges are randomly connected, or random paths on the various branches are studied. Researchers introduced a new combinatorial approach to model and analyse network diffusion processes based on enumerating all possible shorter trajectories between a source node and a destination node. Once all possible shorter trajectories have been enumerated, each trajectory is assigned a probability that the diffusion process will follow that trajectory. The probability of a trajectory depends on several factors, including the trajectory's length, the network's structure, and the nodes' properties.

This combinatorial approach has several advantages over traditional approaches for analysing diffusion processes over networks (Mazzilli & Radicchi, 2021). First, it can capture the complexity of the network, considering all the possible paths that the deployment can follow. Second, it can accurately estimate the likelihood of spread following a particular trajectory. To analyse how an infection spreads through a network, one can enumerate all the shortest trajectories from a source node to a destination node. Each trajectory will have an associated probability based on the network structure and node topology.

Program 108 Trajectory

```
import networkx as nx
import matplotlib.pyplot as plt
def create_graph():
# Create a random graph
G = nx.erdos_renyi_graph(n=10, p=0.3) # 10 nodes, connection probability 0.3
return G
def find_shortest_paths(G, source, destination):
```

```
# Find all shortest paths from source to destination
return list(nx.all_shortest_paths(G, source=source, target=destination))
def assign_probabilities(paths):
# Assign a uniform probability to each path
total_paths = len(paths)
probabilities = [1 / total_paths for _ in range(total_paths)] # Uniform probability
return probabilities
def visualize_graph(G, source, destination, paths, probabilities):
pos = nx.spring_layout(G) # Node positions
plt.figure(figsize=(10, 6))
# Draw the graph
nx.draw(G, pos, with_labels=True, node_color='lightgrey', node_size=700, font_size=12,
font_color='black', edge_color='gray')
# Highlight the source and destination nodes
nx.draw_networkx_nodes(G, pos, nodelist=[source], node_color='red', label='Source')
nx.draw_networkx_nodes(G, pos, nodelist=[destination], node_color='green', label='Destina-
tion')
# Highlight the shortest paths
for i, path in enumerate(paths):
edges = list(zip(path[:-1], path[1:])) # Convert zip to list
nx.draw_networkx_edges(G, pos, edgelist=edges, edge_color='black', width=3)
# Add annotations for paths and probabilities
path_str = " -> ".join(map(str, path))
x = pos[path[0]][0]
y = pos[path[0]][1] + (0.1 * (i + 1)) # Incremental offset based on index
plt.text(x, y, f"{path_str}\nP: {probabilities[i]:.2f}", fontsize=10, color='black',
ha='center')
plt.title('Graph Representation with Shortest Paths')
plt.legend()
plt.show()
def main():
# Create the graph
G = create_graph()
# Display the graph
print("Graph:")
print(G.edges())
# Define the source and destination nodes
source = 0
destination = 9 # Ensure a path exists
# Find the shortest paths
shortest_paths = find_shortest_paths(G, source, destination)
# Print the paths
print("\nShortest Paths:")
for path in shortest_paths:
print(path)
# Assign probabilities to each path
probabilities = assign_probabilities(shortest_paths)
# Print the probabilities
print("\nProbabilities associated with the paths:")
```

```
for path, prob in zip(shortest_paths, probabilities):
print(f"Path: {path} - Probability: {prob:.2f}")
# Visualize the graph with paths and probabilities
visualize_graph(G, source, destination, shortest_paths, probabilities)
if __name__ == "__main__":
main()
```

Figure 1. Trajectories

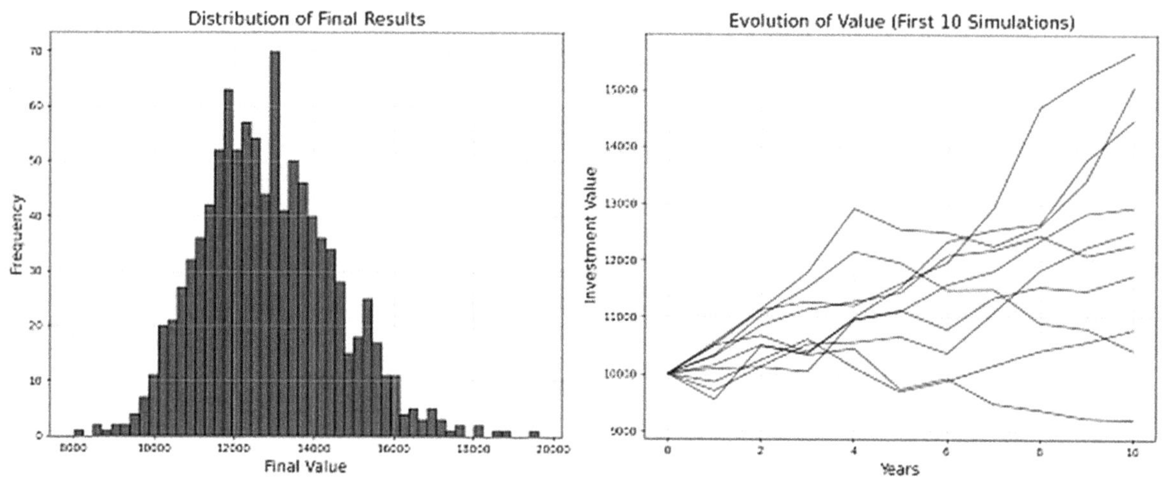

Other examples can be found in the generation of cryptographic keys used to encrypt and decrypt data, as well as in many random generation algorithms based on mathematical algorithms that utilise combinatorial structures, such as matrices or graphs, to generate random values. The Mersenne Twister algorithm, one of the most widely used random number generators, employs a combinatorial matrix to generate random values. The process of generating a new pseudorandom value can be viewed as a combinatorial enumeration process, as the algorithm enumerates all possible combinations for adjacent elements of a 624-element array and then selects a random combination.

This program generates ten random numbers:

Program 109 Random numbers

```
class MersenneTwister:
def __init__(self, seed=5489):
# Initialize the parameters and state of the generator
self.state = [0] * 624
self.f = 1812433253
self.m = 397
self.u = 11
self.s = 7
self.b = 0x9D2C5680
self.t = 15
self.c = 0xEFC60000
```

```
self.l = 18
self.index = 624
self.lower_mask = (1 << 31) - 1
self.upper_mask = 1 << 31
# Initialize the generator with the seed
self.state[0] = seed
for i in range(1, 624):
self.state[i] = self.int_32(self.f * (self.state[i-1] ^ (self.state[i-1] >> 30)) + i)
def twist(self):
# Twist function to mix the state
for i in range(624):
temp = self.int_32((self.state[i] & self.upper_mask) + (self.state[(i+1) % 624] & self.
lower_mask))
temp_shift = temp >> 1
if temp % 2 != 0:
temp_shift = temp_shift ^ 0x9908b0df
self.state[i] = self.state[(i + self.m) % 624] ^ temp_shift
self.index = 0
def get_random_number(self):
# Generate a random number between 0 and 1
if self.index >= 624:
self.twist()
y = self.state[self.index]
y = y ^ (y >> self.u)
y = y ^ ((y << self.s) & self.b)
y = y ^ ((y << self.t) & self.c)
y = y ^ (y >> self.l)
self.index += 1
return self.int_32(y) / float(2**32)
def int_32(self, number):
# Mask to ensure a 32-bit value
return int(0xFFFFFFFF & number)
if __name__ == "__main__":
mt = MersenneTwister(123)
for _ in range(10):
print(mt.get_random_number())
```

The method provides a good balance between computational efficiency and the quality of the generated pseudorandom values. The algorithm has a periodicity of $2^{19937}-1$, which means that the generated numbers do not repeat until after a considerable number of iterations. However, it is unsuitable for cryptographic contexts since it does not guarantee security against predictive attacks.

Examples could be multiplied by referring to the simulation of rare or complex events, such as the behaviour of subatomic particles or the propagation of infectious diseases.

An interesting application refers to the description of the evolution of populations. The population evolution is a random combinatorial problem because it involves choosing a combination of genes from a set of possibilities. Each individual in the population has a genotype, a sequence of 0s and 1s. The parents' genotypes are combined during breeding to create the child's genotype. This combination process is random, as each bit of the child's genotype is randomly chosen by one of the two parents. The random aspect of the problem makes it difficult to predict population behaviour over time. Simulating the population's behaviour and observing how it varies over time is possible.

Let us look at an example of a program that graphically shows population trends:

Program 110 population

```
import random
import matplotlib.pyplot as plt
class Population:
def __init__(self, n, k):
self.n = n # Number of individuals
self.k = k # Length of the genotype
self.genotypes = [[random.randint(0, 1) for _ in range(k)] for _ in range(n)]
def fitness(self):
# Calculate the fitness of each individual
return [sum(individual) for individual in self.genotypes]
def select_parents(self):
# Selection based on fitness (roulette wheel selection)
fitness = self.fitness()
total_fitness = sum(fitness)
probabilities = [f / total_fitness for f in fitness]
parents = random.choices(self.genotypes, weights=probabilities, k=2)
return parents
def crossover(self, parent1, parent2):
crossover_point = random.randint(1, self.k - 1)
child = parent1[:crossover_point] + parent2[crossover_point:]
return child
def mutate(self, individual):
# Mutation with a probability of 0.01
mutation_probability = 0.01
return [gene if random.random() > mutation_probability else 1 - gene for gene in individu-
al]
def reproduce(self):
new_population = []
for _ in range(self.n):
parents = self.select_parents()
child = self.crossover(parents[0], parents[1])
child = self.mutate(child)
new_population.append(child)
self.genotypes = new_population
def main():
n = 100 # Number of individuals
k = 10 # Length of the genotype
generations = 100
population = Population(n, k)
average_fitness = []
for _ in range(generations):
population.reproduce()
fitness = population.fitness()
average_fitness.append(sum(fitness) / len(fitness)) # Calculate the average fitness
```

```
plt.plot(average_fitness)
plt.title("Population Development")
plt.xlabel("Generation")
plt.ylabel("Average Fitness")
plt.show()
if __name__ == "__main__":
main()
```

Let us try to take the financial analyst's example, thanks to a program that simulates the return on an investment using the Monte Carlo method (Carlo, 2001). This method is a statistical approach to modelling uncertainty and the trend of random variables. This approach is useful for modelling the uncertainty related to returns, which can vary annually in an unpredictable way.

Program 111 Financial analysis

```
import numpy as np
import matplotlib.pyplot as plt
def simulate_investment(initial_amount, average_return, std_deviation, years, simula-
tions):
# Convert average return and standard deviation to decimal values
average_return /= 100
std_deviation /= 100
# Prepare an array for final results
final_values = []
investment_traces = []
for _ in range(simulations):
# Generate random annual returns
annual_returns = np.random.normal(average_return, std_deviation, years)
# Calculate the investment value year by year
investment_value = initial_amount
yearly_values = [investment_value] # List to track the progression
for annual_return in annual_returns:
investment_value *= (1 + annual_return)
yearly_values.append(investment_value)
# Store the final result and the trace
final_values.append(investment_value)
investment_traces.append(yearly_values)
return np.array(final_values), np.array(investment_traces)
def plot_simulation_results(final_values, investment_traces):
"""Plot the simulation results (histogram + progression)."""
plt.figure(figsize=(15, 6))
# Plot 1: Histogram of final results
plt.subplot(1, 2, 1)
plt.hist(final_values, bins=50, color='black', alpha=0.7, edgecolor='black')
plt.title("Distribution of Final Results", fontsize=16)
plt.xlabel("Final Value", fontsize=14)
plt.ylabel("Frequency", fontsize=14)
plt.grid(True, alpha=0.4)
```

```
# Plot 2: Scatter plot of the evolution of investment over time
plt.subplot(1, 2, 2)
for trace in investment_traces[:10]: # Limit to 10 simulations for clarity
plt.plot(trace, color='black', alpha=0.6)
plt.title("Evolution of Value (First 10 Simulations)", fontsize=16)
plt.xlabel("Years", fontsize=14)
plt.ylabel("Investment Value", fontsize=14)
plt.grid(True, alpha=0.4)
plt.tight_layout()
plt.show()
# Function to get input from the user
def get_input():
try:
initial_amount = float(input("Enter the initial amount (in euros): "))
average_return = float(input("Enter the annual average return (in percentage): "))
std_deviation = float(input("Enter the annual standard deviation (in percentage): "))
years = int(input("Enter the number of years of the investment: "))
simulations = int(input("Enter the number of Monte Carlo simulations: "))
return initial_amount, average_return, std_deviation, years, simulations
except ValueError:
print("Error: Please enter valid numerical values.")
return None
# Get parameters from the user
parameters = get_input()
if parameters:
initial_amount, average_return, std_deviation, years, simulations = parameters
# Perform the simulation
final_values, investment_traces = simulate_investment(initial_amount, average_return, std_
deviation, years, simulations)
# Print some statistics
print("\nStatistics of the results:")
print(f"Average final value: {np.mean(final_values):.2f} euros")
print(f"Median final value: {np.median(final_values):.2f} euros")
print(f"Minimum final value: {np.min(final_values):.2f} euros")
print(f"Maximum final value: {np.max(final_values):.2f} euros")
# Plot the results
plot_simulation_results(final_values, investment_traces)
```

The simulation illustrates how different performance trajectories can yield varying results, even with the same initial parameters. Because the returns are randomly generated, each program run will produce a slightly different distribution of the final results. The program generates two different types of charts. The histogram allows us to see the range of final values in the results. The scatter chart, on the other hand, represents the evolution of the investment over time, showing the impact of annual fluctuations on returns, which allows us to better understand the volatility and variability of investment results.

These are the graphs obtained from the following data:

Initial amount (in euros): 10000
Average annual return (in percentage): 2.5
Annual standard deviation (in percentage): 4

Number of years of investment: 10
Number of Monte Carlo simulations: 1000

Figure 2. Financial analysis

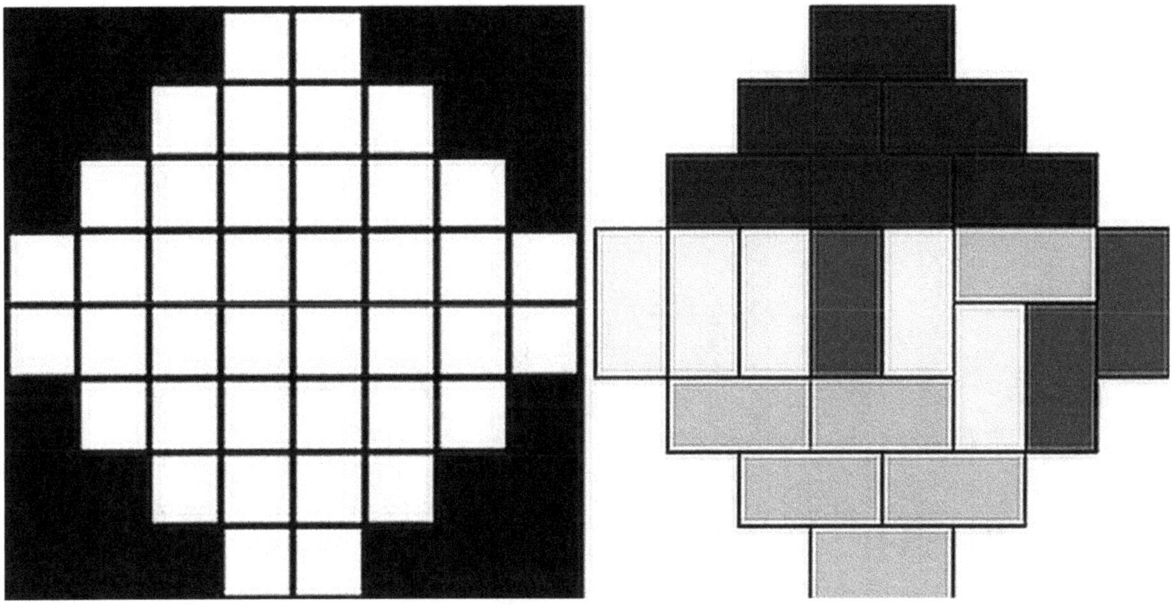

The summary indicators are:

Average final value: 12920.39 euros
Median final value: 12829.02 euros
Minimum final value: 8226.43 euros
Final maximum value: 18842.39 euros

Over the last twenty years, random matrix theory has evolved as a tool for solving combinatorial problems (Novak, 2009; Baik, 2016; Izenman, 2021). An interesting example is the random tilings of the Aztec diamond. It concerns the number of ways in which a particular region of the plane can be covered with domino-shaped tiles (Elkies et al., 1992; Stéphan, 2021). The Aztec diamond is a diamond-shaped grid of squares. Precisely, if the length of the diamond side is n, it consists of a total of $2n^2$-n unit squares arranged in a diamond shape. The diamond can be covered with domino pieces. "random tiling" refers to a process in which one covers an Aztec diamond using a probabilistic or random method. As n increases, the Aztec diamond's random tilings show a shape known as the "boundary shape." The formula represents the exact number of possible tilings. This result is a consequence of advanced combinatorial studies $2^{n(n+1)/2}$.

Figure 3. Empty and filled Aztec diamond

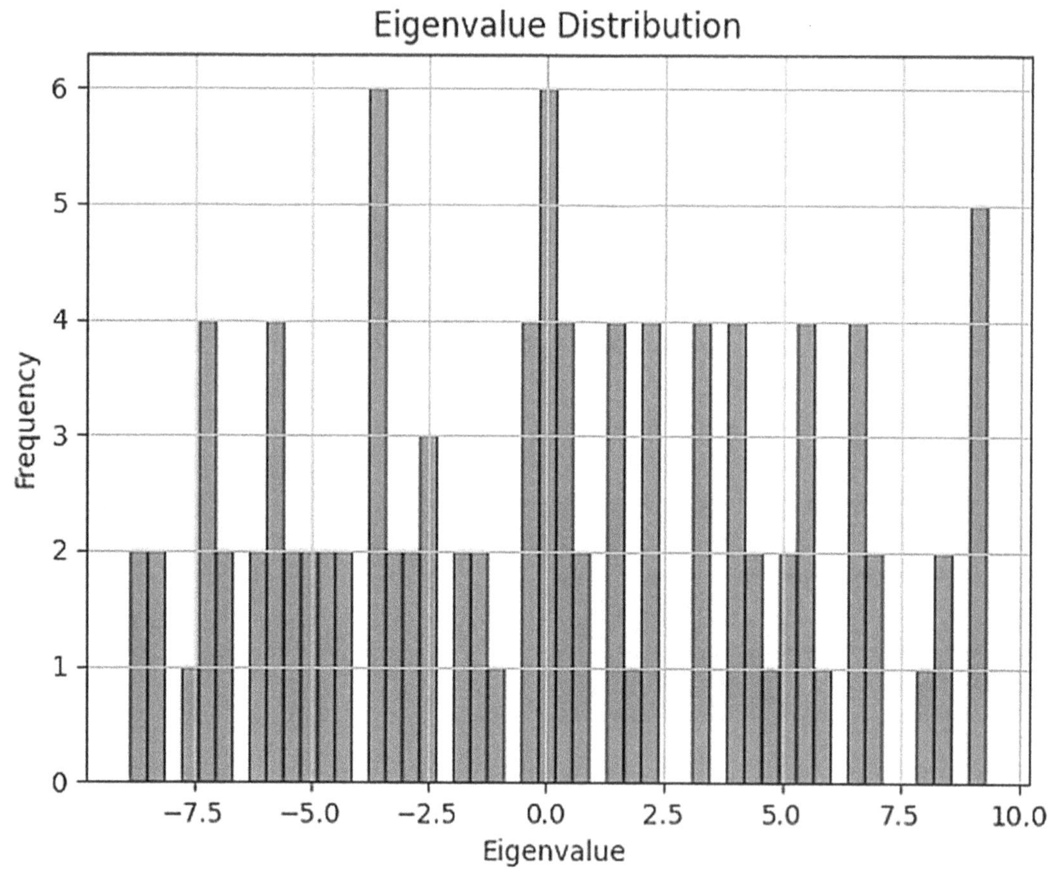

For a size 4 Aztec diamond, the program below calculates 1024 different ways to cover the diamond with domino-shaped tiles. For size 40, the order of magnitude is huge: about 10^{256}.

Program 112 Aztec diamond

```
# Aztec Diamond
def count_tiles(n):
return 2**(n * (n + 1) // 2)
n = int(input("Enter the number of rows/columns in the Aztec Diamond: "))
print(count_tiles(n))
```

If we want to provide a different scenario, we can refer to the analysis of the eigenvalue moments' results. They can be used as a significant example of random matrix theory.

Program 113 Eigenvalues

```
import numpy as np
import matplotlib.pyplot as plt
# Function to generate a large random matrix
def generate_random_matrix(n):
return np.random.randn(n, n)
# Function to compute the eigenvalues of a matrix
def compute_eigenvalues(matrix):
eigenvalues, _ = np.linalg.eig(matrix)
return eigenvalues
# Function to calculate the moments of eigenvalues
def compute_moments(eigenvalues, k):
return np.mean(eigenvalues**k).real
# Function to plot the histogram of eigenvalue distribution
def plot_eigenvalue_distribution(eigenvalues, num_bins=50):
plt.hist(eigenvalues.real, bins=num_bins, alpha=0.75, edgecolor='black')
plt.title('Eigenvalue Distribution')
plt.xlabel('Eigenvalue')
plt.ylabel('Frequency')
plt.grid(True)
plt.show()
# Parameters
matrix_size = 100 # Size of the matrix
num_moments = 4 # Number of moments to calculate
# Generate the random matrix
random_matrix = generate_random_matrix(matrix_size)
# Compute the eigenvalues
eigenvalues = compute_eigenvalues(random_matrix)
# Calculate and print the moments of the eigenvalues
moments = [compute_moments(eigenvalues, k) for k in range(1, num_moments + 1)]
for i, moment in enumerate(moments, start=1):
print(f'Moment {i}: {moment:.3f}')
# Plot the eigenvalue distribution
plot_eigenvalue_distribution(eigenvalues)
```

The program generates a random matrix, calculates its eigenvalues, and analyses its distribution through the first four moments.

A possible output might be as follows:

Moment 1 (average): 0.124 slight deviation from the perfect symmetry of the distribution
Moment 2 (variance): 2,578 eigenvalues fairly dispersed with respect to their mean
Moment 3 (skewness): -24.025 major tail on the left side
Moment 4 (kurtosis): 179,898 distributions with very heavy tails

Figure 4. Distribution of eigenvalues

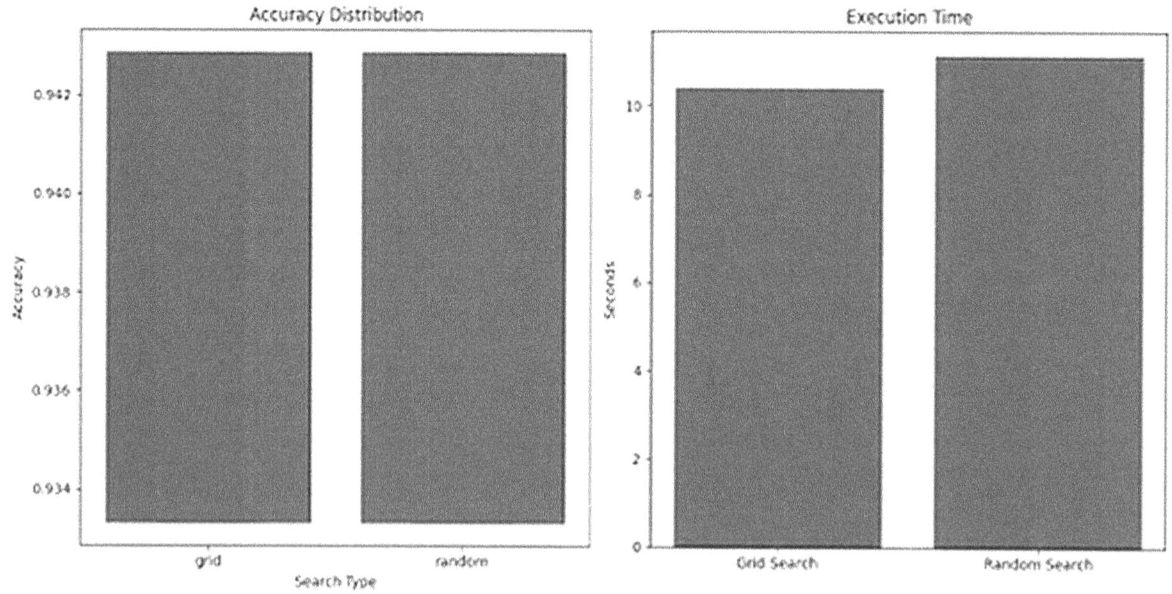

Hyperparameter optimization is a crucial process in machine learning. Its goal is to find the best configuration that maximizes a model's performance (Agrawal, 2024).

Hyperparameter optimization techniques can be associated with random combinatorics in various contexts. Hyperparameter optimization can be seen as a selection problem, which can be treated using random combinatorics in some cases:

- *Hyperparameters*: They are variables that are not learned during the model training process but are set before training (such as the learning rate, the depth of a decision tree, the neurons in a neural network, etc.). There are several methods of optimizing hyperparameters.
- *Grid search*: This method explores the hyperparameter space by evaluating all possible combinations. It is close to random combinatorics since, while not "random" in the true sense, it explores all combinations of hyperparameters defined in a specific range. For example, if we want to optimize two parameters, such as the learning rate and the number of trees, each with three values, the research explores nine combinations.
- *Random search*: Randomly selects hyperparameter values from a defined range instead of exploring all of space. This approach is related to random combinatorics, as it selects combinations of hyperparameters in a non-deterministic manner (Barrett, 2022).
- *Genetic algorithms*: These algorithms attempt to optimise hyperparameters by exploring the space like natural selection. These algorithms create new combinations of hyperparameters through crossing, mutation, and selection, exploring space randomly but with an evolutionary strategy. These methods can be considered an advanced form of random combinatorics, where combinations of parameters evolve (Naseri & Koffas, 2020).

These types of approaches have several advantages, such as simplicity and efficiency. However, their limitations should not be underestimated, as they are random, which means the search may not find the optimal combination or take a long time to do so. Additionally, random search can be ineffective if the hyperparameter space is too large and complex.

Random combinatorics is one of the foundations of many hyperparameter optimisation techniques, especially in contexts where exhaustive exploration is unnecessary or when a solution is to be found relatively quickly.

The following program can highlight the comparison between grid and random search using the classic Iris dataset. The Grid search explores all possible combinations, while the Random Search function randomly selects a few parameter combinations and trains the model for each, calculating the accuracy. Both measure the time taken to complete the search, and ultimately, the optimal parameters, accuracy, and execution time for both techniques are printed.

Program 114 Method comparison

```
import numpy as np
from sklearn.ensemble import RandomForestClassifier
from sklearn.datasets import load_iris
from sklearn.model_selection import train_test_split, cross_val_score
from sklearn.metrics import accuracy_score
import seaborn as sns
import pandas as pd
import matplotlib.pyplot as plt
import random
import time
# Data Preparation
data = load_iris()
X = data.data
y = data.target
X_train, X_test, y_train, y_test = train_test_split(X, y, test_size=0.3, random_state=42)
# Parameters
n_trees = [50, 100, 200]
max_depth = [None, 5, 10]
min_samples_split = [0.01, 0.1, 0.5]
n_iterations = len(n_trees) * len(max_depth) * len(min_samples_split)
def grid_search():
start_time = time.time()
results = []
for n in n_trees:
for depth in max_depth:
for split in min_samples_split:
model = RandomForestClassifier(n_estimators=n, max_depth=depth,
min_samples_split=split, random_state=42)
scores = cross_val_score(model, X_train, y_train, cv=5)
results.append({
'search_type': 'grid',
'n_trees': n,
'max_depth': depth,
'min_samples_split': split,
'accuracy': np.mean(scores)
})
return results, time.time() - start_time
def random_search():
start_time = time.time()
results = []
for _ in range(n_iterations):
```

```
n = random.choice(n_trees)
depth = random.choice(max_depth)
split = random.choice(min_samples_split)
model = RandomForestClassifier(n_estimators=n, max_depth=depth,
min_samples_split=split, random_state=42)
scores = cross_val_score(model, X_train, y_train, cv=5)
results.append({
'search_type': 'random',
'n_trees': n,
'max_depth': depth,
'min_samples_split': split,
'accuracy': np.mean(scores)
})
return results, time.time() - start_time
# Perform the searches
grid_results, grid_time = grid_search()
random_results, random_time = random_search()
# Combine results
all_results = pd.DataFrame(grid_results + random_results)
# Visualization
plt.figure(figsize=(12, 6))
plt.subplot(1, 2, 1)
sns.boxplot(data=all_results, x='search_type', y='accuracy')
plt.title('Accuracy Distribution')
plt.xlabel('Search Type')
plt.ylabel('Accuracy')
plt.subplot(1, 2, 2)
plt.bar(['Grid Search', 'Random Search'], [grid_time, random_time])
plt.title('Execution Time')
plt.ylabel('Seconds')
plt.tight_layout()
plt.show()
# Print best results
times = {'grid': grid_time, 'random': random_time}
for search_type in ['grid', 'random']:
subset_results = all_results[all_results['search_type'] == search_type]
best_result = subset_results.loc[subset_results['accuracy'].idxmax()]
print(f"\nBest results from {search_type} search:")
print(f"Parameters: n_trees={best_result['n_trees']}, "
f"max_depth={best_result['max_depth']}, "
f"min_samples_split={best_result['min_samples_split']}")
print(f"Accuracy: {best_result['accuracy']:.3f}")
print(f"Execution time: {times[search_type]:.2f} seconds")
# Final comparison
print("\nPerformance Comparison:")
print(f"Accuracy difference (grid - random): "
f"{all_results[all_results['search_type']=='grid']['accuracy'].max() - all_results[all_re-
sults['search_type']=='random']['accuracy'].max():.3f}")
print(f"Time difference (grid - random): {grid_time - random_time:.2f} seconds")
```

Best results from grid search:
Parameters: n_trees=100, max_depth=nan, min_samples_split=0.01

Accuracy: 0.943
Execution time: 10.40 seconds

Best results from random search:
Parameters: n_trees=100, max_depth=5.0, min_samples_split=0.5

Accuracy: 0.943
Execution time: 11.13 seconds

Performance Comparison:

Accuracy difference (grid - random): 0.000
Time difference (grid - random): -0.73 seconds

Figure 5. Method comparison

	I	II	III	IV	V	VI	VII	VIII
2	96	22	141	41	105	122	11	30
3	32	6	128	63	146	46	134	81
4	69	95	158	13	153	55	110	24
5	40	17	113	85	161	2	159	100
6	148	74	163	45	80	97	36	107
7	104	157	27	167	154	68	118	91
8	152	60	171	53	99	133	21	127
9	119	84	114	50	140	86	169	94
10	98	142	42	156	75	129	62	123
11	3	87	165	61	135	47	147	33
12	54	130	10	103	28	37	106	5

	I	II	III	IV	V	VI	VII	VIII
2	70	121	26	9	112	49	109	14
3	117	39	126	56	174	18	116	83
4	66	139	15	132	73	58	145	79
5	90	176	7	34	67	160	52	170
6	25	143	64	125	76	136	1	93
7	138	71	150	29	101	162	23	151
8	16	155	57	175	43	168	89	172
9	120	88	48	166	51	115	72	111
10	65	77	19	82	137	38	149	8
11	102	4	31	164	144	59	173	78
12	35	20	108	92	12	124	44	131

REFERENCES

Agrawal, P. (2024, March). A Survey on Hyperparameter Optimization of Machine Learning Models. In *2024 2nd International Conference on Disruptive Technologies (ICDT)* (pp. 11-15). IEEE.

Baik, J., Deift, P., & Suidan, T. (2016). *Combinatorics and random matrix theory* (Vol. 172). American Mathematical Soc. DOI: 10.1090/gsm/172

Barrett, J. (2022). *Combinatorial and algorithmic approaches to random trees and graphs* (Doctoral dissertation, McGill University (Canada)).

Bollobás, B. (1981). Degree sequences of random graphs. *Discrete Mathematics, 33*(1), 1–19. DOI: 10.1016/0012-365X(81)90253-3

Burroughs, W. S. (1960). *Minutes to Go: Pieces by William Burroughs in Minutes to Go.* Two Cities Editions.

Calvino, I. (1994). *Il castello dei destini incrociati* [The castle of crossed destinies]. Mondadori.

Cameron, P. J. (1997). The random graph. *The Mathematics of Paul Erdös, II*, 333–351.

. Carlo, Q. M. (2001). *Monte Carlo methods in financial engineering.*

Clarke, A. C. (1967). *The nine billion names of God.* Harcourt.

Deng, Y. (2022). Random permutation set. *International Journal of Computers, Communications & Control, 17*(1). Advance online publication. DOI: 10.15837/ijccc.2022.1.4542

Elkies, N., Kuperberg, G., Larsen, M., & Propp, J. (1992). Alternating-sign matrices and domino tilings (Part I). *Journal of Algebraic Combinatorics, 1*(2), 111–132. DOI: 10.1023/A:1022420103267

Frieze, A., & Karoński, M. (2016). *Introduction to random graphs.* Cambridge University Press.

Izenman, A. J. (2021). Random matrix theory and its applications. *Statistical Science, 36*(3), 421–442. DOI: 10.1214/20-STS799

Mazzilli, D., & Radicchi, F. (2021). Combinatorial approach to spreading processes on networks. *The European Physical Journal B, 94*(1), 1–13. DOI: 10.1140/epjb/s10051-020-00029-z

Naseri, G., & Koffas, M. A. (2020). Application of combinatorial optimization strategies in synthetic biology. *Nature Communications, 11*(1), 2446. DOI: 10.1038/s41467-020-16175-y PMID: 32415065

Novak, J. I. (2009). *Topics in combinatorics and random matrix theory* (Doctoral dissertation, Queen's University).

Queneau, R. (1961). *Cent mille milliards de poèmes* [One hundred thousand billion poems]. Gallimard.

Stéphan, J. M. (2021). Extreme boundary conditions and random tilings. *SciPost Physics Lecture Notes*, 026.

Chapter 11
Combinatorics and Probability

ABSTRACT

The chapter explores the intersection of combinatorics and probability through various engaging problems and real-world applications. It delves into classic mathematical puzzles such as the Monty Hall problem, the birthday paradox, and unique scenarios like Mozart's musical dice game. The text examines probability calculations in diverse contexts, including smartphone unlock patterns, board games like Nine Men's Morris, and gambling games like Win for Life and Keno. Each example illustrates how combinatorial methods can be used to calculate possible combinations and determine probabilities. The chapter analyses Key mathematical concepts through practical examples, such as smartphone unlock codes, Dyck walks, Catalan numbers, and Schröder-Hipparchus numbers. The chapter highlights how seemingly simple counting problems can reveal complex mathematical relationships and provide insights into randomness, pattern recognition, and probability estimation across different domains. The narrative emphasizes that combinatorics and probability are powerful tools for understanding uncertainty.

INTRODUCTION

When you have eliminated all which is impossible, then whatever remains, however improbable, must be the truth
Arthur Conan Doyle

Combinatorics and probability are two fundamental branches of mathematics. Although distinct, they are deeply intertwined in their application and study. Combinatorics examines the counting and arrangement of objects according to specific rules, while probability focuses on measuring uncertainty and predicting events. Together, these fields offer powerful tools for addressing various challenges.

A simple example can serve as an introduction: for each wheel of the lottery, five numbers are drawn from 90. How many different combinations can be drawn? What is the probability of guessing the five? Combinatorial calculus tells us that $C_{90,5}$ = 43,949,268 so the probability is 1/43,949,268, which is extremely low.

Wanting to summarize the possible combinations, these are the results:

Two numbers: 4,005 combinations
Three numbers: 117,480 combinations
Four numbers: 2,555,190 combinations
Five numbers: 43,949,268 combinations

Of course, there are many other versions of this game; the American Powerball lottery is based on the choice of six numbers: five between 1 and 69, one between 1 and 26. Consider another simple problem: We want to calculate the probability of getting exactly 3 "heads" by flipping a coin 5 times. Combinatorics provides us with the solution, the number of ways in which we can get 3 "heads" and 2 "tails" in 5 tosses is given by:

$$C_{5,3} = \frac{5!}{3!(5-3)!} = 10$$

DOI: 10.4018/979-8-3373-3089-1.ch011

Let us calculate the probability of each combination. Since each outcome (heads or tails) has a probability of 1/2, the probability of getting exactly 3 "heads" is:

$$P(3) = C_{5,3}\left(\frac{1}{2}\right)^5 = \frac{5}{16}$$

Let us now examine some more or less well-known situations.

MONTY HALL PROBLEM

The first case considered, which is essentially related to probability, is known as the Monty Hall problem (Meylani, 2023), derived from a television game called "Let's Make a Deal" conducted by Monty Hall, pseudonym of Monte Halperin (1921-2017). Here is how it worked:

1. There are three doors. Behind one is a Machine, while behind the other two are Goats.
2. The contestant chooses a door, and the presenter, who knows what is behind each door, opens one of the other two doors, which has a goat behind it.
3. Monty then asks the contestant if he wants to change his initial choice to the other door.

The question is, should the competitor change his choice?

The Monty Hall problem can also be viewed in combinatorial terms because the contestant must consider all possible combinations of initial choices and intermediate outcomes. For example, there are six ways in which the prizes can be arranged behind the three doors distinguishing the two goats (MG_1G_2, MG_2G_1, G_1MG_2, G_2MG_1, G_1G_2M, G_2G_1M), and after the handler opens a door, some of these combinations are eliminated. All this can be represented as follows by supposing that the contestant chooses the second door and Monty Hall opens the door with a goat:

Table 1. Possible combinations and choices

MG_1G_2	MG_2G_1	G_1**MG2**	G_2**MG1**	G_1G_2**M**	G_2G_1**M**

One wins the car by changing the initial choice in two out of three cases (vos Savant, 1990).

PROBLEM OF BIRTHDAYS

An interesting example of combinatorics and probability is the birthday problem, which concerns the probability that at least two people share the exact date of birth in a randomly chosen group (DasGupta, 2005; Borja & Haigh, 2007). The paradox of the birthday was defined in 1939 by Richard von Mises (1883-1953).

For simplicity, ignore leap years and assume that every day of the year is equally likely as a date of birth. If we have only one person, there are 365 possible birthdays, so the probability that no one else has the same birthday is 1. If we add a second person, there are now 364 days remaining, so the probability that this second person will not share the first person's date of birth is 364/365. If we add a third person, the probability that they will not share the date of birth of the first two people is 363/365, and so on.

Thus, the probability p that *no one* in the group of n people shares a date of birth is:

$$p = \frac{365}{365} \cdot \frac{364}{365} \cdot \frac{363}{365} \cdot \ldots \cdot \frac{365 - n + 1}{365}$$

The probability *p'* that *at least* two people share the exact date of birth is, therefore, p' = 1 - p.

For a group of only 23 people, the probability is 0.51.

Combinatorics is the study of how objects can be combined in specific ways. In the birthday problem, we use combinatorics to count how dates of birth can be distributed among a group of people. If we have *n* people, there are 365^n ways in which their birth dates can fall on any day of the year. Each person has 365 possible choices for his or her date of birth; therefore, the total number of combinations is 365 • 365 • ... • 365 (n times), which is 365^n. However, not all of these cases are distinct. Some involve at least two people sharing the exact date of birth. So, to calculate the probability that at least two people will share the exact date of birth, we must first calculate the ways in which all dates of birth can be distinguished. This is given by 365 • 364 • ... • (365 – n + 1), the number of n-day permutations chosen from 365. So, the probability that at least two people share the exact date of birth is 1 - (365 • 364 • ... • (365 – n + 1)) / 365^n, precisely the previous formula.

The paradox has several practical applications in various fields. For example, it is used in cryptanalysis and cybersecurity problems. When choosing the size of a block to encrypt, it is important to consider the probability that two blocks have the same hash value (a hash function completely breaks down the input data and converts it into a sequence of characters of fixed length).

This program calculates the estimated probability that in a group of *n* people, two will share the same birthday:

Program 115 Birthday

```
import numpy as np
def birthday_simulation(n, days=365, simulations=10000):
multiple_people_count = 0
for _ in range(simulations):
# Generate random birthdays for the group
birthdays = np.random.randint(0, days, n)
# Check if there's at least one pair with the same birthday
if len(birthdays) != len(set(birthdays)):
multiple_people_count += 1
# Calculate the probability
probability = multiple_people_count / simulations
return probability
n = int(input("Number of people in the group: "))
simulations = 10000 # Number of simulation runs
probability = birthday_simulation(n, simulations=simulations)
print(f"Estimated probability that at least one pair in a group of {n} people share a
birthday: {probability:.6f}")
```

MAGIC SQUARE

As seen above, when talking about magic squares, in a 3 x 3 matrix, the natural numbers from 1 to 9 are randomly inserted. We can ask ourselves two questions: How many “combinations” are possible? What is the probability that the numbers entered will form a magic square?

We can use the concept of permutations to determine the total of possible combinations to insert the integer from 1 to 9 into a 3 x 3 matrix. This matrix has 9 cells, and we have to fill them with the values 1, 2, ..., 9 without repetitions. The total quantity of ways we can arrange these numbers is 9! = 362,880. It is known that there are precisely eight distinct magic squares, so the probability will be P = 8 / 362880 = 1 / 45360 = 0.000022.

Program 116 Magic Square 3x3

```
import itertools
import numpy as np
def is_magic_square(matrix):
# Check rows, columns, and diagonals
row_sum = sum(matrix[0])
for i in range(3):
if sum(matrix[i]) != row_sum:
return False
if sum(matrix[:, i]) != row_sum:
return False
if sum(np.diag(matrix)) != row_sum:
return False
if sum(np.diag(np.fliplr(matrix))) != row_sum:
return False
return True
def magic_square_probability():
numbers = list(range(1, 10))
total_combinations = 0
magic_squares = 0
# Generate all permutations of numbers from 1 to 9
for permutation in itertools.permutations(numbers):
matrix = np.array(permutation).reshape(3, 3)
total_combinations += 1
if is_magic_square(matrix):
magic_squares += 1
probability = magic_squares / total_combinations
return total_combinations, probability
if __name__ == "__main__":
total_combinations, probability = magic_square_probability()
print(f"Total number of possible combinations: {total_combinations}")
print(f"Probability of forming a magic square: {probability:.10f}")
```

The results can be printed using the following program:

Program 117 Magic Square Solutions

```
import tkinter as tk
from itertools import permutations
def is_magic_square(square):
a, b, c, d, e, f, g, h, i = square
return (a + b + c == 15 and d + e + f == 15 and g + h + i == 15 and
a + d + g == 15 and b + e + h == 15 and c + f + i == 15 and
a + e + i == 15 and c + e + g == 15)
# Generate all possible permutations of numbers from 1 to 9
all_permutations = permutations(range(1, 10))
```

```
# Find all magic squares
magic_squares = []
for perm in all_permutations:
if is_magic_square(perm):
magic_squares.append(perm)
# Create the main window
root = tk.Tk()
root.title("3x3 Magic Squares")
# Function to display a magic square
def display_square(square, frame):
for i in range(3):
for j in range(3):
value = square[i * 3 + j]
label = tk.Label(frame, text=str(value), borderwidth=1, relief="solid", width=5, height=2)
label.grid(row=i, column=j, padx=2, pady=2)
# Create a frame for each magic square and display them
for index, square in enumerate(magic_squares):
sub_frame = tk.Frame(root, borderwidth=2, relief="solid", padx=5, pady=5)
sub_frame.grid(row=index//2, column=index%2, padx=10, pady=10)
display_square(square, sub_frame)
# Start the GUI window
root.mainloop()
```

THE GAME OF MUSICAL DICE

Mozart (1756-1791) was responsible for constructing the game of musical dice (Hedges, 1978; Ruttkay, 1997). The great musician demonstrated that it was possible to construct a minuet (which is composed of 16 bars) in a completely random way, using a combinatorial scheme. To play, a table with 11 rows and 16 columns is used that shows the number of the bar to be used at the intersection.

Figure 1. Mozart table

One has to roll two dice to generate a random number between 2 and 12. This result corresponds to a particular bar that constitutes the beginning of the minuet. If one rolls six, the piece numbered 148 is chosen. By repeating the procedure sixteen times, the remaining pieces are gradually queued. Mozart's ability, in this case, was to write 176 bars that could, in any case, logically be concatenated while respecting, in each column, the harmonic base. With this system, 45,949,729,863,572,161 minuets can be generated, data from 11^{16}.

The problem is that the results of rolling two dice are different; the seven is more likely to come out than the twelve. We can overcome the obstacle by using a twelve-sided die, also known as a dodecahedron.

SMARTPHONE PASSWORDS

One relevant problem for smartphone users is protection from malicious individuals; therefore, Android users are asked to choose a PIN or password (Ramadhani, 2021) or select and remember a graphic pattern on a 3 x 3 grid. The template must consist of at least four points.

Since Android devices block access after 20 incorrect attempts, an attacker must guess the unlock code within this timeframe. What are the most commonly used unlocking patterns by users? The theoretical pattern space calculated by the researchers is 389,112, but the actual pattern space chosen by the user is significantly smaller (Cho et al., 2017).

The interesting aspect is that the starting points were strongly oriented towards the highest point on the left, which has a probability of use of 63.5%, and that the most used segment was the 2—3, while some diagonal lines were never used. In the Figure, there are the 20 most used codes based on a study on the subject:

Figure 2. Most used codes

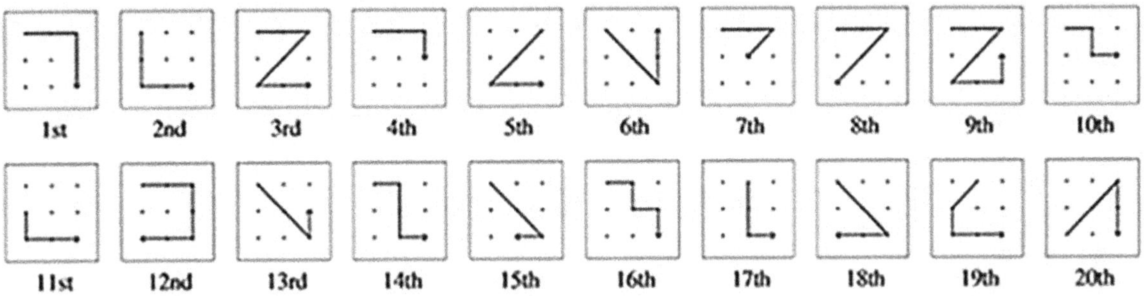

What is interesting is to examine if the points are touched with different probabilities; the Table shows the probability of use for each of the nine points (Drivet, 2020):

Table 2. Probability of using the 9 points

0.117	**0.114**	**0.113**
0.104	0.120	0.105
0.109	0.108	0.110

A simplified version of the problem that is limited to determining paths with lines connecting three points can be as follows:

Program 118 Smartphone

```
import matplotlib.pyplot as plt
import matplotlib.animation as animation
import numpy as np
def get_paths(grid_size=3):
directions = [(0, 1), (1, 0), (0, -1), (-1, 0), (1, 1), (1, -1), (-1, 1), (-1, -1)]
paths = []
seen_paths = set()
def is_valid(x, y):
return 0 <= x < grid_size and 0 <= y < grid_size
def add_path(path):
sorted_path = tuple(sorted(path))
if sorted_path not in seen_paths:
paths.append(path)
seen_paths.add(sorted_path)
for x in range(grid_size):
```

```
for y in range(grid_size):
for dx1, dy1 in directions:
nx1, ny1 = x + dx1, y + dy1
if not is_valid(nx1, ny1):
continue
for dx2, dy2 in directions:
nx2, ny2 = nx1 + dx2, ny1 + dy2
if not is_valid(nx2, ny2):
continue
start = (y + 0.5, x + 0.5)
middle = (ny1 + 0.5, nx1 + 0.5)
end = (ny2 + 0.5, nx2 + 0.5)
if start != middle and middle != end and start != end:
path = (start, middle, end)
add_path(path)
return paths
def draw_grid(ax):
for x in range(4):
ax.plot([x, x], [0, 3], color='black', linewidth=0.5)
for y in range(4):
ax.plot([0, 3], [y, y], color='black', linewidth=0.5)
def draw_points(ax, grid_size=3):
for x in range(grid_size):
for y in range(grid_size):
ax.plot(y + 0.5, x + 0.5, 'ko', markersize=10)
def update(frame, paths, ax):
ax.clear()
draw_grid(ax)
draw_points(ax)
if paths:
path = paths[frame]
ax.plot([path[0][0], path[1][0], path[2][0]], [path[0][1], path[1][1], path[2][1]], col-
or='blue', linewidth=2)
ax.text(0.5, 3.1, f'Path: {frame + 1}/{len(paths)}', fontsize=12, ha='center', va='bot-
tom', color='red')
ax.set_xlim(0, 3)
ax.set_ylim(0, 3)
ax.set_xticks(np.arange(0, 4, 1))
ax.set_yticks(np.arange(0, 4, 1))
ax.set_xticklabels([])
ax.set_yticklabels([])
ax.invert_yaxis()
def main():
paths = get_paths()
fig, ax = plt.subplots()
ani = animation.FuncAnimation(fig, update, frames=len(paths), fargs=(paths, ax), inter-
val=500, repeat=False)
plt.show()
if __name__ == "__main__":
```

```
main()
```

NINE MEN'S MORRIS

An interesting game is the Nine Men's Morris game (Loewer, 2020). It is one of the oldest games known. The diagram of its board has been engraved on stone since Egyptian times. Shakespeare, in the play A Midsummer Night's Dream, writes in the second act "The nine men's morris is filled up with mud". The game consists of a board with 24 intersections, and the two players have nine pieces each, which are placed in turn on one of the intersections.

Figure 3. The board

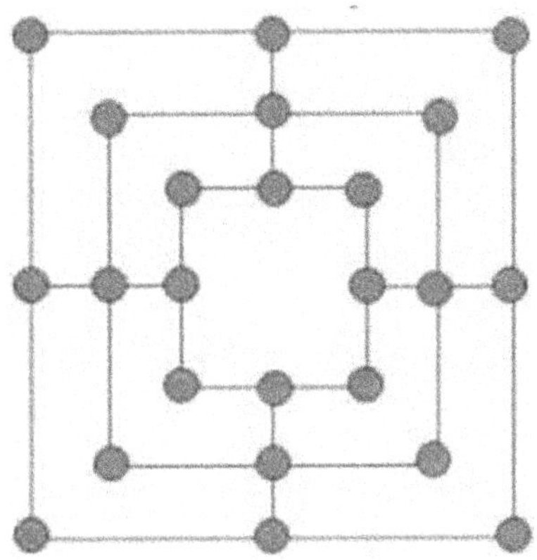

In turn, each player moves a checker to an adjacent free spot. When the player creates an uninterrupted row of three checkers of the same colour, he forms a so-called "mill" and can eliminate a checker of his opponent's choice from the game. When a player has only three checkers, he can jump them; that is, he can move the checker to any free spot each turn. If a player cannot move any checkers or has only two checkers, he loses the game.

Since there are 24 positions on the board, the possible configurations are:

$$n = \sum_{w=0}^{9}\sum_{b=0}^{9}\frac{24!}{w!b!(24-w-b)!} = 143,122,694,691$$

In reality, there are very few states based on the game's rules.

DICK'S WALK

The Dyck Walk is an important concept in the theory of walks in combinatorial mathematics (Deutsch, 1999). It is a sequence of steps on a Cartesian plane in which one moves up (u) or down (d) one step without ever descending below the x-axis. These walks are called "Dyck's" in honour of the English mathematician Walther Dyck, who studied them in 1882.

The rules of the Dyck Walk can be counted using combinatorial methods and are connected to Catalan numbers, a sequence of natural values appearing in combinatorics' various counting problems. This formula defines them:

$$C_n = \frac{1}{n+1}\binom{2n}{n} = \frac{(2n)!}{(n+1)|n|}$$

Recursive program to generate some Catalan numbers:

Program 119 Catalan

```
def catalan(n):
if n <= 1:
return 1
res = 0
for i in range(n):
res += catalan(i) * catalan(n - i - 1)
return res
for i in range(10):
print(catalan(i), end=" ")
```

Catalan numbers count the Dick's paths of length *2n*. In other words, C_n represents the number of Dick's paths of length *2n* that start from the origin and end at the x = 2n abscissa without ever going below the x-axis. For example, for n = 3, there are precisely 5 Dick paths of length 6. A possible path can be the following:

Figure 4. Dick's path

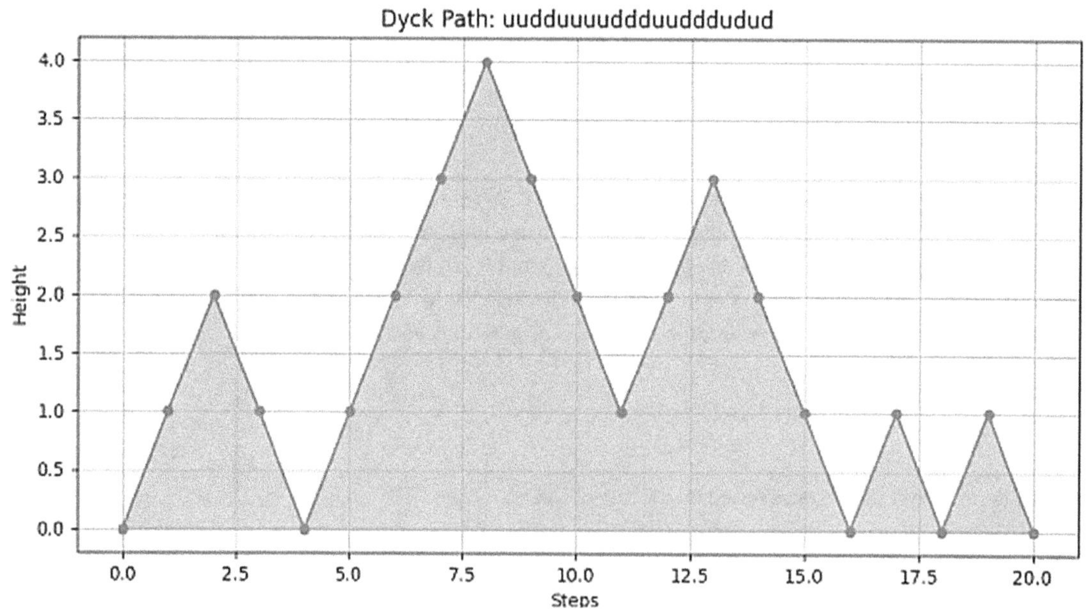

Generated by the following program:

Program 120 Dick's Path

```
import matplotlib.pyplot as plt
import numpy as np
def generate_dyck_path(n):
path = ['u'] * n + ['d'] * n
while True:
np.random.shuffle(path)
if is_dyck_path(path):
return path
def is_dyck_path(path):
height = 0
for step in path:
if step == 'u':
height += 1
else:
height -= 1
if height < 0:
return False
return height == 0
def plot_dyck_path(path):
x = []
y = []
current_x, current_y = 0, 0
# Add the starting point
x.append(current_x)
y.append(current_y)
for step in path:
if step == 'u':
current_y += 1
else:
current_y -= 1
current_x += 1
x.append(current_x)
y.append(current_y)
plt.figure(figsize=(10, 5))
plt.plot(x, y, 'grey', marker='o', markersize=5) # Plot the path
plt.fill_between(x, y, color='grey', alpha=0.5) # Fill the area under the path
plt.title(f'Dyck Path: {"".join(path)}')
plt.xlabel('Steps')
plt.ylabel('Height')
plt.grid(True)
plt.show()
n = 10 # Number of steps up and down
path = generate_dyck_path(n)
plot_dyck_path(path)
```

A Dyck walk is a sequence of steps that can be interpreted in various mathematical contexts, particularly in combinatorics and probability theory. One fundamental interpretation involves the matching of parentheses: each "up" step corresponds to an opening parenthesis, and each "down" step to a closing parenthesis. The condition for a valid Dyck walk mirrors the condition for well-formed parentheses: each opening must be matched by a closing, and no closing parenthesis may occur before its corresponding opening.

Another application can be the division of a convex polygon into triangles. It is possible to apply probability concepts to the walk; for example, one can think of the Dyck walk as a random process. Suppose we have a sequence of *2n* random steps, where each step can be "up" or "down" with equal probability (1/2). We aim to determine the probability that this sequence of steps constitutes a Dyck walk. Not all random sequences of *n* "up" and *n* "down" steps are Dyck walks because they can lie below the x-axis. Knowing the combinations given by Catalan's formula and dividing them by the possible combinations $\binom{2n}{n}$, we obtain $P(walk) = \frac{1}{n+1}$.

This result elegantly connects combinatorics with probability, highlighting the structure inherent in random walks constrained by boundary conditions.

A concrete example can clarify the concept: What is the probability that a random walk of length 6 (with three steps up and three steps down) is a Dyck Walk?

1. The total number of sequences is $\binom{6}{3} = 20$.
2. The number of Dyck walks of length six is given by 5.

So, the probability is:

P(walk) = 1/4.

SCHRÖDER HIPPARCHUS NUMBERS

An interesting variant applicable to paths, but also to the subdivisions of a polygon or the characteristics of a tree, refers to a particular type of number. Let us imagine a square grid of size n × n. The number of Schröder Hipparchus (Qi & Guo, 2017) indicates the total number of possible paths, assuming we can only move vertically, horizontally, or diagonally to the right. In this case, the sequence is: 1, 1, 3, 11, 45, 197, 903, 4,279, 20,793, 103,049, ...

The n_{th} term of the sequence represents how many ways we can divide a polygon of n+1 sides into smaller polygons using the diagonals of the polygon itself.

Figure 5. Subdivision of a polygon

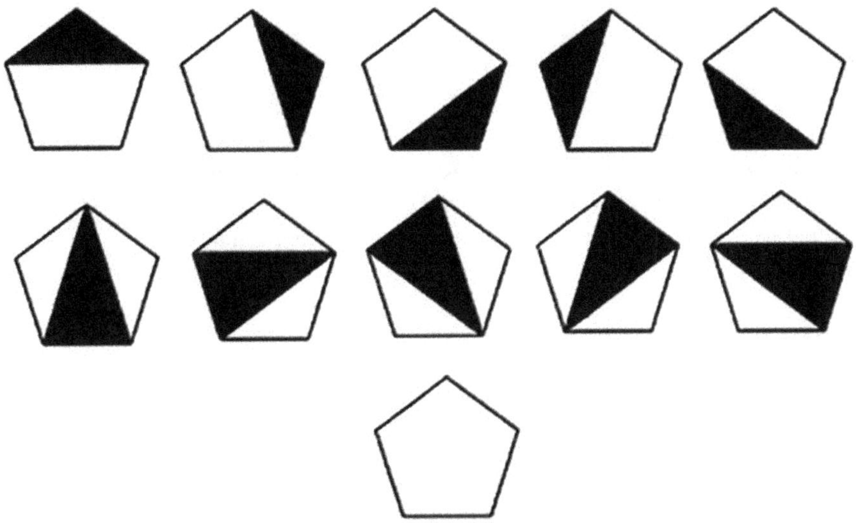

There is a recursive formula that can generate a sequence, like the famous Fibonacci one:

$$S_n = \frac{1}{n}((6n-9)\cdot S_{n-1} - (n-3)\cdot S_{n-2}$$

Program 121 Schroder_Hipparchus

```
def Schröder_Hipparchus(n):
if n == 1 or n == 2:
return 1
return ((6 * n - 9) * Schröder_Hipparchus(n - 1) - (n - 3) * Schröder_Hipparchus(n -
2))//n
for i in range(1, 11):
print(f"Il {i}-esimo numero di Schröder-Hipparchus è:", Schröder_Hipparchus(i))
```

SNOW WHITE AND THE SEVEN DWARFS

Suppose this event occurs three times in a row. Is it a fortuitous event, or was there a choice?

Since it is possible to assume that Snow White always stands at the head of the table, the first dwarf has seven free chairs available, the second 6, etc., while the last can only occupy the last remaining chair. In practice, the positions are 7 • 6 •... • 1. It is the factorial; therefore, the seven diners can arrange themselves around the table in 7! = 5040 different ways. Let us now examine the position of Bashful: he can sit to the left or right of Snow White, and the other dwarfs can occupy the six positions left free. This leads to 2 • 6! = 1440 different ways favourable to the phenomenon's occurrence.

So far, we have determined the number of positions; the next step will be to calculate the probability of the event, which is given by the cases favourable to proximity divided by the total number of cases; therefore, it can be said that p = 1440 / 5040 = 0.286. It is a non-negligible probability, but the probability that the phenomenon will repeat itself three times is given by p • p • p = 0.286 • 3 = 0.023. The probability that it is a coincidence is just over 2%, which is low enough to suggest it was merely a fortuitous case.

WIN FOR LIFE

Win for life is one of the many betting games available in Italy, in other countries there are similar games (Set for Life in the UK, Cash4Life in the USA, etc.). Playing is effortless: choose at least 10 numbers out of 20 plus the "Numerone" in the purchased card. By guessing 10 or 0 numbers plus the "Numerone", you win the special prize, up to €3,000 per month for 20 years. Otherwise, one can win smaller amounts. One does not win by rolling 4, 5 or 6.

Figure 6. Win for life

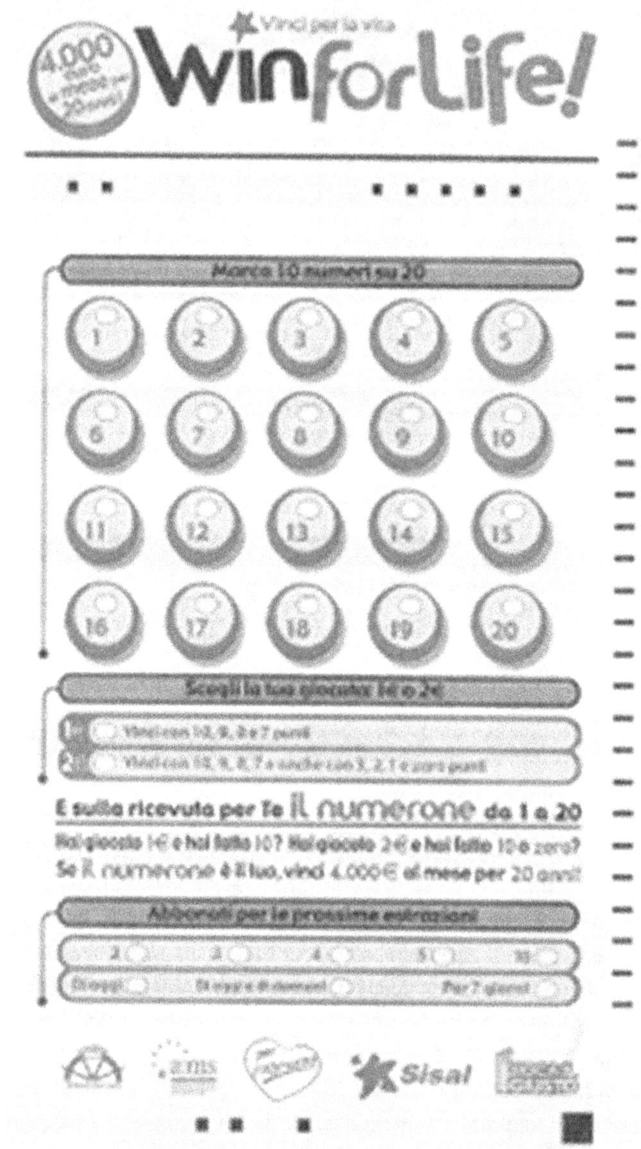

Like all betting games, there is a link with the calculation of probabilities and, above all, with the calculation of combinations necessary to estimate probability. The law of probability that governs the outcome of a play ("Numerone" excluded) is the hypergeometric distribution. The hypergeometric distribution is a discrete probability distribution that describes the extraction of a number, card, ball, etc., without replacement.

$$P(k) = \frac{\binom{h}{k}\binom{n-h}{r-k}}{\binom{n}{r}}$$

where $n = 20$ are the available numbers, $h = 10$ are chosen by the player, $r = 10$ are drawn, and P(k) is the probability of guessing k exactly. These are fairly low odds. The probability of winning the largest prize is 1/1,847,560. Since we are talking about gambling, this is equivalent to having 1,847,560 tickets (each 22 cm long) on the road that leads from Flagstaff to Las Vegas (405 km) and choosing the winning ticket by taking one at random.

KENO

Keno is a game of chance, similar to a lottery, that is popular in many casinos and online platforms (Bollman, 2018). Players select a set of values from a larger pool, hoping their choices match the drawn results. In this game, participants select a specific quantity from a total of 80. For example, a player might choose ten options, after which 20 are randomly drawn from the set of 80.

The combination formula determines the possible ways a player can select k elements from a total of n. For instance, if a player picks 10, the possible selections are:

$$C_{80,10} = \frac{80!}{10!(80-10)!} = 1,646,492,110,120$$

A similar calculation can be made for the 20 numbers drawn:

$$C_{80,20} = \frac{80!}{20!(80-20)!} = 3,535,316,142,212,174,320$$

The winnings depend on how many of the player's chosen values match the ones drawn. Suppose a player chooses 10 (k = 10) and wants to calculate the probability of guessing exactly three correct numbers (m = 3) with 20 numbers drawn (t = 20). $C_{10,3}$ are the way to choose three correct values from the 10 chosen, and $C_{70,17}$ is the way to choose the remaining 17. The choice of the numbers drawn is the one already seen, $C_{80,20}$. The probability of guessing exactly m numbers is given by the following formula:

$$P(m) = \frac{C_{k,m} \cdot C_{80-k,20-m}}{C_{80,20}} = \frac{C_{10,3} \cdot C_{70,17}}{C_{80,20}} = 0.27$$

The figure shows the 22 prime numbers present on the board. What is the probability of winning by betting on the first 20 (therefore excluding 73 and 79)?

There is an extremely low probability, $2.83 \cdot 10^{-19}$. This means it is improbable that one will guess all 20 prime numbers exactly and, in general, any choice of 20 values.

Figure 7. Prime numbers play in Keno

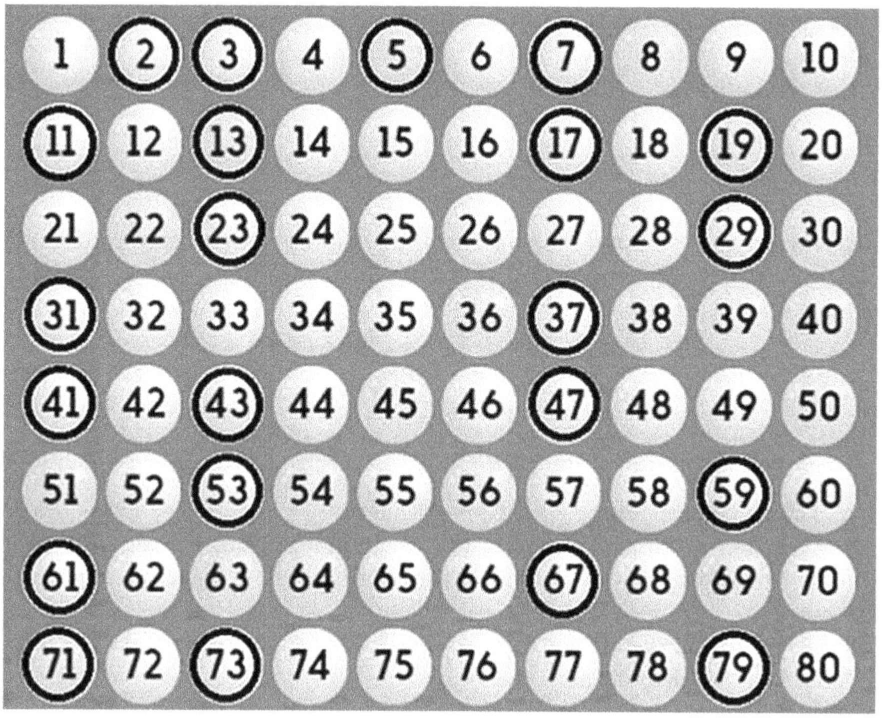

CLUEDO

In Cluedo, a popular deduction game, combinatorics plays a crucial role in determining probabilities and solving the mystery. Each player's move progressively reduces the number of possibilities by excluding combinations. Analysing the game from a combinatorial perspective allows us to understand better how probabilities evolve and how we can make more informed decisions during the game.

The goal is to guess who the murderer is, what weapon he used, and in which room the crime took place. So, one has to answer three questions: Who did it? With which weapon? Where?

The board represents the floor plan of the house, divided into rooms and corridors. Each room has a door that allows access through the corridors. The house has nine rooms and various spaces that connect them. Players roll a die to determine how many spaces they can move. Each player can move in a straight line along the corridors but cannot pass through walls or block the passage of other players.

Three cards are removed and left face down: one for each category (a suspect, a weapon, a room). It is the solution to the crime. The other cards are shuffled and secretly dealt to each player.

If a player arrives at the door of a room during their turn, they can decide to enter. When a player enters a room, they can guess the murderer. It is a crucial step in the game because it allows them to gather information.

Each game involves a combination of three elements that together determine the assassin. The initial configuration of the game can be seen as a combination problem, where all possible murder configurations must be calculated. In Cluedo, in the original version, there are:

6 suspects (Miss Scarlett, Prof. Plum, Mrs. Peacock, Col. Mustard, Mrs. White, Rev. Green)

6 weapons (Rope, Dagger, Wrench, Revolver, Candlestick, Lead Pipe)
9 rooms (Kitchen, Dining Room, Lounge, Hall, Study, Library, Billiard Room, Conservatory, Ballroom)

The combinatorial formula to calculate the total number of possible murder combinations in Cluedo is the product of the possible choices for each category:

Total combinations = 6(suspects)×6(weapons)×9(rooms) = 324

This number represents the total of possibilities that players must reduce as the game progresses, based on information gathered during guesses and card exchanges.

During the game, whenever a player makes a guess (for example, "I think Colonel Mustard committed the knife murder in the library"), the other players must secretly show the guesser cards that contradict the guess (for example, a Colonel Mustard card, a Revolver card, or a Library card).

This stage of the game can be analysed in terms of exclusionary combinations. For example, if a player discovers that Colonel Mustard is not the murderer, all combinations involving him are eliminated, drastically reducing the possibilities remaining. At this point, the chances of finding the real culprit are concentrated on the other combinations of suspects, weapons, and rooms. In combinatorial terms, we can say that each new hypothesis decreases the number of "possible combinations" left to explore.

Combinatorics in Cluedo is not limited to calculating initial combinations; it also extends to handling conditional probability. Conditional probability is the probability that an event will occur, given that another event has already occurred. In the context of Cluedo, this probability changes based on the information a player acquires during the game. Suppose a player has ruled out a specific weapon. The probability of each of the remaining combinations changes, and the player can update his guesses based on this new information.

To calculate the probability of a hypothesis based on the available information, one can use the conditional probability formula:

$$P(A|B) = \frac{P(A \cap B)}{P(B)}$$

Where:

P(A|B) is the probability that event A (the murder in a specific room) occurs, given that event B (the exclusion of a particular weapon) is already known.

P(A∩B) is the joint probability that A and B occur.

P(B) is the probability that B occurs.

If a player excludes a weapon (e.g., the Dagger), the solution can only be among the remaining 5 weapons. Thus, the number of still possible combinations becomes 6 • 5 • 9 = 270 and the probability P(B) = 270/324 = 0.83. Now, we want to calculate the probability that a specific weapon (e.g., the Rope) is the murder weapon, given that the Dagger has been excluded. Since there are still 5 possible weapons: P(A∩B) = 1/5 = 0.2. At this point P(A | B) = 0.2/0.83 = 0.24.

This approach enables players to adapt their strategies dynamically, updating the probabilities of remaining combinations as new information becomes available.

Regarding strategy, Cluedo players must make smart choices about which guesses to make to get the most information. In a game, it is possible to make guesses about suspects, weapons, and rooms, allowing the possibilities to be quickly narrowed down. For example, suppose a player has two cards in a category. In that case, they know the solution cannot involve those two suspects, which immediately eliminates many potential combinations. As mentioned above, the total of possible combinations at the start is 324.

Now, let us say a player has two suspect cards. The secret solution can only include the remaining four suspects. So, the new number of possible combinations becomes 4 • 6 • 9 = 216, excluding 108 combinations. If the player has two weapons, he could narrow down the possibilities: 4 • 4 • 9 = 144. The more cards one has of the same category, the faster the possible combinations are reduced, making the deduction much more efficient!

Ultimately, combinatorics is at the heart of Cluedo's strategy: progressively reducing the number of possible combinations through exclusion and probability management is essential to arrive at the solution. Understanding combinatorics helps players think in terms of probability and make more informed decisions, turning the game into a mathematical challenge and a simple game of deduction.

REFERENCES

Bollman, M. (2018). *Mathematics of keno and lotteries*. Chapman and Hall/CRC. DOI: 10.1201/9781315192840

Borja, M. C., & Haigh, J. (2007). The birthday problem. *Significance*, *4*(3), 124–127. DOI: 10.1111/j.1740-9713.2007.00246.x

Cho, G., Huh, J. H., Cho, J., Oh, S., Song, Y., & Kim, H. (2017, May). Syspal: System-guided pattern locks for android. In *2017 IEEE Symposium on security and privacy (SP)* (pp. 338-356). IEEE. DOI: 10.1109/SP.2017.61

DasGupta, A. (2005). The matching, birthday and the strong birthday problem: A contemporary review. *Journal of Statistical Planning and Inference*, *130*(1-2), 377–389. DOI: 10.1016/j.jspi.2003.11.015

Deutsch, E. (1999). Dyck path enumeration. *Discrete Mathematics*, *204*(1-3), 167–202. DOI: 10.1016/S0012-365X(98)00371-9

Drivet, A. (Ed.). (2020). *Examining an Operational Approach to Teaching Probability*. IGI Global.

Hedges, S. A. (1978). Dice music in the eighteenth century. *Music & Letters*, *59*(2), 180–187. DOI: 10.1093/ml/59.2.180

Loewer, W. (2020). The Effects of Rule Variations on Perfect Play Databases for Nine Men's Morris.

Meylani, R. (2023, October). THREE DOORS, ENDLESS POSSIBILITIES: UNPACKING THE MONTY HALL PROBLEM ACROSS DISCIPLINES. In *The Sixteenth International Congress of Educational Research* (p. 91).

Qi, F., & Guo, B. N. (2017). Explicit and recursive formulas, integral representations, and properties of the large Schröder numbers. *Kragujevac J. Math*, *41*(1), 121–141. DOI: 10.5937/KgJMath1701121F

Ramadhani, F., Ahmad, E., Fahroza, M. F., Syahfitri, A., & Fadhlurohhman, D. (2021). Implementation Of Combinatorial in Determining Password Technology. *Journal of Computer Science. Information Technology and Telecommunication Engineering*, *2*(1), 150–154.

Ruttkay, Z. (1997). Composing Mozart variations with dice. *Teaching Statistics*, *19*(1), 18–19. DOI: 10.1111/j.1467-9639.1997.tb00313.x

vos Savant, M. (1990, September 2). Ask Marilyn Column. *Parade Magazine*, p. 14.

Chapter 12
Combinatorics and Artificial Intelligence

ABSTRACT

The chapter explores the relationship between combinatorics and artificial intelligence (AI), shedding light on their historical roots and points of convergence. The historical roots are traced from Ramon Llull to deep learning, showing points of convergence and applications, such as optimization, graph theory, and natural language processing (NLP). The text also presents combinatoric problems, such as the travelling salesman problem and the knapsack problem, demonstrating how combinatorial methods are used to find efficient solutions. Furthermore, the text highlights the role of combinatorics in machine learning, optimization, and scientific discovery, underlining its importance in developing advanced AI algorithms. Finally, the text presents several promising avenues for future research, including advanced search algorithms, graph neural networks, generative models, computer defense, game theory, and the Internet of Things, underlining the potential for further integration between combinatorics and AI.

INTRODUCTION

Artificial intelligence will be the last invention made by humanity

Nick Bostrom

The relationship between combinatorics and artificial intelligence (AI) is a fascinating and increasingly important area of research. Although these fields come from distinct domains, they have become closely intertwined over the years. Combinatorics provided the tools to represent and analyse search spaces, while AI developed efficient algorithms to explore such spaces.

The historical roots and points of convergence between combinatorics and artificial intelligence offer a compelling perspective on the evolution of both disciplines (Russell & Norvig, 2016); Briganti & Le Moine, 2020; Gawlikowski et al., 2023).

In this context, we could define Ramon Llull (or Ramon Llull) as a precursor of the intertwining of combinatorics and artificial intelligence. Llull, a 13th-century philosopher, theologian, and logician, developed his "Ars Magna," a combinatorial system for generating knowledge through the manipulation of symbols and concepts. He also devised a machine to combine ideas and concepts systematically and designed a universalistic alphabet where basic elements could be mechanically combined, anticipating concepts found in modern neural networks.

DOI: 10.4018/979-8-3373-3089-1.ch012

Figure 1. Llull's machine

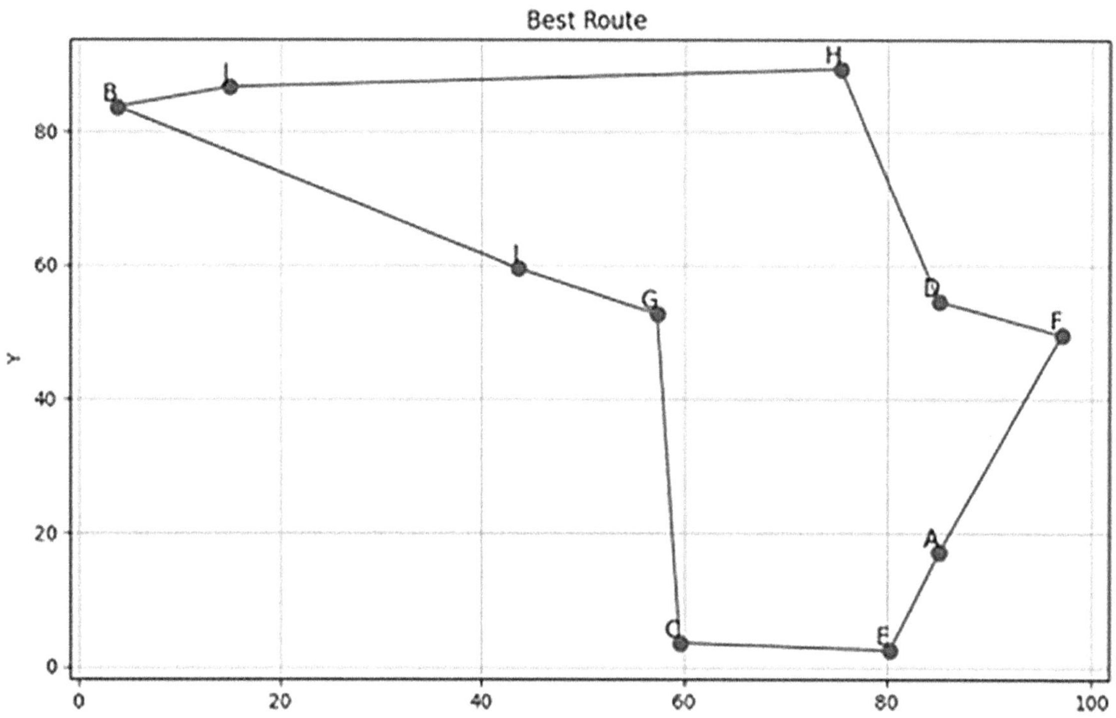

In this sense, we could consider Llull, if not a direct "progenitor", at least a forerunner of some modes of thought that are reflected today in computational sciences and AI. It is a historical example of combinatorial thinking that, even in a quite different philosophical and religious context, anticipates some ideas of symbolic manipulation and knowledge generation that we find in AI.

The real meeting between these fields can be traced back to the dawn of AI. In the 1950s, games such as chess were used as a test bed for the development of search and reasoning algorithms (Shannon, 1950).

In the 1960s, figures such as Allen Newell and Herbert Simon developed programs, including the General Problem Solver (GPS), which employed combinatorial search techniques to tackle logical challenges. Developed in 1957, GPS was one of the first artificial intelligence systems designed to address a wide range of tasks. It introduced concepts such as problem decomposition and heuristic search.

The first concept is based on the idea that solving simpler subproblems can lead to a more comprehensive overall solution. Heuristics, on the other hand, are strategies that, while not always guaranteeing an optimal outcome, often yield acceptable results within a reasonable timeframe.

In the 1970s and 1980s, the emergence of logic programming (e.g., with the Prolog language) provided a new framework in which combinatorics intersected with AI through automatic deduction and the discovery of solutions in combinatorial spaces (Clocksin & Mellish, 1984). Based on rules and representations of knowledge, expert systems have represented an important line of research in AI in which combinatorics has played a fundamental role in the representation and manipulation of knowledge. Unfortunately, expert systems have shown limitations, including the difficulty of accurately representing knowledge and the lack of flexibility in addressing new and unforeseen problems.

In the 1990s, the introduction of genetic algorithms (Wirsansky, 2020) and other techniques revolutionized the approach to solving combinatorial optimization problems. These methods, inspired by natural evolutionary processes, have proved remarkably effective.

The rise of deep learning and neural networks in the 2000s, fuelled by increased computing power and vast data sets, further transformed artificial intelligence (Wu et al., 2020; Li et al., 2021; Dong et al., 2021) . Although the origins can be traced back several decades earlier with the first experiments of neural networks, it is only in recent decades that the field has experienced a real explosion, thanks mainly to technological factors.

Intensive use of a programming language can make a significant contribution to understanding the subject. In this case, Python was chosen because it is known for its clear, simple, and readable syntax. This makes the code easily understandable, even for those without specific training in computer science. Python can also have numerous libraries that simplify the construction of AI models and the application of combinatorial techniques. It should be noted that the contribution of the LLMs has been fundamental in implementing the programs presented in the text.

Therefore, combinatorics is a fundamental branch of mathematics with a crucial role in the field of AI. It offers several powerful tools for solving complex problems. For example, optimization, knowledge representation, and planning models often benefit from combinatorial techniques to find efficient solutions.

One of the most interesting applications of combinatorics in AI is its use in graph theory (Bondy & Murty, 1976). Graph theory is famously used to represent relationships between objects, where nodes represent objects and edges represent relationships. This approach is particularly useful in applications such as neural networks and pathfinders.

Another field where combinatorics tools are helpful is that of enumeration techniques, which are fundamental for solving optimization problems.

The relationship between combinatorics and AI is particularly evident when managing vast spaces of possibilities, a feature common to many AI applications, such as finding optimal outcomes, planning actions, and recognizing patterns. Techniques like *branch and bound* are used to systematically explore the search space, while enumeration algorithms are applied to resolve constraint satisfaction problems. Approaches such as branch and bound are used to systematically explore the solution space, while enumeration algorithms are used to find solutions for constraint satisfaction problems. The concept is central to AI. Often, an AI challenge can be described as a set of possible states or configurations to identify one that meets specific criteria, such as an optimisation target or a satisfiability condition.

Another important area where combinatorics intersects with artificial intelligence is natural language processing (NLP). This interdisciplinary discipline combines computer science, AI, and linguistics to facilitate interaction between humans and machines through natural language. Recent advances in deep learning have significantly enhanced the ability of machines to comprehend and produce natural language. This discipline has seen significant advances thanks to the integration of deep learning and machine learning techniques (Goodfellow et al., 2016; Carleo et al., 2019; Kaur et al., 2020; Bengio et al. 2021; Karimi-Mamaghan et al., 2022), which have significantly improved the ability of machines to understand, generate, and interact in human language, allowing for much more effective human-machine communication. Combinatorial methods, combined with machine learning, have improved NLP models, from feature selection to optimizing algorithm performance.

In this regard, many other topics could be mentioned, such as game theory or computer vision. Machine learning is a rapidly growing field that focuses on developing algorithms to learn from data. Combinatorial methods can reduce the search space size when learning complex models or automatically identify the most informative features in a dataset. In machine learning, combinatorics identifies the most relevant variables to use in predictive models and explores different hyperparameter configurations to improve model performance.

Reinforcement learning plays a crucial role in exploring combinatorial spaces (Mazyavkina et al., 2021; Yang & Whinston, 2023). Think of problems such as global optimization or improving research by applying strategies that optimize the selection of actions to be taken to explore ever-better solutions, as in the case of sequences of actions. Some examples include planning and scheduling or solving combinatorial games.

A classic example in artificial intelligence is the travelling salesman problem (TSP), in which the goal is to find the shortest route that visits a set of cities and returns to the origin. Tackling the TSP requires navigating a vast combinatorial space, and algorithms such as brute force, genetic methods, or *simulated annealing* are commonly used to approximate optimal routes. Brute force methods, while accurate, are impractical for large-scale problems due to the exponential growth of permutations. In contrast, genetic and other heuristic approaches efficiently explore space by creating new solutions through the combination and mutation of existing ones, which is characteristic of combinatorial methods.

An example of a genetic algorithm:

Program 122 genetic algorithm

```
import numpy as np
import random
import matplotlib.pyplot as plt
class City:
def __init__(self, x, y, name):
self.x, self.y, self.name = x, y, name
def distance(self, city):
return np.hypot(self.x - city.x, self.y - city.y)
def __repr__(self):
return f"{self.name} ({self.x:.2f}, {self.y:.2f})"
class Fitness:
def __init__(self, path):
self.path = path
self.distance = sum(path[i].distance(path[i+1]) for i in range(-1, len(path)-1))
self.fitness = 1 / self.distance
def create_population(size, cities):
return [random.sample(cities, len(cities)) for _ in range(size)]
def rank_population(population):
return sorted(((i, Fitness(p).fitness) for i, p in enumerate(population)), key=lambda x:
x[1], reverse=True)
def selection(ranked, elite_size):
selection_probs = np.cumsum([f for _, f in ranked]) / sum(f for _, f in ranked)
selected = [ranked[i][0] for i in range(elite_size)]
selected += [ranked[np.searchsorted(selection_probs, random.random())][0] for _ in
range(len(ranked) - elite_size)]
return selected
def crossover(parent1, parent2):
geneA, geneB = sorted(random.sample(range(len(parent1)), 2))
child = parent1[geneA:geneB] + [c for c in parent2 if c not in parent1[geneA:geneB]]
return child
def mutate(path, mutation_rate):
for i in range(len(path)):
if random.random() < mutation_rate:
j = random.randint(0, len(path) - 1)
path[i], path[j] = path[j], path[i]
return path
def evolve_population(population, elite_size, mutation_rate):
ranked = rank_population(population)
selected = selection(ranked, elite_size)
mating_pool = [population[i] for i in selected]
children = mating_pool[:elite_size] + [crossover(mating_pool[i], mating_pool[-i-1]) for i
in range(len(mating_pool) - elite_size)]
return [mutate(child, mutation_rate) for child in children]
def genetic_algorithm(cities, pop_size, elite_size, mutation_rate, generations):
population = create_population(pop_size, cities)
print(f"Initial distance: {1 / rank_population(population)[0][1]:.2f}")
```

```
for _ in range(generations):
population = evolve_population(population, elite_size, mutation_rate)
best = rank_population(population)[0]
print(f"Final distance: {1 / best[1]:.2f}")
return population[best[0]]
def plot_path(path):
x, y = [c.x for c in path] + [path[0].x], [c.y for c in path] + [path[0].y]
plt.figure(figsize=(10, 6))
plt.plot(x, y, 'o-', markersize=8)
for i, city in enumerate(path + [path[0]]):
plt.text(city.x, city.y, city.name, fontsize=12, ha='right', va='bottom')
plt.title('Best Path')
plt.xlabel('X')
plt.ylabel('Y')
plt.grid(True)
plt.show()
# Parameters and execution
n = 10
city_list = [City(random.uniform(0, 100), random.uniform(0, 100), chr(65+i)) for i in
range(n)]
best_path = genetic_algorithm(city_list, pop_size=100, elite_size=20, mutation_rate=0.01,
generations=500)
plot_path(best_path)
```

Figure 2. Traveling salesman path

An equally important case can be traced back to one of the classic problems, filling a backpack (Knapsack Problem). The statement is quite simple: if we give a backpack with a specific capacity and we give *N* objects, each of which is characterized by a weight and a value, the problem is to choose which of these objects to put in the backpack to obtain the greatest value without exceeding the capacity *W* of the backpack itself. It is a classic example of combinatorial optimization, widely studied in computer science and algorithm theory (Martello & Toth, 1987). It comes in several variants, among which the best known is the 0/1 Knapsack Problem, where the goal is, as mentioned, to maximize the total value of the selected objects without exceeding a pre-established weight limit.

The interesting aspect is that there are practical applications in various fields, including logistics, which helps to optimize the loading of goods, and in managing company resources, which allows one to decide which projects to finance to maximize profit.

Techniques for solving the Knapsack Problem include *brute force* approaches, which examine all possible combinations of objects, and more efficient methods such as *dynamic programming*. The latter takes advantage of overlapping subproblems to reduce the number of calculations required by storing the intermediate results in a table. The temporal complexity of this approach is $O(N \times W)$.

An example of a dynamic algorithm is as follows:

Program 123 Knapsack

```
class Object:
def __init__(self, weight, value):
self.weight = weight
self.value = value
self.ratio = value / weight
def __repr__(self):
return f"(Weight: {self.weight}, Value: {self.value}, Ratio: {self.ratio:.2f})"
def knapSack(W, wt, val, n):
# Scale weights and capacity to handle them as integers
scaling_factor = 100 # Use a scale to convert floats to integers
W = int(round(W * scaling_factor))
wt = [int(round(w * scaling_factor)) for w in wt]
# Base case: no items or zero capacity
if n == 0 or W == 0:
return 0, []
# Store results for subproblems
dp = [[0] * (W + 1) for _ in range(n + 1)]
# Table to track included items
inclusion = [[False] * (W + 1) for _ in range(n + 1)]
# Populate the dp table
for i in range(1, n + 1):
for w in range(1, W + 1):
if wt[i-1] <= w:
if dp[i-1][w] < val[i-1] + dp[i-1][w - wt[i-1]]:
dp[i][w] = val[i-1] + dp[i-1][w - wt[i-1]]
inclusion[i][w] = True
else:
dp[i][w] = dp[i-1][w]
else:
dp[i][w] = dp[i-1][w]
```

```
# Retrieve the included items
selected_items = []
w = W
for i in range(n, 0, -1):
if inclusion[i][w]:
selected_items.append((wt[i-1] / scaling_factor, val[i-1], val[i-1] / (wt[i-1] / scaling_
factor)))
w -= wt[i-1]
# Convert the maximum value back to float
max_value = dp[n][W] / scaling_factor
return max_value, selected_items
def input_data():
n = int(input("Enter the number of items: "))
values = []
weights = []
for i in range(n):
v = float(input(f"Enter the value of item {i+1}: "))
w = float(input(f"Enter the weight of item {i+1}: "))
values.append(v)
weights.append(w)
W = float(input("Enter the capacity of the knapsack: "))
return W, weights, values, n
if __name__ == '__main__':
W, weights, values, n = input_data()
max_value, selected_items = knapSack(W, weights, values, n)
# Print maximum value
print(f"The maximum obtainable value is: {max_value * 100:.2f}")
print("The selected items are:")
for i, (w, v, r) in enumerate(selected_items, start=1):
print(f"Item {i}: (Weight: {w}, Value: {v}, Ratio: {r:.2f})")
```

For example, if the backpack is five in size and the objects are four, with utilities (5, 3, 3, 1) and weights (4, 1, 3, 2), the maximum value that can be obtained is eight by inserting the first two objects in the backpack.

A whole category is related to bin packing problems (Malkevitch, 1991). Ronald Graham (1935-2020) proposed a singular problem: how many containers with a capacity of 524 kg we need for 33 packages that have the weights 9, 9, 10, 10, 10, 10, 10, 10, 12, 12, 12, 37,37, 46, 84, 85, 106, 106, 106, 106, 127, 127, 127, 127, 252, 252, 252, 252, 252, 252, 442. The answer is 7; the surprising aspect is that we will need one more if we remove the weight from 46 kg.

This version of the program uses *First Fit Decreasing* (FFD) to solve Graham's bin packing problem. The code also generates two graphs with the expected packages.

Program 124 bin packing

```
import matplotlib.pyplot as plt
def first_fit_decreasing(packages, capacity):
# Sort packages in descending order
packages.sort(reverse=True)
# Initialize bins
bins = []
```

```
for package in packages:
placed = False
for bin in bins:
if sum(bin) + package <= capacity:
bin.append(package)
placed = True
break
if not placed:
bins.append([package])
return bins
def draw_bins(bins, capacity):
fig, ax = plt.subplots(figsize=(10, len(bins) * 0.5))
bottom = 0
for i, bin in enumerate(bins):
for package in bin:
ax.barh(i, package, left=bottom, color='grey', edgecolor='black')
bottom += package
bottom = 0
ax.set_xlabel('Weight (kg)')
ax.set_ylabel('Bin')
ax.set_title('Package Distribution Across Bins')
ax.set_xlim(0, capacity)
ax.set_yticks(range(len(bins)))
ax.set_yticklabels([f'Bin {i+1}' for i in range(len(bins))])
plt.tight_layout() # Adjust layout to prevent overlap
plt.show()
def main():
packages = [9, 9, 10, 10, 10, 10, 10, 10, 12, 12, 12, 37, 37, 46, 84, 85, 106, 106, 106,
106, 127, 127, 127, 127, 127, 252, 252, 252, 252, 252, 252, 252, 442]
capacity = 524
# Test with all packages
bins = first_fit_decreasing(packages, capacity)
print(f"Number of bins needed with all packages: {len(bins)}")
draw_bins(bins, capacity)
# Test without the 46 kg package
packages_without_46 = [package for package in packages if package != 46]
bins_without_46 = first_fit_decreasing(packages_without_46, capacity)
print(f"Number of bins needed without the 46 kg package: {len(bins_without_46)}")
draw_bins(bins_without_46, capacity)
if __name__ == "__main__":
main()
```

For the specific problem, the exact number of possible combinations is challenging to calculate directly without an exhaustive enumeration algorithm, which would be computationally prohibitive. However, the number of possible combinations is huge and grows very rapidly with the amount of packets. The number of packet permutations is n! This represents all possible sequences in which parcels can be considered. Each parcel permutation can be distributed in many ways in the containers, but not all distributions are valid due to capacity restrictions. For this reason, the bin packing is considered NP-hard.

Natural language processing is a field of artificial intelligence that deals with the interaction between computers and human language. It involves breaking text into smaller units, called tokens, which can be words, phrases, or characters.

A useful tool is the *Term Frequency-Inverse Document Frequency* (TF-IDF). This algorithm should not be confused with the creation of a Word Cloud. In the latter case, it is a visual representation of the words present in a text. Words are arranged randomly but with a size that reflects their frequency or importance in the text. TF-IDF is a measure that assesses the importance of a word within a document compared to a collection or corpus of documents (Carvalho & Guedes, 2020; Al-Obaydy et al., 2022). It is a textual analysis technique used to extract significant features from texts. The result is the product of the frequency of a word in the document (TF):

TF(t,d) = occurrences of *t* in text *d* / words in text *d*

and the measure of how rare a word is in the corpus documents (IDF):

$$IDF(t,D) = \log\left(\frac{N}{|\{d \in D: t \in d\}|}\right)$$

The following program generates this image based on the TF-IDF scores of the words in the text Alice.txt, taken from the book Alice's Adventures in Wonderland.

Figure 3. Word cloud generated with TF-ITF

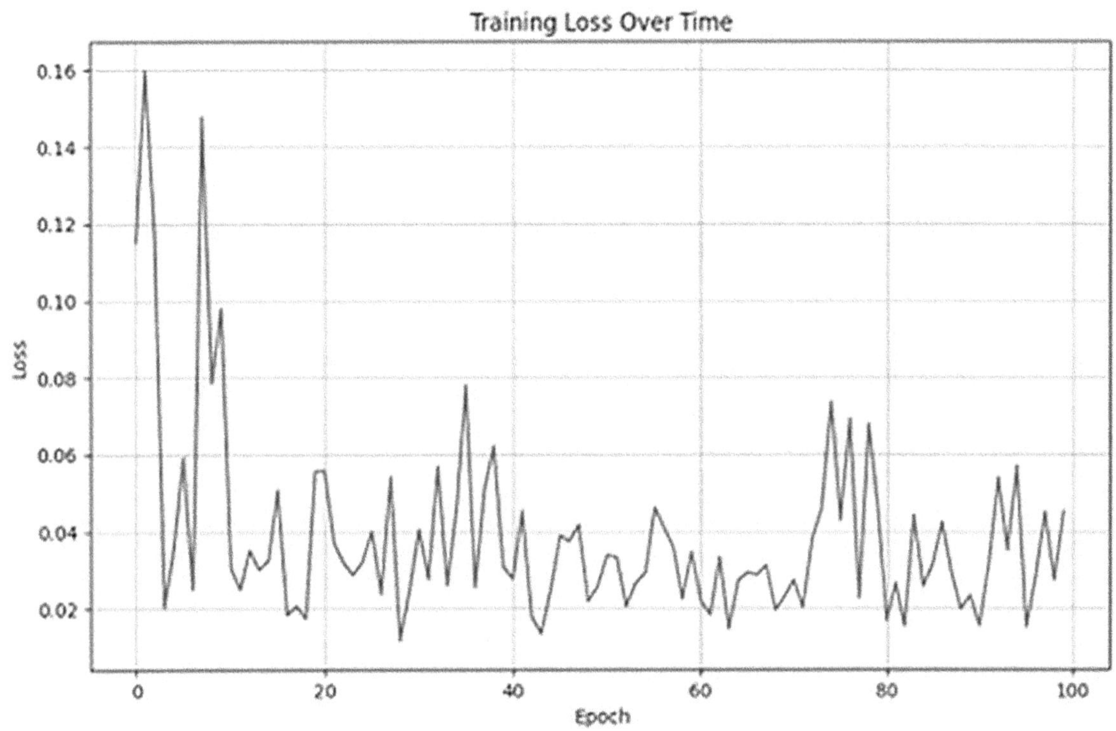

Program 125 TF-ITF

```
from sklearn.feature_extraction.text import TfidfVectorizer
from wordcloud import WordCloud
import matplotlib.pyplot as plt
import os
# Function to read the content of the file with a specified encoding
def read_file(file_path, encoding='utf-8'):
with open(file_path, 'r', encoding=encoding) as file:
return file.read()
# Function to filter short words
def filter_short_words(text, min_length=4):
words = text.split()
filtered_words = [word for word in words if len(word) >= min_length]
return ' '.join(filtered_words)
# Ask the user to enter the file name
file_name = input("Enter the name of the text file (in the same directory): ")
# Add the .txt extension if it's not present
if not file_name.endswith('.txt'):
file_name += '.txt'
# Check if the file exists
if not os.path.isfile(file_name):
print(f"The file {file_name} was not found. Please ensure the file exists in the same di-
rectory as the script.")
else:
try:
# Attempt to read the file content using UTF-8 encoding
try:
text = read_file(file_name, encoding='utf-8')
except UnicodeDecodeError:
# If it fails, try reading the file content using ISO-8859-1 encoding
text = read_file(file_name, encoding='ISO-8859-1')
# Filter short words
filtered_text = filter_short_words(text)
# Use TfidfVectorizer to calculate TF-IDF
vectorizer = TfidfVectorizer(stop_words='english') # Use predefined stopwords
tfidf_matrix = vectorizer.fit_transform([filtered_text])
feature_names = vectorizer.get_feature_names_out()
tfidf_scores = tfidf_matrix.toarray()[0]
# Combine words with their TF-IDF scores
tfidf_dict = {feature_names[i]: tfidf_scores[i] for i in range(len(feature_names))}
# Generate the word cloud
wordcloud = WordCloud(
width=800,
height=400,
background_color='white',
max_words=30,
contour_color='steelblue',
contour_width=2,
min_font_size=10
).generate_from_frequencies(tfidf_dict)
```

```
# Display the generated image
plt.figure(figsize=(10, 5))
plt.imshow(wordcloud, interpolation='bilinear')
plt.axis('off')
plt.show()
except FileNotFoundError:
print(f"The file {file_name} was not found. Please ensure the file exists in the same directory as the script.")
```

Although the TF-IDF is not a combinatorial tool in the strict sense of the term, it uses combinatorial concepts to calculate its metrics. It can be integrated into algorithms that address combinatorial problems. Frequency of terms (TF) is a combinatorial measure that counts how many times a specific word appears in a document. It is an example of combinatorial counting. The IDF measures how rare or common a word is in a collection of documents. This is the logarithm of the ratio of the total number of documents to the number of documents that contain the word. It is a typical case of combinatorial counting, where the distribution of an element (the word) in a larger set is considered.

The combination of these two measures leading to the calculation of the TF-IDF involves data from multiple documents. In this sense, we can consider the process an example of a combinatorial operation, as it involves aggregating information from different sources to obtain a final value.

To examine the field of machine learning and its connection to combinatorics, one can begin with an extensive database. Let us consider the videogame.xlsx file that contains information about Gender, Age, and Preference for a particular type of video game. We can use this program to determine which video game is recommended by knowing the person's gender and age.

Program 126 Combinatorial video games

```
import pandas as pd
from sklearn.tree import DecisionTreeClassifier
from networkx import Graph, shortest_path
import random
# Define the combinatory rules for prediction
rules = {
"Male_Young": ["shooter", "fighting"],
"Male_Adult": ["management", "fantasy"],
"Female_Young": ["fantasy", "platform"],
"Female_Adult": ["management", "platform"]
}
# Load the dataset
players = pd.read_excel('videogame.xlsx')
# Print the dataset columns to verify their names
print("Dataset columns:", players.columns)
# Remove any whitespace from column names
players.columns = players.columns.str.strip()
# Ensure the required columns exist in the dataset
required_columns = ['Gender', 'Age', 'videogame']
for col in required_columns:
if col not in players.columns:
raise ValueError(f"The dataset must contain the column '{col}'.")
# Add necessary columns to the dataset
```

```
players['GenderAge'] = players['Gender'].astype(str) + "_" + (players['Age'] // 10 * 10).
astype(str)
players['CategoryAgeGroup'] = players.apply(
lambda row: ("Male" if row['Gender'] == 1 else "Female") + "_" + ("Young" if row['Age'] <
18 else "Adult"), axis=1
)
# Define prediction functions
def decision_tree_prediction(gender, age):
gender_age = f"{gender}_{int(age / 10) * 10}"
category_age_group = ("Male" if gender == 1 else "Female") + "_" + ("Young" if age < 18
else "Adult")
X = players[['GenderAge', 'CategoryAgeGroup']]
y = players['videogame']
# Encode categorical variables
X_encoded = pd.get_dummies(X)
model = DecisionTreeClassifier()
model.fit(X_encoded, y)
# Prediction
X_pred = pd.DataFrame([[gender_age, category_age_group]], columns=['GenderAge', 'Catego-
ryAgeGroup'])
X_pred_encoded = pd.get_dummies(X_pred)
X_pred_encoded = X_pred_encoded.reindex(columns=X_encoded.columns, fill_value=0)
prediction = model.predict(X_pred_encoded)
return prediction[0]
def combinatory_rules_prediction(gender, age):
category = "Male" if gender == 1 else "Female"
category += "_Young" if age < 18 else "_Adult"
prediction = rules[category]
return random.choice(prediction)
def optimized_combinatory_prediction(gender, age):
graph = create_compatibility_graph(players)
player_node = f"Player_{gender}_{age}"
# Check if the player's node exists in the graph
if player_node not in graph.nodes:
return None
# Get all videogames directly connected to the player's node
possible_games = list(graph.neighbors(player_node))
if not possible_games:
return None
# Return a random videogame from the possible options
return random.choice(possible_games)
def is_videogame(node):
return node not in players.apply(lambda row: f"Player_{row['Gender']}_{row['Age']}",
axis=1).values
def create_compatibility_graph(players):
graph = Graph()
for i, player in players.iterrows():
player_node = f"Player_{player['Gender']}_{player['Age']}"
graph.add_node(player_node)
```

```
graph.add_edge(player_node, player['videogame'])
for videogame in players['videogame'].unique():
graph.add_node(videogame)
return graph
def main():
gender = int(input("Enter the player's gender (1 for Male, 0 for Female): ").strip())
age = int(input("Enter the player's age: ").strip())
print("\nPrediction using a decision tree:")
print(decision_tree_prediction(gender, age))
print("\nPrediction using combinatory rules:")
print(combinatory_rules_prediction(gender, age))
print("\nOptimized combinatory prediction:")
print(optimized_combinatory_prediction(gender, age))
if __name__ == "__main__":
main()
```

The primary function collects user input (Gender and Age), calculates three different predictions (decision tree, combinatorial rules, optimized combinatorics) and prints them.

Let us examine how predictions are formed.

1. *Decision tree*: It separates features (X) and targets (y) from the dataset, builds and trains a decision tree classification model on X and y, predicts the player using the model and combinatorial variables, and returns the predicted video game type.
2. *Combinatorial rules*: It determines the combinatorial category of the player; retrieves a list of video game genres suggested by the combinatorial category in the rules; Randomly selects a video game from the suggested list and returns the selected video game type.
3. *Optimized combinatorics*: This approach constructs a compatibility network, where nodes represent players and video games, and edges indicate meaningful associations (e.g., based on preferences or traits). A node is created for the player within this structure. The algorithm then searches for the shortest path from the player node to any node in the video game. If a valid path exists, the genre of the destination game is returned. If no such path exists, the output is None.

As the world of research and technology constantly evolves, it is crucial to move in promising and innovative directions. New and fascinating areas of application are emerging at the intersection of combinatorics and artificial intelligence (AI). For example, the application of combinatorial concepts to optimise quantum algorithms in *quantum machine learning* represents fertile ground for discoveries and applications.

A fascinating field where combinatorics and AI intersect is that of automated scientific discovery. Artificial intelligence algorithms can increasingly exploit combinatorial techniques to analyse large amounts of scientific data, discovering hidden patterns and relationships between variables that may not be immediately obvious to scientists. This approach is particularly beneficial in physics, genomics, computational chemistry, and the general life sciences, as well as in complex systems.

The combined use of AI and combinatorial techniques enables the development of increasingly accurate predictive models and the execution of simulations that bring science closer to a level of precision previously unimaginable. In genomics, combinatorics-based algorithms can be used to analyse complex genetic sequences and predict the behaviour of specific genes in response to environmental or therapeutic variables. In computational chemistry, similar techniques can be applied to simulate molecular interactions and discover new substances with desired properties.

A concrete example of applying a combinatorial algorithm in a scientific context is the automation of formulating mathematical models. In particular, it can be used to automatically generate descriptive equations of physical phenomena, starting from empirical data collected through experiments or simulations. These models, in turn, can be used to predict the behaviour of complex systems, speeding up the process of scientific discovery and reducing the need for direct human intervention.

In this program, a combinatorial algorithm automatically generates a descriptive equation of a phenomenon.

Program 127 scientific discovery

```
import itertools
import numpy as np
def generate_data():
x = np.linspace(0, 10, 100)
m = 5 # slope
y = m * x
y += np.random.normal(0, 0.5, y.shape)
return x, y
variables = ['x']
operators = ['+', '*']
constants = [1, 2, 5, 10]
def generate_equations(max_length=3):
components = variables + [str(c) for c in constants]
for length in range(1, max_length + 1):
for eq in itertools.product(components, operators, repeat=length):
yield ''.join(eq[:-1]) # Exclude the last operator
def evaluate_equation(equation, x):
try:
result = eval(equation, {"x": x}) # Use a safe environment for eval
if np.any(np.isinf(result)) or np.any(np.isnan(result)):
return np.full_like(x, np.inf)
return result
except:
return np.full_like(x, np.inf)
def calculate_mse(y_true, y_pred):
diff = y_true - y_pred
if np.any(np.isinf(diff)) or np.any(np.isnan(diff)):
return np.inf
return np.mean(np.square(diff))
def discover_law():
x, y = generate_data()
best_equation = ''
best_mse = float('inf')
evaluated_equations = []
for eq in generate_equations():
y_pred = evaluate_equation(eq, x)
mse = calculate_mse(y, y_pred)
evaluated_equations.append((eq, mse))
if mse < best_mse:
best_mse = mse
best_equation = eq
# Sort equations by MSE in ascending order
```

```
evaluated_equations.sort(key=lambda x: x[1])
# Print all evaluated equations
print("All evaluated equations (sorted by ascending MSE):")
for eq, mse in evaluated_equations:
print(f"y = {eq:10s} MSE: {mse:.3f}")
print("\nFinal Results:")
print(f"Best equation found: y = {best_equation}")
print(f"MSE: {best_mse:.3f}")
if best_mse < np.inf:
r2 = 1 - (best_mse / np.var(y))
print(f"R^2: {r2:.3f}")
else:
print("Unable to calculate R^2 due to infinite or NaN values.")
discover_law()
```

The program generates simulated data, based on a known law that we assume we do not know, with a bit of added noise. From this data, the program generates various equations to describe the dataset, using combinations of mathematical operators and variables. Then evaluate these equations and select the one with the lowest mean squared error (MSE) as the best approximation of the physical law. The result indicates that the program has "rediscovered" Ohm's law (V = IR).

In machine learning, combinatorial optimization is crucial for selecting the most suitable parameters for models, identifying the best structure of neural networks, and optimizing game strategies in multi-agent systems. In many applications, the parameters (or hyperparameters) of a model determine the system's final performance. Therefore, combinatorial optimisation becomes an indispensable tool for exploring and selecting the most effective combination of hyperparameters, thereby increasing the model's efficiency.

In neural networks, the space of possible configurations is vast and complex, with many combinations, including the number of layers, the activation function, the number of neurons per layer, and other variables. Combinatorial analysis enables us to study the structure of this space, facilitating an understanding of the relationships between different configurations and reducing the space of possible solutions. This way, one can explore the combinations that lead to the best results in a more efficient and targeted way. In the context of deep learning, it is essential to understand and optimize the structure of the neural network itself. For example, the architecture design of a *convolutional neural network* (CNN) for image processing can vary greatly. One can decide how many layers there should be, the size of the filters to apply, the number of neurons in each layer, and how these layers are connected. Combinatorial optimization helps to effectively explore these variables, guiding the construction of architectures that maximize performance in the specific application.

Many AI algorithms require searching vast and complex solution spaces, where the right combination of features can make a significant difference. Combinatorial heuristic techniques, such as local search, genetic algorithms, or gradient descent, are essential for reducing the time it takes to find an acceptable result. These techniques enable us to explore and select the best combinations of variables in a reasonable amount of time, even when the space of possible solutions is enormous.

A practical example can be drawn from a classification model for medical diagnosis. In these contexts, one can have thousands of characteristics to describe a patient's health status, such as vital signs, lab test results, diagnostic images, and more. However, not all these features are necessarily relevant to the diagnosis. Some may need to be more relevant, useful, or even misleading, increasing the risk of overfitting, which is the model's tendency to learn noise from data instead of valuable patterns.

The following program can highlight this aspect.

Program 128 Medical diagnosis

```
from itertools import combinations
```

```
class SimulatedPatient:
def __init__(self, fever, cough, fatigue, headache, nausea):
self.symptoms = {
'fever': fever,
'cough': cough,
'fatigue': fatigue,
'headache': headache,
'nausea': nausea
}
def overfitting_diagnosis(patient):
if patient.symptoms['fever'] and patient.symptoms['cough'] and patient.symptoms['fa-
tigue']:
return "Flu"
elif patient.symptoms['fever'] and patient.symptoms['headache'] and not patient.symptom-
s['nausea']:
return "Viral infection"
elif patient.symptoms['cough'] and patient.symptoms['fatigue'] and patient.symptoms['head-
ache']:
return "Bronchitis"
elif patient.symptoms['nausea'] and patient.symptoms['headache'] and not patient.symptom-
s['fever']:
return "Food poisoning"
else:
return "Unclear diagnosis"
def simplify_model(patients, k=2):
combination_counts = {}
for patient in patients:
for comb in combinations(patient.symptoms.items(), k):
key = frozenset(comb)
if key not in combination_counts:
combination_counts[key] = {'count': 0, 'diagnoses': set()}
combination_counts[key]['count'] += 1
combination_counts[key]['diagnoses'].add(overfitting_diagnosis(patient))
simplified_rules = {}
for comb, info in combination_counts.items():
if len(info['diagnoses']) == 1:
simplified_rules[comb] = list(info['diagnoses'])[0]
return simplified_rules
def simplified_diagnosis(patient, rules):
for symptoms, diagnosis in rules.items():
if all(patient.symptoms[s] == v for s, v in symptoms):
return diagnosis
return "Unclear diagnosis"
# Create a dataset of patients
patients = [
SimulatedPatient(True, True, True, False, False),
SimulatedPatient(True, False, False, True, False),
SimulatedPatient(False, True, True, True, False),
SimulatedPatient(False, False, True, True, True),
```

```
SimulatedPatient(True, True, False, False, True)
]
# Simplify the model
simplified_rules = simplify_model(patients)
print("Comparison between overfitting model and simplified model:")
print("-" * 60)
for i, patient in enumerate(patients, 1):
overfit_diag = overfitting_diagnosis(patient)
simplified_diag = simplified_diagnosis(patient, simplified_rules)
print(f"Patient {i}:")
print(f"Symptoms: {patient.symptoms}")
print(f"Overfitting diagnosis: {overfit_diag}")
print(f"Simplified diagnosis: {simplified_diag}")
print("-" * 60)
print("\nSimplified rules:")
for symptoms, diagnosis in simplified_rules.items():
print(f"{dict(symptoms)} -> {diagnosis}")
```

The output for each patient shows symptoms, diagnosis of the overfitting model, and diagnosis of the simplified model. The overfitting model gives specific diagnoses based on precise combinations of symptoms. The simplified model more often generates an "Unclear diagnosis," but when it gives a diagnosis, it is based on more general and robust rules.

A fascinating and constantly evolving field in artificial intelligence is image recognition, particularly facial recognition. Combinatorics plays a crucial role in the analysis of this problem since it defines both the problem itself and the space of possible solutions. In facial recognition, the variety of configurations a face can take is enormous. Each change in expression, lighting, viewing angle, or posture represents a new configuration to analyse. Combinatorics helps us understand the vastness of this space and develop more efficient algorithms to explore it, allowing for accurate matches to be found even in the presence of significant variations.

The distinguishing features of a face, such as the distance between the eyes, the shape of the nose, the type of chin, and other unique characteristics, are collectively referred to as facial features. Selecting the best features is a complex task, as it involves choosing the optimal subset of features from a larger set without compromising essential information or introducing noise.

Combinatorial optimization helps precisely in this: by exploring the possible combinations of features, we can identify those that best represent a person's identity, improving the performance of the facial recognition system.

Machine learning algorithms used in facial recognition often rely on combinatorial optimization techniques to identify optimal model parameters, such as weights in a neural network or values associated with each feature. However, despite this technology's enormous potential, facial recognition has several limitations and issues that hinder its widespread adoption and raise important questions, both ethical and substantive.

One of the most apparent problems concerns privacy. Facial recognition systems can collect and store vast amounts of biometric data, raising concerns about protecting personal information and the risk of abuse.

Another significant issue is cognitive bias: facial recognition algorithms are sensitive to training data. The system may produce discriminatory results if the dataset contains bias or is biased concerning certain ethnicities. Studies have shown that facial recognition systems tend to be less accurate at recognizing the faces of dark-skinned people than those with fair skin.

In addition, there is the problem of false positives, i.e., the inability of systems to avoid errors, such as the incorrect identification of a person. A notable case of facial recognition failure occurred in 2020, when an innocent man in Detroit was wrongly arrested due to a positive match with a facial recognition system. Although video evidence proved his innocence, the system had misidentified his face as that of a thief. Another significant example concerns an episode reported by Gerd Gigerenzer: In 2017, during the UEFA Champions League final between Real Madrid and Juventus, the Cardiff police used a facial recognition system to examine the faces of 170,000 fans. The system recorded as many as 2470 matches with a criminal database, but 93% (2297) were false alarms (Gigerenzer, 2022).

These examples highlight the reliability and accuracy issues plaguing facial recognition systems.

Another area in which combinatorics plays a fundamental role is probabilistic reasoning in artificial intelligence, particularly in *Bayesian networks*, *Hidden Markov models*, and *Dependent Markov chains*. To determine the optimal configuration of a Bayesian network, the structure of the network requires exploring a combinatorial space of possible graphs. In addition, learning the network parameters, i.e., conditional probabilities, can be seen as a combinatorial optimization problem, especially when the data is incomplete or missing.

A concrete application of combinatorics in the field of artificial intelligence concerns Bayesian filters, a class of machine learning algorithms used for classification, as in the case of spam filtering in emails. The principle behind these filters is based on Bayes' theorem, which allows probabilities to be updated a posteriori based on observations. The spam filter must decide whether an email is "spam", or "non-spam" (ham) based on a set of characteristics, i.e., words or phrases contained in the text.

The filtering process takes place through various steps:

1. Data collection: The frequency of words in each category is extracted using a set of emails labelled as spam or non-spam.
2. Calculation of a priori probability: The basic probabilities of a given email being spam (P(spam) or not spam P(non-spam) are calculated.
3. Conditional probability calculation: Calculate the probabilities that a specific word will appear in a spam email P(word|spam) or a non-spam email P(word|non-spam).

However, what is the link with combinatorics? The fundamental problem is that every email contains a combination of words, and it is not enough to consider a single word to determine whether the email is spam or not. It is the combination of several words that becomes indicative of spam. For example, the presence of the word "offer" may not be enough, but the combination of "offer", "free", and "click here" can make the decision lean towards spam.

Bayesian filters, therefore, must consider all possible combinations of words in the email. It is a remarkably complex situation, and the solution requires using mathematical techniques such as logarithms, which allow the avoidance of multiplicative underflow during probability calculations.

A simple example of using a Bayesian spam filter would be analysing a dataset of emails labelled as spam or non-spam, verifying how a sentence such as "Congratulations! You won a free iPhone, reply now!" would be classified as spam compared to another email with a more innocent phrase such as "You have an unclaimed refund; please complete the form to receive it." This example illustrates the effectiveness and complexity of Bayesian filters in addressing classification problems that involve combinations of characteristics.

Program 129 Spam filter

```
import re
from collections import defaultdict
from math import log
# Example dataset: 20 emails classified as "spam" or "non-spam"
emails = [
("unmissable offer, buy now!", "spam"),
("new promotion: buy immediately", "spam"),
("interesting job proposal", "non-spam"),
("subscribe to our newsletter", "non-spam"),
("free offer, click here!", "spam"),
("you have a new email message", "non-spam"),
```

```
("50% discount, only for today!", "spam"),
("account update", "non-spam"),
("you've won an exclusive prize, reply now!", "spam"),
("the best deals just for you", "spam"),
("new message from your boss", "non-spam"),
("invitation to attend our conference", "non-spam"),
("try our premium service for free", "spam"),
("customer service: your issue has been resolved", "non-spam"),
("buy now and receive a discount voucher", "spam"),
("reminder: appointment scheduled for tomorrow", "non-spam"),
("payment request, act immediately", "spam"),
("your subscription is expiring, renew now", "non-spam"),
("exclusive investment opportunity", "spam"),
("annual report of your company", "non-spam")
]
# Function to tokenize a sentence into keywords
def tokenize(text):
# Convert to lowercase and remove special characters
return re.findall(r'\b\w+\b', text.lower())
# Keyword counts in spam and non-spam emails
spam_word_counts = defaultdict(int)
non_spam_word_counts = defaultdict(int)
# Global counts of spam and non-spam emails
spam_count = 0
non_spam_count = 0
# Train the filter
for email, label in emails:
words = tokenize(email)
if label == "spam":
spam_count += 1
for word in words:
spam_word_counts[word] += 1
else:
non_spam_count += 1
for word in words:
non_spam_word_counts[word] += 1
# Function to calculate the probability of spam using Bayes' theorem
def calculate_spam_probability(email):
words = tokenize(email)
# Prior probabilities
p_spam = spam_count / len(emails)
p_non_spam = non_spam_count / len(emails)
# Logarithms to avoid multiplicative underflow
log_prob_spam = log(p_spam)
log_prob_non_spam = log(p_non_spam)
# Compute conditional probabilities for each word
for word in words:
# Probabilities P(word|spam) and P(word|non-spam)
p_word_given_spam = (spam_word_counts[word] + 1) / (sum(spam_word_counts.values()) + 1)
```

```
p_word_given_non_spam = (non_spam_word_counts[word] + 1) / (sum(non_spam_word_counts.val-
ues()) + 1)
log_prob_spam += log(p_word_given_spam)
log_prob_non_spam += log(p_word_given_non_spam)
# Compare the calculated probabilities
if log_prob_spam > log_prob_non_spam:
return "spam"
else:
return "non-spam"
# Input a new email from the user
new_email = input("Enter the text of the new email to classify: ")
result = calculate_spam_probability(new_email)
# Print the classification result
print(f"The email: '{new_email}' is classified as: {result}")
```

Hidden Markov models (HMMs) are statistical models in which the system is viewed as a Markov process with unobservable (hidden) states (Cheng et al., 2021; Xue & Liu, 2021). The space of possible states in an HMM is combinatorial. There are Nt possible sequences of states for a model with *N* possible states and *t* time steps. The use of these probabilistic models in AI is pervasive. HMM is used in speech recognition and genetic sequence analysis, Bayesian networks in previously examined medical diagnoses, and computer vision.

A concrete example of how an HMM is applied in practice is speech recognition, where the Viterbi algorithm determines the most probable sequence of hidden states. Based on the observed data, the Viterbi algorithm evaluates all possible states and selects the one with the highest likelihood. Of course, implementing such an algorithm in a real-world environment would typically require libraries such as *PyAudio* or *Sounddevice* to record audio from a microphone, followed by converting the audio signal into a numerical format, a process that can be pretty complex. However, to illustrate how it works, one could use a set of basic phonemes (distinct sound units) and assign the necessary transition and emission probabilities to those phonemes, simulating the process without dealing with the full complexity of real audio processing.

In this simplified example, the combinatorial nature of the problem remains evident: the number of possible phoneme sequences increases exponentially with the length of the voice input, and Viterbi's algorithm helps navigate this vast space to identify the most likely interpretation of the audio signal.

In this case, it may be enough to give an idea of the algorithm by using a few simple phonemes and assigning the necessary probabilities.

Program 130 Viterbi algorithm

```
import numpy as np
def viterbi(observations, states, start_prob, trans_prob, emit_prob):
# Number of observations and states
n_observations = len(observations)
n_states = len(states)
# Initialize the probability matrix (dynamic programming table)
viterbi_table = np.zeros((n_states, n_observations))
# Initialize the matrix to keep track of the best paths
path = np.zeros((n_states, n_observations), dtype=int)
# Initialization step: calculate probabilities for the first state
for s in range(n_states):
viterbi_table[s, 0] = start_prob[s] * emit_prob[s][observations[0]]
# Iterate through observations to fill the table
```

```
for t in range(1, n_observations):
for s in range(n_states):
# Maximize over all previous states
max_prob = -1
best_state = 0
for prev_s in range(n_states):
prob = viterbi_table[prev_s, t-1] * trans_prob[prev_s][s] * emit_prob[s][observations[t]]
if prob > max_prob:
max_prob = prob
best_state = prev_s
viterbi_table[s, t] = max_prob
path[s, t] = best_state
# Find the path with the highest probability in the last column
best_final_state = np.argmax(viterbi_table[:, n_observations-1])
best_path = [best_final_state]
# Backtrace to find the optimal path
for t in range(n_observations-1, 0, -1):
best_final_state = path[best_final_state, t]
best_path.insert(0, best_final_state)
return best_path, np.max(viterbi_table[:, n_observations-1])
# Example HMM for speech recognition:
# States: could be phonemes or words
states = ['silence', 'a', 'b', 'c']
# Observations (could be codes corresponding to audio features like MFCC)
observations = [0, 1, 2, 2, 1] # For example, these might be acoustic features
# Initial probabilities for each state (start probabilities)
start_prob = [0.6, 0.1, 0.1, 0.2]
# Transition probabilities between states
trans_prob = [
[0.1, 0.3, 0.2, 0.1], # from "silence"
[0.2, 0.2, 0.3, 0.1], # from "a"
[0.3, 0.3, 0.3, 0.5], # from "b"
[0.2, 0.4, 0.2, 0.3], # from "c"
]
# Emission probabilities
# Rows represent states and columns represent observations
emit_prob = [
[0.6, 0.3, 0.1], # silence
[0.1, 0.2, 0.7], # a
[0.3, 0.4, 0.3], # b
[0.5, 0.3, 0.7], # c
]
# Run the Viterbi algorithm
best_path, best_prob = viterbi(observations, states, start_prob, trans_prob, emit_prob)
# Print the results
decoded_states = [states[state] for state in best_path]
print("The most probable state path is:", decoded_states)
print("The maximum probability is:", best_prob)
```

The connection between dependent Markov chains and combinatorics is profound and manifests differently. In a language model, Markov chains can be used to calculate the probability of a given sequence of words, which involves counting the possible combinations of words. Another situation involves these chains being used to model DNA and protein sequences. To help understand the problem, we can formulate a simple example of a dependent Markov chain related to weather forecasting. Let us imagine we want to model the weather. Instead of just considering whether it is sunny or rainy today (as in a simple Markov chain), we also want to consider what the weather was like before by defining the possible states:

- State 1: Sunny today and sunny yesterday
- State 2: Sunny today and rainy yesterday
- State 3: Rainy today and sunny yesterday
- State 4: Rainy today and rainy yesterday

The probability of going from state 1 (sunny today and sunny yesterday) to state 2 (sunny today and rainy yesterday) could be low because it is more likely that a sunny day will be followed by another sunny day. We will then have a 4x4 transition matrix, where each element represents the probability of going from one state to another. It is a dependent Markov chain because the probability of a sunny or rainy day today depends on yesterday's weather and the weather of the past few days. The weather example can be seen as a combinatorial problem for several reasons; even simplified weather can take on different combinations of states (sunny, rainy, foggy, snowy, cloudy, etc.).

When we consider the dependence on the past, the number of possible sequences of states grows rapidly. Considering only two states (sunny and rainy) and a dependence on the previous day, we have $2^2 = 4$ possible states. If we extend the dependence to two previous days, we have $2^3 = 8$ states, and so on. Combinatorics comes into play in counting these possible combinations of states and how they follow one another in time.

Combinatorics in logic and automatic reasoning involves solving problems that involve manipulating and processing discrete sets of rules, facts, and logical deductions. In particular, it deals with identifying and analysing the possible combinations of elements in a given formal system to derive conclusions or solve complex problems. A central area in which combinatorics plays a fundamental role is the automatic proof of theorems, which deals with the search for formal proof starting from a set of axioms and logical rules.

Given a set of axioms and rules of inference, is it possible to prove that a specific statement is true? Proving a theorem involves exploring many sequences of deductions and combinations of rules, making the problem computationally complex. In particular, the process of finding evidence requires generating and exploring a tree of possibilities that can grow exponentially as the complexity of logical formulas increases.

One of the main methods for the automatic proof of theorems is the solving method, which looks for a contradiction. This approach is based on unifying logical formulas, i.e., the search for substitutions for variables that make two logical expressions identical. The unification process requires a combinatorial analysis of possible substitutions, exploring all combinations of values for variables. Since the number of combinations can grow rapidly with the complexity of the expressions, the efficiency of the unification algorithms becomes a crucial factor. The resolution algorithm must be able to effectively explore the space of possible deductions to avoid an exponential growth in execution time.

Combinatorial circuits represent another significant example of how combinatorics interfaces with artificial intelligence. These circuits are a type of logic circuit in which the outputs depend solely on the current inputs without storing information about previous states (unlike sequential circuits). In other words, the output of a combinatorial circuit is determined directly by the combinations of input values through specific logical relationships without the intervention of time memory.

Each combinatorial circuit can be described by a set of logic gates, which operate on inputs and produce outputs following certain logic rules. To fully understand the circuit's behaviour, it is necessary to evaluate it for all possible combinations of the inputs, which for a circuit with two inputs can be, for example, the four combinations (0,0), (0,1), (1,0), and (1,1).

This combinatorial analysis allows us to determine the function of each logic gate and, consequently, optimize the entire circuit to improve performance, such as speed and power consumption.

Program 131 Combinatorial circuits

```
from sympy import symbols, And, Or, Not
# Define input variables
A, B = symbols('A B')
# Define the logic gates
AND_gate = And(A, B)
NOT_gate = Not(A)
OR_gate = Or(AND_gate, NOT_gate)
# Define the combinational circuit
circuit = OR_gate
# Function to evaluate the circuit and print intermediate steps
def evaluate_circuit(A_val, B_val):
# Substitute input values into the logic gates
and_result = AND_gate.subs({A: A_val, B: B_val})
not_result = NOT_gate.subs({A: A_val, B: B_val})
or_result = OR_gate.subs({A: A_val, B: B_val})
# Print intermediate steps
print(f"A={A_val}, B={B_val}")
print(f" AND(A, B) = {bool(and_result)}")
print(f" NOT(A) = {bool(not_result)}")
print(f" OR(AND(A, B), NOT(A)) = {bool(or_result)}")
print(f"Output = {bool(or_result)}\n")
# Evaluate the circuit with all possible input combinations
inputs = [(0, 0), (0, 1), (1, 0), (1, 1)]
for A_val, B_val in inputs:
evaluate_circuit(A_val, B_val)
```

Combinatorics and artificial intelligence (AI) are vast and evolving research fields with many potential future research directions. Here are some promising areas:

1. *Advanced search algorithms*: These are combinatorial techniques for optimizing the search for complex solutions, such as global optimization and the search for solutions in large or discontinuous spaces.
2. *Graphical neural networks*: GNNs represent an innovative area in which combinatorics on graphs is used to develop new convolutional algorithms. These algorithms can be more computationally efficient and generalize to large data sets, enabling applications in various fields, from chemistry to social analysis.
3. *Generative models*: The development of generative models based on graphs and combinatorial structures can be used to generate music, images, text, and other types of content (Balasubramanian, 2021).
4. *Cyber defence*: In cybersecurity, AI is becoming critical for developing models that use combinatorial techniques to identify anomalies in data and detect suspicious activity and for generating cyberattacks, simulating the actions of a hacker, and evaluating the resilience of systems.
5. *Game theory*: Although there have been many advances in this field, there are still scenarios that can be explored further, where the integration of combinatorial and machine learning techniques promises to develop more sophisticated artificial intelligence systems capable of solving complex problems. For example, machine learning can be used to approximate Nash equilibria in complex games that feature a large number of players and strategies. A typical scenario is the use of auto-trading systems for pricing.
6. *Internet of Things* (IoT): This fast-growing field promises to transform many aspects of our lives. In industrial sectors, IoT can monitor the status of machinery and, thanks to AI, predict when maintenance is required, prevent-

ing breakdowns and reducing repair costs. Combinatorial algorithms help plan and predict how action is needed, reducing costs and increasing efficiency. Smart homes can learn from the habits of the inhabitants to automatically adjust lights, thermostats, appliances, and security systems. Combinatorics is used to analyse and find the best combinations of resources to optimize energy efficiency and home automation.

Among all these directions, we report two cases.

The first is an interesting example of GNN related to the simulation of road traffic (Jiang & Luo, 2022) in a grid situation such as Manhattan.

Program 132 traffic simulation

```
import numpy as np
import networkx as nx
import torch
import torch.nn as nn
import torch.nn.functional as F
import matplotlib.pyplot as plt
from matplotlib.colors import LinearSegmentedColormap
class LayerGNN(nn.Module):
def __init__(self, in_features, out_features):
super(LayerGNN, self).__init__()
self.linear = nn.Linear(in_features, out_features)
def forward(self, x, adjacency_matrix):
support = torch.mm(adjacency_matrix, x)
output = self.linear(support)
return output
class TrafficGNN(nn.Module):
def __init__(self, num_node_features, hidden_channels):
super(TrafficGNN, self).__init__()
self.conv1 = LayerGNN(num_node_features, hidden_channels)
self.conv2 = LayerGNN(hidden_channels, hidden_channels)
self.conv3 = LayerGNN(hidden_channels, 1)
def forward(self, x, adjacency_matrix):
x = F.relu(self.conv1(x, adjacency_matrix))
x = F.dropout(x, p=0.5, training=self.training)
x = F.relu(self.conv2(x, adjacency_matrix))
x = F.dropout(x, p=0.5, training=self.training)
x = self.conv3(x, adjacency_matrix)
return x
class TrafficSimulation:
def __init__(self, num_intersections=16):
self.num_intersections = num_intersections
self.grid_size = int(np.sqrt(num_intersections))
self.graph = self._create_road_network()
self.adjacency_matrix = nx.adjacency_matrix(self.graph).todense()
self.adjacency_matrix = torch.FloatTensor(self.adjacency_matrix)
self.adjacency_matrix = self.adjacency_matrix + torch.eye(num_intersections)
deg = torch.diag(torch.sum(self.adjacency_matrix, dim=1).pow(-0.5))
```

```
self.adjacency_matrix = torch.mm(torch.mm(deg, self.adjacency_matrix), deg)
self.model = TrafficGNN(num_node_features=4, hidden_channels=16)
def _create_road_network(self):
G = nx.grid_2d_graph(self.grid_size, self.grid_size)
return nx.convert_node_labels_to_integers(G)
def generate_traffic_data(self):
node_features = torch.zeros((self.num_intersections, 4))
for node in range(self.num_intersections):
node_features[node, 0] = torch.rand(1)
node_features[node, 1] = torch.rand(1)
node_features[node, 2] = torch.rand(1)
node_features[node, 3] = torch.randint(1, 4, (1,)) / 3.0
return node_features
def train_model(self, num_epochs=100):
optimizer = torch.optim.Adam(self.model.parameters(), lr=0.01)
criterion = nn.MSELoss()
losses = []
print("Starting training...")
for epoch in range(num_epochs):
self.model.train()
optimizer.zero_grad()
features = self.generate_traffic_data()
out = self.model(features, self.adjacency_matrix)
target = torch.zeros((self.num_intersections, 1))
for node in range(self.num_intersections):
neighbors = list(self.graph.neighbors(node))
if neighbors:
target[node] = features[neighbors, 0].mean()
loss = criterion(out, target)
loss.backward()
optimizer.step()
losses.append(loss.item())
if (epoch + 1) % 10 == 0:
print(f'Epoch {epoch+1}/{num_epochs}, Loss: {loss.item():.4f}')
return losses
def predict_traffic(self):
self.model.eval()
with torch.no_grad():
features = self.generate_traffic_data()
predictions = self.model(features, self.adjacency_matrix)
return predictions.numpy().flatten()
def plot_training_history(self, losses):
plt.figure(figsize=(10, 6))
plt.plot(losses)
plt.title('Loss Progress During Training')
plt.xlabel('Epoch')
plt.ylabel('Loss')
plt.grid(True)
plt.show()
```

```
def visualize_network(self, predictions=None):
try:
# Close all existing figures
plt.close('all')
# Create a new figure
fig, ax = plt.subplots(figsize=(10, 10))
# Create grid layout
pos = {i: (i % self.grid_size, self.grid_size - 1 - i // self.grid_size)
for i in range(self.num_intersections)}
if predictions is None:
predictions = self.predict_traffic()
# Normalize predictions more control
norm_predictions = (predictions - predictions.min()) / (predictions.max() - predictions.
min())
# Limit the range of grays for better visibility
norm_predictions = 0.2 + (norm_predictions * 0.6)
# Create a grayscale color map
node_colors = ['lightgray' if v < 0.5 else 'gray' for v in norm_predictions]
# Draw the graph
nx.draw(self.graph, pos,
node_color=node_colors,
node_size=1000,
width=2,
edge_color='black',
with_labels=True,
font_size=8,
font_color='black',
font_weight='bold',
ax=ax)
# Add labels with the values, shifting them a bit lower
labels = {i: f"{predictions[i]:.2f}" for i in range(self.num_intersections)}
# Add a vertical shift (modify 'y' in positions)
offset = - 0.2
shifted_pos = {k: (v[0], v[1] - offset) for k, v in pos.items()}
nx.draw_networkx_labels(self.graph, shifted_pos, labels,
font_size=8,
font_color='black')
# Add color bar
sm = plt.cm.ScalarMappable(cmap='gray_r', norm=plt.Normalize(vmin=predictions.min(), vmax-
=predictions.max()))
sm.set_array([])
cbar = plt.colorbar(sm, ax=ax)
cbar.set_label('Traffic Level')
plt.title("Traffic Simulation (Grayscale)")
plt.axis('on')
plt.grid(True)
# Show the plot
plt.show()
except Exception as e:
```

```
print(f"Error during visualization: {str(e)}")
def main():
try:
print("Initializing the simulation...")
simulation = TrafficSimulation(num_intersections=16)
print("\nStarting model training...")
losses = simulation.train_model(num_epochs=100)
print("\nVisualizing the training progress...")
simulation.plot_training_history(losses)
print("\nGenerating traffic predictions...")
predictions = simulation.predict_traffic()
print("\nVisualizing the traffic network...")
simulation.visualize_network(predictions)
print("\nTraffic predictions for all intersections:")
for intersection in range(simulation.num_intersections):
print(f"Intersection {intersection}: Traffic level = {predictions[intersection]:.3f}")
except Exception as e:
print(f"Error during execution: {str(e)}")
if __name__ == "__main__":
main()
```

After a training phase, the program reports the traffic values for each node. The training phase is crucial for monitoring whether the model is learning and improving over time. In this case, the loss function used in the code is a *Mean Squared Error Loss* (MSELoss), which measures the difference between the model's predictions and the target. This loss should decrease during training as the model becomes more adapted to the data. The graph shows the Loss values, so one can check for a decrease over time, indicating that the model is improving.

Figure 4. Loss chart

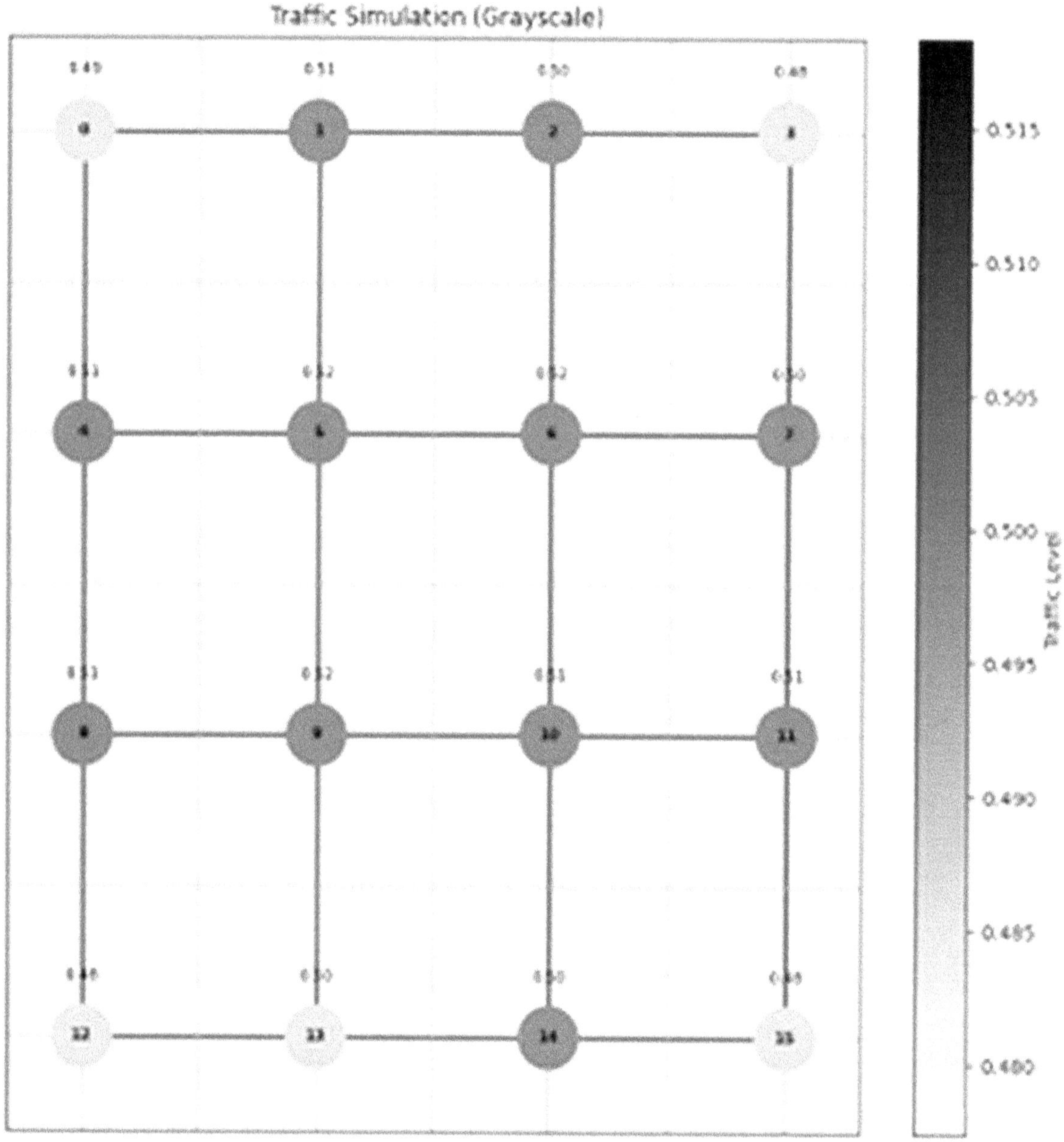

The second phase involves printing the grid with the attribution of the estimated traffic value.

Figure 5. Road traffic simulation

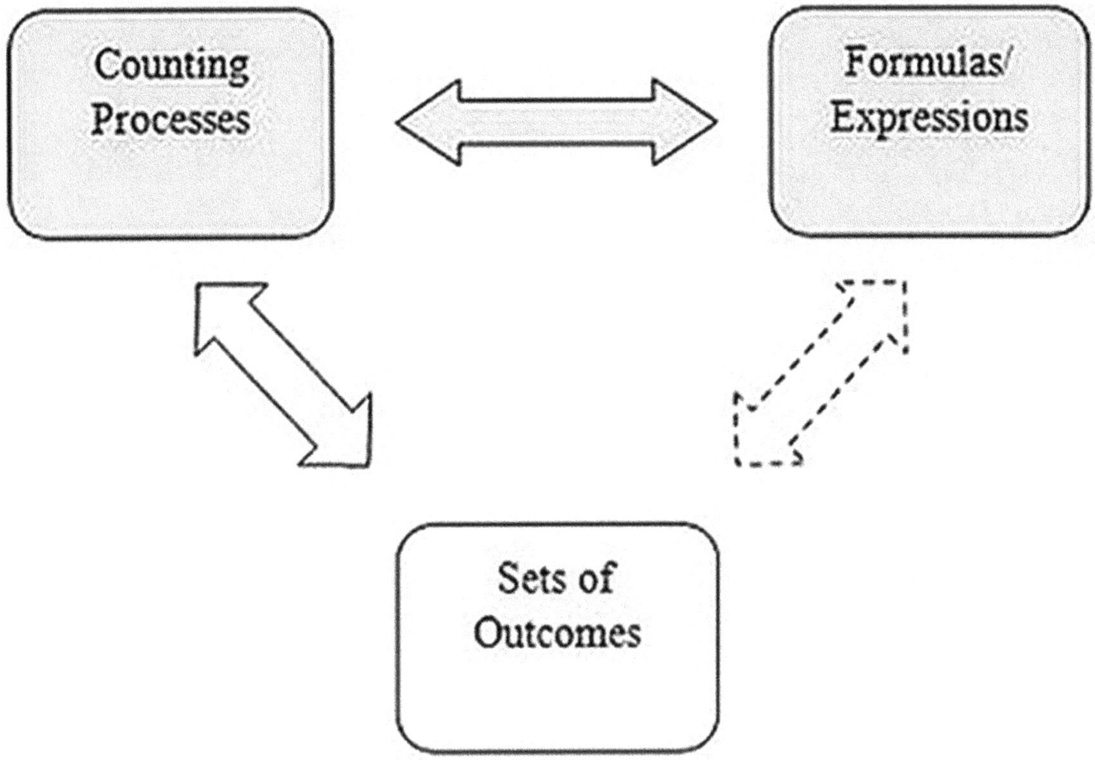

This program combines combinatorial techniques and Artificial Intelligence, particularly machine learning. It can be interpreted as an example of how these two areas interact to solve complex problems, such as traffic simulation in a network of road intersections.

Combinatorics is present in several aspects. A grid of crossings creates a structure of connected nodes that can be studied combinatorially. The network of intersections is an example of how connections between nodes can be combined in different ways, and this brings out combinatorial problems such as calculating minimum distances, finding the nearest neighbours, or analysing traffic flows on different routes. Although the GNN model does not explicitly solve the TSP, it can be applied to situations where a solution depends on how different traffic combinations are distributed across the road network. Regarding artificial intelligence, we highlight that GNN is an advanced AI technique suitable for working with structured data such as graphs (in this case, the road network). The model is trained using a supervised learning process. Real (target) traffic data is compared to the model's predictions to optimize the loss function that measures the difference between the prediction and the true value.

The second simulates an auto-negotiation system for pricing.

Program 133 Price negotiation

```
import random
import numpy as np
class Agent:
```

```
def __init__(self, name, initial_price, strategy='random'):
self.name = name
self.initial_price = initial_price
self.current_price = initial_price
self.strategy = strategy
def negotiate(self, other_agents):
if self.strategy == 'random':
return self._negotiate_random(other_agents)
elif self.strategy == 'average':
return self._negotiate_average(other_agents)
elif self.strategy == 'competitive':
return self._negotiate_competitive(other_agents)
def _negotiate_random(self, other_agents):
variation = random.uniform(-0.1, 0.1)
self.current_price *= (1 + variation)
return self.current_price
def _negotiate_average(self, other_agents):
other_prices = [agent.current_price for agent in other_agents]
average_price = np.mean(other_prices)
self.current_price = (self.current_price + average_price) / 2
return self.current_price
def _negotiate_competitive(self, other_agents):
other_prices = [agent.current_price for agent in other_agents]
max_price = max(other_prices)
self.current_price = min(self.current_price * 1.05, max_price)
return self.current_price
def negotiation_simulation(num_agents=5, num_rounds=10):
strategies = ['random', 'average', 'competitive']
agents = [
Agent(f'Agent_{i}',
random.uniform(50, 100),
random.choice(strategies))
for i in range(num_agents)
]
price_history = []
for round in range(num_rounds):
round_prices = []
for agent in agents:
other_agents = [a for a in agents if a != agent]
new_price = agent.negotiate(other_agents)
round_prices.append(new_price)
price_history.append(round_prices)
return agents, price_history
def main():
agents, price_history = negotiation_simulation()
# Print price evolution across rounds
print("Price evolution across rounds:")
for round, prices in enumerate(price_history):
print(f"Round {round + 1}: {[f'{p:.2f}' for p in prices]}")
```

```
# Print final results
print("\nNegotiation Results:")
for agent in agents:
print(f"{agent.name}: Strategy {agent.strategy}, Final price {agent.current_price:.2f}")
if __name__ == "__main__":
main()
```

This program presents a negotiation model with the following characteristics: three negotiation strategies (random, medium, and competitive), multiple simulation rounds occur, and agents are assigned prices.

In summary, combinatorics provides essential tools and techniques for addressing numerous complex challenges in artificial intelligence (AI). The intersection of combinatorics and AI is crucial for developing more efficient algorithms and innovative solutions in various fields, from solving logical problems to designing intelligent systems. However, it is essential to note that, despite significant progress, AI still faces substantial challenges, particularly in solving complex mathematical problems, including those involving combinatorics.

Some of the significant hurdles in integrating combinatorics with AI include:

1. *Scalability* – Many combinatorial challenges are NP-complete, meaning no efficient solutions exist for large instances. A key difficulty is developing algorithms that can optimally handle these cases. The exponential growth of possible outcomes complicates optimization and the search for practical answers.
2. *Uncertainty* – In real-world applications, data is often incomplete, noisy, or uncertain, adding complexity to the use of combinatorial techniques. Additionally, an AI model trained to solve a specific combinatorial task may become overly adapted to its training data, making it difficult to generalize to new variants. This is particularly problematic in combinatorial contexts, where adaptability is crucial. Developing algorithms that effectively manage uncertainty is a significant challenge, particularly in fields such as natural language processing, pattern recognition, and robotics.
3. *Interpretability* – Many AI algorithms, particularly those based on neural networks and deep learning, function as "black boxes," making their decisions difficult to understand or justify. This poses a problem in applications where transparency and trust are essential, such as healthcare and the justice system. Enhancing the interpretability of combinatorial methods is a priority for AI developers.
4. *Data Bias* – AI algorithms can reflect and amplify biases present in their training data. It is especially concerning when applying combinatorial techniques, as AI-generated solutions may not be neutral but instead mirror historical and cultural inequalities. Bias often originates from training datasets, which inherently carry past societal patterns. Since machine learning algorithms do not create new knowledge but rather reproduce existing trends, addressing bias is a critical challenge (Dastin, 2022).

In summary, combinatorics remains a key pillar in the evolution of AI, providing tools to tackle complex challenges and optimise algorithms. Its ability to explore and manage large volumes of data makes it both a theoretical and practical point of reference, essential for enhancing the performance of AI systems. As technologies and methodologies advance, the possibilities for integration will continue to expand, leading to significant advancements in various industries.

Combinatorics in logic can contribute to developing more efficient algorithms for solving decision-making and constrained problems.

In the context of optimization, combinatorial techniques can reduce computational time and increase the accuracy of results.

In circuit design, combinatorics can be applied to optimize the connectivity of electronic components, improving energy efficiency and performance.

Game theory offers innovative approaches to analysing strategies and behaviours, helping to develop more precise and predictive models.

In addition, the integration of combinatorics with other areas, such as machine learning and deep learning, promises to bring even more advanced innovations. Making AI systems more robust, efficient, and adaptable.

REFERENCES

Al-Obaydy, W. I., Hashim, H. A., Najm, Y. A., & Jalal, A. A. (2022). Document classification using term frequency-inverse document frequency and K-means clustering. *Indonesian Journal of Electrical Engineering and Computer Science, 27*(3), 1517–1524. DOI: 10.11591/ijeecs.v27.i3.pp1517-1524

Balasubramanian, K. (2021). Symmetry, combinatorics, artificial intelligence, music and spectroscopy. *Symmetry, 13*(10), 1850. DOI: 10.3390/sym13101850

Bengio, Y., Lodi, A., & Prouvost, A. (2021). Machine learning for combinatorial optimization: A methodological tour d'horizon. *European Journal of Operational Research, 290*(2), 405–421. DOI: 10.1016/j.ejor.2020.07.063

Bondy, J. A., & Murty, U. S. R. (1976). *Graph theory with applications* (Vol. 290). Macmillan. DOI: 10.1007/978-1-349-03521-2

Briganti, G., & Le Moine, O. (2020). Artificial intelligence in medicine: Today and tomorrow. *Frontiers in Medicine, 7*, 509744. DOI: 10.3389/fmed.2020.00027 PMID: 32118012

Carleo, G., Cirac, I., Cranmer, K., Daudet, L., Schuld, M., Tishby, N., Vogt-Maranto, L., & Zdeborová, L. (2019). Machine learning and the physical sciences. *Reviews of Modern Physics, 91*(4), 045002. DOI: 10.1103/RevModPhys.91.045002

Carvalho, F., & Guedes, G. P. (2020). TF-IDFC-RF: a novel supervised term weighting scheme. arXiv preprint arXiv:2003.07193.

Cheng, X., Huang, B., & Zong, J. (2021). Device-free human activity recognition based on GMM-HMM using channel state information. *IEEE Access : Practical Innovations, Open Solutions, 9*, 76592–76601. DOI: 10.1109/ACCESS.2021.3082627

Clocksin, W. F., & Mellish, C. S. (1984). *Programming in Prolog*. Springer.

Dastin, J. (2022). Amazon scraps secret AI recruiting tool that showed bias against women. In *Ethics of data and analytics* (pp. 296–299). Auerbach Publications. DOI: 10.1201/9781003278290-44

Dong, S., Wang, P., & Abbas, K. (2021). A survey on deep learning and its applications. *Computer Science Review, 40*, 100379. DOI: 10.1016/j.cosrev.2021.100379

Gawlikowski, J., Tassi, C. R. N., Ali, M., Lee, J., Humt, M., Feng, J., Kruspe, A., Triebel, R., Jung, P., Roscher, R., Shahzad, M., Yang, W., Bamler, R., & Zhu, X. X. (2023). A survey of uncertainty in deep neural networks. *Artificial Intelligence Review, 56*(S1, Suppl 1), 1513–1589. DOI: 10.1007/s10462-023-10562-9

Gigerenzer, G. (2022). *How to stay smart in a smart world: Why human intelligence still beats algorithms*. MIT Press.

Goodfellow, I., Bengio, Y., & Courville, A. (2016). *Deep Learning*. MIT Press.

Jiang, W., & Luo, J. (2022). Graph neural network for traffic forecasting: A survey. *Expert Systems with Applications, 207*, 117921. DOI: 10.1016/j.eswa.2022.117921

Karimi-Mamaghan, M., Mohammadi, M., Meyer, P., Karimi-Mamaghan, A. M., & Talbi, E. G. (2022). Machine learning at the service of meta-heuristics for solving combinatorial optimization problems: A state-of-the-art. *European Journal of Operational Research, 296*(2), 393–422. DOI: 10.1016/j.ejor.2021.04.032

Kaur, H., Nori, H., Jenkins, S., Caruana, R., Wallach, H., & Wortman Vaughan, J. (2020, April). Interpreting interpretability: understanding data scientists' use of interpretability tools for machine learning. In *Proceedings of the 2020 CHI conference on human factors in computing systems* (pp. 1-14). DOI: 10.1145/3313831.3376219

Li, Z., Liu, F., Yang, W., Peng, S., & Zhou, J. (2021). A survey of convolutional neural networks: Analysis, applications, and prospects. *IEEE Transactions on Neural Networks and Learning Systems, 33*(12), 6999–7019. DOI: 10.1109/TNNLS.2021.3084827 PMID: 34111009

Malkevitch, J. (1991). Implementing the Standards: Bin Packing: A Mathematical Application. *Mathematics Teacher, 84*(1), 29–31. DOI: 10.5951/MT.84.1.0029

Martello, S., & Toth, P. (1987). Algorithms for knapsack problems. *North-Holland Mathematics Studies, 132*, 213–257. DOI: 10.1016/S0304-0208(08)73237-7

Mazyavkina, N., Sviridov, S., Ivanov, S., & Burnaev, E. (2021). Reinforcement learning for combinatorial optimization: A survey. *Computers & Operations Research, 134*, 105400. DOI: 10.1016/j.cor.2021.105400

Russell, S. J., & Norvig, P. (2016). *Artificial intelligence: a modern approach.* Pearson.

Shannon, C. E. (1950). XXII. Programming a computer for playing chess. *The London, Edinburgh and Dublin Philosophical Magazine and Journal of Science, 41*(314), 256–275. DOI: 10.1080/14786445008521796

Wirsansky, E. (2020). *Hands-on genetic algorithms with Python: applying genetic algorithms to solve real-world deep learning and artificial intelligence problems.* Packt Publishing Ltd.

Wu, Z., Pan, S., Chen, F., Long, G., Zhang, C., & Philip, S. Y. (2020). A comprehensive survey on graph neural networks. *IEEE Transactions on Neural Networks and Learning Systems, 32*(1), 4–24. DOI: 10.1109/TNNLS.2020.2978386 PMID: 32217482

Xue, T., & Liu, H. (2021, July). Hidden Markov Model and its application in human activity recognition and fall detection: A review. In *International conference in communications, signal processing, and systems* (pp. 863-869). Singapore: Springer Singapore.

Yang, Y., & Whinston, A. (2023, July). A survey on reinforcement learning for combinatorial optimization. In 2023 *IEEE World Conference on Applied Intelligence and Computing* (AIC) (pp. 131-136). IEEE. DOI: 10.1109/AIC57670.2023.10263956

Chapter 13
Methodological Choices

ABSTRACT

The chapter explores combinatorics teaching through a structured methodological approach, considering theoretical and practical aspects. Jean Piaget's idea that combinatorial thinking emerges during the formal operational stage (ages 11-12) is discussed, although later studies have challenged this theory. Vergnaud's distinction between conceptualization and symbolization highlights the importance of an interpretative model based on operational invariants. Lockwood's model identifies three essential components of combinatorics: mathematical formulas, counting processes, and sets of outcomes. It emphasizes the need for a seamless transition between these elements for a deeper understanding. The suggested teaching approach includes progressiveness, diversification, and cooperation, stressing the use of visual and digital tools such as diagrams and interactive software to facilitate learning. The chapter demonstrates how combinatorics can be applied to real-world contexts through practical examples, such as password problems, sports tournaments.

METHODOLOGICAL CHOICES

What matters is not the amount of things you do, but the efficiency of each individual action

Tim Ferriss

Didactic experimentation in combinatorics requires a careful and structured methodological approach. Many mathematicians, psychologists, and pedagogues have contributed (Biggs, 2002; Anderson, 2012; Tucker, 2012; Zazkis & Liljedahl, 2019; Mazur, 2022; Sandefur et al., 2022), and it is not easy to move forward with the formulations and suggestions, sometimes being opposed. In his studies, the well-known developmental psychologist Jean Piaget theorized that combinatorial thinking could not be formed before a stage of growth indicated by the scholar as the formal operative stage (11-12 years), an approach later criticized based on further studies. Several studies that are not specific to combinatorics can be easily included.

Consider the following statement by Vergnaud, father of conceptual field theory: "The distinction between conceptualizing and symbolizing is essential, up to the point where the understanding of words and sentences by different persons, particularly students and teachers, is not simply a binary relationship signifier/signified, but a ternary one with the interpretation privileged by operational invariants". The distinction between conceptualizing and symbolizing is fundamental. Students and teachers' understanding of combinatorial problems and solutions is not reduced to a simple binary relationship between the signifier (e.g., a combinatorial formula) and the meaning (the underlying mathematical concept). Rather, it is a ternary relationship that also includes interpretation based on operational invariants, i.e., the conceptual structures individuals use to deal with combinatorial situations (Vergnaud, 1992). For example, when considering the formula for combinations, we have: the signifier, which is the formula, the meaning that is enclosed in the concept of selecting r elements from a set of n without repetition and without considering the order, the operational invariants represented by the mental strategies (and operations) that students use to interpret and apply the concept. This ternary perspective is particularly relevant in combinatorics, where students often struggle to connect symbolic formulas to underlying concepts and concrete situations.

DOI: 10.4018/979-8-3373-3089-1.ch013

Another interesting model of students' combinatorial thinking states that combinatorial reasoning (English, 2005) is based on three distinct but related components: formulas, counting processes, and sets of outcomes. In this respect, we can cite the model developed by Elise Lockwood, a professor of mathematics education. She is a researcher known for her work on students' learning of combinatorics and mathematical reasoning. Her studies focus on how people understand and solve combinatorial problems, emphasising counting processes and using visual and computational representations (Lockwood, 2013; Lockwood, 2014)).

Figure 1. Lockwood model

The model focuses on three interrelated aspects of combinatorial thinking: the mathematical formulas and notations used to represent combinatorial problems, the procedures and methods employed to systematically count possibilities, and the set of all possible outcomes of a given combinatorial scenario.

The key idea of the model is that these three aspects are closely linked and that students should be able to switch between them to develop a deep understanding of combinatorics. The well-known formula permits calculating combinations of n elements taken k at a time. However, it is not enough to memorize this formula. Students need to understand its meaning and derivation and see if there is a different point of view. An example clarifies this last aspect: a password is made up of ten capital letters. How many of these 10-letter passwords contain at least 3 P? The problem is complex, and we must consider many cases. Another effective approach is to subtract those that contain 0, 1, or 2 P from all passwords. This approach could encourage students to explain the connections between formulas, processes, and outcomes.

Field research has identified that the model followed by the students follows this process:

Figure 2. Process

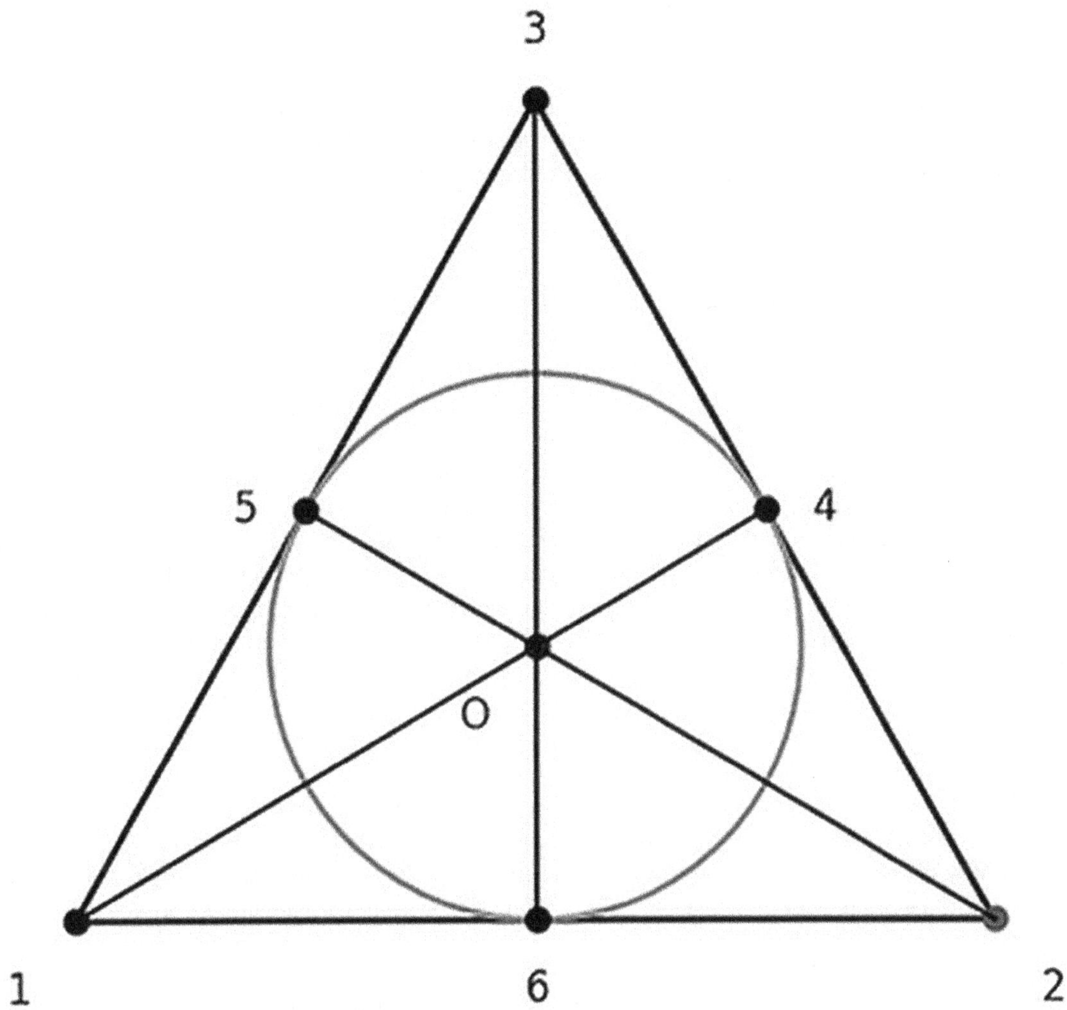

More empirical research, as well as the construction of theoretical models, would be desirable. For example, what could be the most significant methodological indications?

First, it is necessary to distinguish between two possible levels of interpretation: the first centred on the theoretical aspects and the other on the more properly didactic aspects.

The first represents the traditional approach, oriented towards understanding and resolving combinatorial problems using mathematical tools. The principal methodologies at this level include:

1. *Enumerative approach*: This approach involves directly counting the number of possible configurations using basic principles such as the multiplication or inclusion-exclusion principle.
2. *Algebraic approach*: Uses factorials, binomial coefficients, and generating series to represent and manipulate combinatorial sequences.
3. *Algorithmic approach*: Thanks to recursion and backtracking, it is possible to explore possible solutions to a problem.
4. *Graphical approach*: Uses graph theory to represent combinatorial problems through paths, cycles and colourations.

5. *Probabilistic approach*: Analyses configurations from a probabilistic perspective by studying the probabilities of specific configurations and utilising sampling techniques.
6. *Experimental approach*: The use of simulations allows for the exploration and verification of difficult-to-treat combinatorial hypotheses.

On the other hand, combinatorics teaching, especially in high schools, focuses on methods that favour students' active learning (Polya, 1945). In this case, methodological choices are more geared towards promoting conceptual understanding and practical skills through a learning process that stimulates critical thinking, reflection, and collaboration.

Teaching methodologies at this level include:

- *Progressiveness*: Teachers should begin with basic concepts and then gradually progress to more complex steps, allowing students to build a solid foundation of knowledge and reducing the risk of frustration while fostering understanding (Semanišinová, 2021). Above all, it is essential to adapt the problems and exercises to the different skill levels of the students, with basic and advanced exercises. We can plan this development by starting with problems based on the application of learned formulas, then introducing situations that require the combined use of several concepts and finally presenting open problems and real-world situations.
- *Diversification*: It is crucial to provide various examples that demonstrate how combinatorial concepts apply in different contexts and disciplines, highlighting the connections between combinatorics and other areas of mathematics, such as algebra and probability. In abstract algebra, group theory frequently employs combinatorial concepts to investigate symmetries and various arrangements of objects. The permutations form a group under the composition operation, and the results of combinatorial calculus can be applied to analyse the properties of these groups. Combinatorial calculus is essential for calculating probabilities in various contexts. For example, when analysing events in a sample space, combinatorial techniques help determine the total number of favourable outcomes compared to the total number of possible outcomes.
- *Cooperation*: Collaboration between students and group discussion are essential for effective learning. Through group work, students can share ideas and discuss different resolution strategies. Discussing different resolution strategies is one of the most significant aspects of cooperation. Each student may have his or her own method for dealing with a combinatorial problem, and open dialogue allows these strategies to be compared. This process helps identify the most effective solutions and encourages critical thinking and analysis of the choices made. Group work can increase student motivation. Collaborating with peers makes learning more enjoyable and challenging, as students feel part of a group.
- *Problem-solving*: Problem-based learning is a methodology that encourages students to tackle real and complex challenges, stimulating critical thinking and creativity. This approach is particularly effective in combinatorics, where tasks can be framed in practical and applied contexts. Providing real-world scenarios and incorporating games that require combinatorial reasoning is essential. An important aspect to consider is that combinatorial exploration fosters divergent thinking, encouraging students to develop original solutions. The skills acquired through problem-solving in combinatorics prepare students to tackle intricate mathematical and logical tasks (Lockwood & Gibson, 2016). As mathematician George Pólya (1887–1985) famously stated: "Problem-solving is a specific feat of intelligence, and intelligence is the specific gift of mankind: problem-solving can be considered as the most characteristic activity of humankind."
- *Use of visual and computer tools*: They are powerful allies in teaching combinatorics. Using diagrams like decision trees or Venn diagrams helps visualise the relationships and possible combinations of the elements. These tools can make concepts such as permutations and combinations clearer. Using physical objects allows students to manipulate the elements and see how the different combinations are formed. This hands-on activity facilitates learning through hands-on experience. Concept maps visually represent the relationships between various combinatorial concepts, helping students logically organize information. All these tools help students visualize problems and better understand not only concepts but also processes. Technology also offers numerous tools to enrich the teaching of combinatorics. Programs and applications like Python, GeoGebra, or Wolfram Alpha can make learning interactive and engaging. Spreadsheets can be used to perform calculations quickly, but also allow

simulations. Numerous websites and educational platforms also offer interactive exercises on combinatorics. These tools allow students to practice independently and receive immediate feedback.

- *Reflection*: It is very formative to encourage students to reflect on which approaches have worked best and why, analysing errors that are quite frequent in this field. It may be helpful to invite students to keep a journal in which they record their learning experiences, and the strategies used to facilitate discussions about problem-solving strategies and thought processes employed.

The theoretical level is essential for a deep understanding of combinatorial structures and counting methods, the didactic level is crucial for engaging students and developing practical and conceptual skills that go beyond simply memorizing formulas. Only integrating the two levels can lead to more solid and lasting learning. The teacher may introduce the basic concepts of combinatorics through practical problems and games and then move on to more formal and theoretical approaches. In this way, the transition from the concrete to the abstract and vice versa fosters a deeper understanding and the acquisition of skills that can also be applied in more advanced contexts.

In conclusion, the methodological choices in the study of combinatorics, whether oriented towards a theoretical or didactic level, directly reflect teaching strategies and approaches to problem-solving. Both levels are crucial but finding the right balance to stimulate curiosity and understanding is important.

The didactic aspects will be addressed with concrete suggestions in the following chapter. However, it is now possible to present some examples that deepen the discussion on the methodological approaches just enunciated.

Example 1: To open the door and access an Airbnb apartment, we must type a four-digit code with nine digits on a numeric keypad. Unfortunately, we do not recall the code. Judging by how worn out the buttons are, the letters that make up the code are 2, 5, 7 and 9. How many possible codes will we need to try to enter? After passing the gate, it is necessary to enter the apartment, and here, too, the same problem arises. The numbers most consumed are 1, 3, and 8, but the number 3 is strongly consumed; therefore, we assume this number is repeated twice. How many possible codes are there to overcome this obstacle as well?

The problem is easily solved by visualizing the situation with a tree and, formally, using the formulas of permutations and permutations with repetition [24, 12].

However, it is interesting to propose the situation concretely. Viewing the keypads and being able to try to open the doors creates an immersive situation.

Below are the two programs that use a visual approach applied to a problematic situation:

Program 134 door code 1

```
import tkinter as tk
from tkinter import messagebox
# Correct code to enter
correct_code = "5297"
# Function to verify the entered code
def verify_code():
entered_code = entry.get()
if entered_code == correct_code:
messagebox.showinfo("Success", "Correct code! Door opened.")
else:
messagebox.showwarning("Error", "Incorrect code. Please try again.")
entry.delete(0, tk.END)
# Function to add the pressed digit to the input field
def add_digit(digit):
current = entry.get()
entry.delete(0, tk.END)
```

```
entry.insert(0, current + digit)
# Creating the main window
root = tk.Tk()
root.title("Numeric Keypad Simulator")
# Creating the code input field
entry = tk.Entry(root, font=("Helvetica", 24), justify="center")
entry.grid(row=0, column=0, columnspan=3)
# Creating the numeric buttons
buttons = [
('1', 1, 0), ('2', 1, 1), ('3', 1, 2),
('4', 2, 0), ('5', 2, 1), ('6', 2, 2),
('7', 3, 0), ('8', 3, 1), ('9', 3, 2),
('0', 4, 1)
]
# Function to create buttons
def create_button(text, row, col):
text_color = "gray" if text in ['2', '5', '7', '9'] else "black"
button = tk.Button(root, text=text, font=("Helvetica", 24), bg="white", fg=text_color,
command=lambda t=text: add_digit(t))
button.grid(row=row, column=col, sticky="nsew")
# Creating the buttons
for (text, row, col) in buttons:
create_button(text, row, col)
# Creating the submit button
submit_button = tk.Button(root, text="#", font=("Helvetica", 24), command=verify_code)
submit_button.grid(row=4, column=0, columnspan=3, sticky="nsew")
# Configuring the grid to expand the buttons
for i in range(5):
root.grid_rowconfigure(i, weight=1)
for i in range(3):
root.grid_columnconfigure(i, weight=1)
# Starting the main application loop
root.mainloop()
```

Program 135 door code 2

```
import tkinter as tk
from tkinter import messagebox
# Correct code to enter
correct_code = "3138"
# Function to verify the entered code
def verify_code():
entered_code = entry.get()
if entered_code == correct_code:
messagebox.showinfo("Success", "Correct code! Door opened.")
else:
messagebox.showwarning("Error", "Incorrect code. Please try again.")
```

```
entry.delete(0, tk.END)
# Function to add the pressed digit to the input field
def add_digit(digit):
current = entry.get()
entry.delete(0, tk.END)
entry.insert(0, current + digit)
# Creating the main window
root = tk.Tk()
root.title("Numeric Keypad Simulator")
# Creating the code input field
entry = tk.Entry(root, font=("Helvetica", 24), justify="center")
entry.grid(row=0, column=0, columnspan=3)
# Creating the numeric buttons
buttons = [
('1', 1, 0), ('2', 1, 1), ('3', 1, 2),
('4', 2, 0), ('5', 2, 1), ('6', 2, 2),
('7', 3, 0), ('8', 3, 1), ('9', 3, 2),
('0', 4, 1)
]
# Function to create buttons
def create_button(text, row, col):
text_color = "gray" if text in ['3', '1', '8'] else "black"
button = tk.Button(root, text=text, font=("Helvetica", 24), bg="white", fg=text_color,
command=lambda t=text: add_digit(t))
button.grid(row=row, column=col, sticky="nsew")
# Creating the buttons
for (text, row, col) in buttons:
create_button(text, row, col)
# Creating the submit button
submit_button = tk.Button(root, text="#", font=("Helvetica", 24), command=verify_code)
submit_button.grid(row=4, column=0, columnspan=3, sticky="nsew")
# Configuring the grid to expand the buttons
for i in range(5):
root.grid_rowconfigure(i, weight=1)
for i in range(3):
root.grid_columnconfigure(i, weight=1)
# Starting the main application loop
root.mainloop()
```

The next step might be to have students solve a similar problem with a different number of digits.

Example 2: Eight teams participate in a school football tournament; we want to create a calendar so that each team can meet the others. How many matches will be played? How do we reach this conclusion?

Each team must play against all the others exactly once. The total of matches is given by the combinations $\binom{n}{2}$ and calculated with the formula $\frac{n(n-1)}{2}$, where n is the number of teams.

There are several possible approaches. Students can imagine themselves as part of one of the teams and write down a list of games they would play; this encourages active involvement, which can be compared with the results of their peers to stimulate a collaborative process. To stimulate visual learning, we can create an 8x8 table, for example (excluding the diagonal and considering only the upper or lower halves to avoid duplicates). Alternatively, we can arrange the squares as the vertices of an octagon and connect them all to calculate how many arcs there are.

Use this program to print the calendar.

Program 136 calendar

```
import matplotlib.pyplot as plt
import numpy as np
def round_robin_italian(teams, rounds=None):
num_teams = len(teams)
rounds = rounds or (num_teams - 1)
half = num_teams // 2
for round in range(rounds):
left = teams[:half]
right = teams[num_teams - half - 1 + 1:][::-1]
matches = zip(left, right)
if round % 2 == 1:
matches = [(y, x) for (x, y) in matches]
teams.insert(1, teams.pop())
yield matches
teams = ["Team 1", "Team 2", "Team 3", "Team 4", "Team 5", "Team 6", "Team 7", "Team 8"]
schedule = list(round_robin_italian(teams))
# Creating the plot
fig, ax = plt.subplots()
# Calculating the coordinates of the octagon's vertices
theta = np.linspace(0, 2 * np.pi, 8, endpoint=False)
x = np.cos(theta)
y = np.sin(theta)
# Drawing the edges between the vertices
for i in range(8):
for j in range(i + 1, 8):
ax.plot([x[i], x[j]], [y[i], y[j]], 'k-', alpha=0.2)
# Drawing the vertices of the octagon
ax.scatter(x, y)
# Labeling the vertices with the team names
for i, txt in enumerate(teams):
ax.annotate(txt, (x[i], y[i]))
# Printing the number of edges
num_edges = len(teams) * (len(teams) - 1) // 2
plt.title(f'Number of edges: {num_edges}')
# Display the plot
plt.show()
# Printing the full schedule
for i, round in enumerate(schedule):
print("Round", i + 1)
for match in round:
print(match[0], "vs", match[1])
```

The previous problem can be articulated in such a way as to highlight the concept of progressivity.

Example 3: Let us change the situation appropriately. Eight teams participate in a school football tournament. We want to create a calendar so that each team can meet all the others. After this first round, the top four teams advance to the next stage, meeting the same criteria. The tournament proceeds with the two best teams competing to determine the winning team. How many matches will be played?

Program 137 Football Tournament 1

```
import itertools
teams = ['Team A', 'Team B', 'Team C', 'Team D', 'Team E', 'Team F', 'Team G', 'Team H']
# Function to calculate round-robin matches
def round_robin(teams):
matches = list(itertools.combinations(teams, 2))
return matches
# Function to organize the final match
def final():
return [("First place Second Round", "Second place Second Round")]
# Calculate the matches of the first round-robin group
first_round_matches = round_robin(teams)
print("First round-robin group matches:")
for match in first_round_matches:
print(f"{match[0]} vs {match[1]}")
# Simulate the top four ranked teams
top_four_teams = teams[:4] # In practice, they are determined by the ranking of the first
round
# Calculate the matches of the second round-robin group
second_round_matches = round_robin(top_four_teams)
print("\nSecond round-robin group matches:")
for match in second_round_matches:
print(f"{match[0]} vs {match[1]}")
# Calculate the final match
final_match = final()
print("\nFinal match:")
for match in final_match:
print(f"{match[0]} vs {match[1]}")
# Total number of matches
total_matches = len(first_round_matches) + len(second_round_matches) + len(final_match)
print(f"\nTotal number of matches: {total_matches}")
```

Example 4: Let us say that the 8 teams are divided into two groups of 4 teams each, and that in each group each team faces all the other teams in the group once. After the group stage, the top two teams in each group move on to the knockout bracket (semi-finals and final). How many games do I need to finish the tournament?

Program 138 Football Tournament 2

```
import itertools
teams = ['Team A', 'Team B', 'Team C', 'Team D', 'Team E', 'Team F', 'Team G', 'Team H']
# Function to calculate round-robin matches
```

```
def round_robin(teams):
matches = list(itertools.combinations(teams, 2))
return matches
# Function to organize the semifinals and the final
def elimination_phase(qualified_teams):
# Semifinals: the top two from each group face off crosswise
semifinals = [(qualified_teams[0], qualified_teams[3]), (qualified_teams[1], qualified_
teams[2])]
# Final
final = [("Winner of Semifinal 1", "Winner of Semifinal 2")]
return semifinals + final
# Divide the teams into two groups
group_1 = teams[:4]
group_2 = teams[4:]
# Calculate the matches for both groups
group_1_matches = round_robin(group_1)
group_2_matches = round_robin(group_2)
print("Group 1 Matches:")
for match in group_1_matches:
print(f"{match[0]} vs {match[1]}")
print("\nGroup 2 Matches:")
for match in group_2_matches:
print(f"{match[0]} vs {match[1]}")
# Simulate the top two ranked teams from each group
top_two_group_1 = group_1[:2]
top_two_group_2 = group_2[:2]
qualified_teams = top_two_group_1 + top_two_group_2
# Calculate the semifinals and the final
elimination_matches = elimination_phase(qualified_teams)
print("\nElimination Phase Matches:")
for match in elimination_matches:
print(f"{match[0]} vs {match[1]}")
# Total number of matches
total_matches = len(group_1_matches) + len(group_2_matches) + len(elimination_matches)
print(f"\nTotal number of matches: {total_matches}")
```

A later addition could predict the result by attributing a value of “strength” to each team, indicative of the probability of victory. This approach makes the tournament more realistic and introduces the ability to use probability to simulate match outcomes, providing students with a deeper understanding of the tournament's dynamics.

Program 139 Football Tournament 3

```
import itertools
teams = ['Team A', 'Team B', 'Team C', 'Team D', 'Team E', 'Team F', 'Team G', 'Team H']
# Function to calculate round-robin matches
def round_robin(teams):
matches = list(itertools.combinations(teams, 2))
return matches
```

```
# Function to organize the semifinals and the final
def elimination_phase(qualified_teams):
# Semifinals: the top two from each group face off crosswise
semifinals = [(qualified_teams[0], qualified_teams[3]), (qualified_teams[1], qualified_
teams[2])]
# Final
final = [("Winner of Semifinal 1", "Winner of Semifinal 2")]
return semifinals + final
# Divide the teams into two groups
group_1 = teams[:4]
group_2 = teams[4:]
# Calculate the matches for both groups
group_1_matches = round_robin(group_1)
group_2_matches = round_robin(group_2)
print("Group 1 Matches:")
for match in group_1_matches:
print(f"{match[0]} vs {match[1]}")
print("\nGroup 2 Matches:")
for match in group_2_matches:
print(f"{match[0]} vs {match[1]}")
# Simulate the top two ranked teams from each group
top_two_group_1 = group_1[:2]
top_two_group_2 = group_2[:2]
qualified_teams = top_two_group_1 + top_two_group_2
# Calculate the semifinals and the final
elimination_matches = elimination_phase(qualified_teams)
print("\nElimination Phase Matches:")
for match in elimination_matches:
print(f"{match[0]} vs {match[1]}")
# Total number of matches
total_matches = len(group_1_matches) + len(group_2_matches) + len(elimination_matches)
print(f"\nTotal number of matches: {total_matches}")
```

From an educational perspective, a diagram showing how teams compete in a round-robin format could be helpful for football tournaments.

Staying within the realm of football, which can involve at least some students, we can tackle a rather complex combinatorial problem. One of the most fascinating applications of combinatorics in football is semi-automatic offside (SAOT), a technology introduced to help referees determine whether a player is in an offside position during a football match. Each player on the pitch is monitored through 29 points representing the position of specific body parts, such as legs, arms and head. These points are tracked in real time to determine each player's exact position. The system uses 12 cameras around the stadium, capturing 50 frames per second. Each frame provides information on the position of the players and the ball, generating a continuous flow of data. Inside the ball is a sensor that sends its position to the system 500 times per second, allowing it to precisely track the moment the ball is kicked. Of course, not all possible combinations of data are analysed. The system selects only the relevant configurations (e.g., the position of 3 key players) to determine if a player is offside.

To recap:

- player data points = 29
- camera data points = 600

- players = 3
- ball positions = 500

    total combinations = $(29^{600})^3 \cdot 500 \approx 10^{2635}$

To better understand the complexity of the problem, we can try to simulate a simplified version of the SAOT system. Suppose we reduce the values:

- player data points = 10
- camera data-points = 100
- players = 3
- ball positions = 50

Even so, the number of combinations is enormous: total combinations = $(10^{100})^3 \cdot 50 \approx 5 \cdot 10^{301}$.

It is an excellent example of how combinatorics and data analysis are applied in authentic contexts such as professional football. The system's ability to handle such a massive number of combinations in real-time is a testament to the power of modern mathematics and technology.

REFERENCES

Anderson, I. (2012). *A first course in discrete mathematics.* Springer Science & Business Media.

Biggs, N. (2002). *Discrete mathematics.* Oxford University Press.

English, L. D. (2005). Combinatorics and the development of children's combinatorial reasoning. In *Exploring probability in school: Challenges for teaching and learning* (pp. 121–141). Springer US. DOI: 10.1007/0-387-24530-8_6

Lockwood, E. (2013). A model of students' combinatorial thinking. *The Journal of Mathematical Behavior, 32*(2), 251–265. DOI: 10.1016/j.jmathb.2013.02.008

Lockwood, E. (2014). A set-oriented perspective on solving counting problems. *For the Learning of Mathematics, 34*(2), 31–37.

Lockwood, E., & Gibson, B. R. (2016). Combinatorial tasks and outcome listing: Examining productive listing among undergraduate students. *Educational Studies in Mathematics, 91*(2), 247–270. DOI: 10.1007/s10649-015-9664-5

Mazur, D. R. (2022). *Combinatorics: a guided tour* (Vol. 55). American Mathematical Society.

Polya, G. (1945). *How to solve it: A new aspect of mathematical method.* Princeton university press. DOI: 10.1515/9781400828678

Sandefur, J., Lockwood, E., Hart, E., & Greefrath, G. (2022). Teaching and learning discrete mathematics. *ZDM Mathematics Education, 54*(4), 753–775. DOI: 10.1007/s11858-022-01399-7 PMID: 35528041

Semanišinová, I. (2021). Multiple-Solution Tasks in Pre-Service Teachers Course on Combinatorics. *Mathematics, 9*(18), 2286. DOI: 10.3390/math9182286

Tucker, A. (2012). *Applied combinatorics.* Wiley Global Education.

Vergnaud, G. (1992). Conceptual fields, problem solving and intelligent computer tools. In *Computer-based learning environments and problem solving* (pp. 287–308). Springer Berlin Heidelberg. DOI: 10.1007/978-3-642-77228-3_14

Zazkis, R., & Liljedahl, P. (2019). *Teaching mathematics as storytelling.* Brill.

Chapter 14
Didactics of Combinatorics

ABSTRACT

The chapter focuses on teaching combinatorics, a challenging mathematical discipline to learn and teach. Its importance lies in developing logical reasoning, problem-solving skills, and applications in computer science, statistics, and probability. The suggested teaching approach emphasizes real-world problems, interdisciplinary connections, and technological tools. Active discovery, using games, simulations, and collaborative activities, is encouraged rather than mere memorization of formulas. Various examples illustrate how to make combinatorics more engaging. These include the "Pyramid Game," based on a dice roll and a triangular structure, and the use of the Fano plane to organize work shifts or optimize lottery strategies. Other problems include the "Tower of London," which enhances logical skills, and the carpooling problem, which explores different group arrangements. The goal is to integrate theory and practice, making combinatorics more accessible and stimulating for students.

DIDACTICS OF COMBINATORICS

A good teacher is a master of simplification and an enemy of simplism

Louis A. Berman

After examining the most relevant aspects of the topic, one can ask what the important elements are for those who have to teach this discipline to students, considering that, for Eizenberg and Zaslavsky (2004), "Combinatorics is considered one of the more difficult mathematical topics to teach and to learn". The study of combinatorics is important for several reasons: it helps to improve combinatorial reasoning (Kenney & Hirsch, 1991; Batanero et al., 1997; Hart & Martin, 2018; Medová et al., 2020; Simamora & Zunaiedy, 2021; Nopriana et al., 2023) and problem-solving skills, is helpful in many fields such as computer science, statistics, and probability theory, and forms a basis for more advanced mathematical concepts useful for achieving certain tasks (Semanišinová, 2021).

Teaching combinatorics is a vast and complex topic that can be approached differently depending on the student's school level, educational goals, and interests (Braithwaite & Goldstone, 2013; Vásquez et al., 2021; Tillema & Burch, 2022; Andriano et al., 2023). Of course, presenting the topic at the elementary school level is different from presenting it at the university level.

As we saw in the previous chapter, a first suggestion is to start with simple and concrete problems so as not to present the topic as a mountain to climb; In some cases, the reference to artifacts can greatly help. Secondly, it effectively demonstrates how combinatorics is connected to statistics, probability, algebra, and geometry, and more broadly, to all the different STEM areas. From a methodological point of view, it is necessary to encourage discovery instead of presenting formulas and rules "ready for use" to develop critical thinking skills through problem-solving activities, games and simulations with group and collaborative work (Mashiach Eizenberg & Zaslavsky, 2004; Lockwood, 2013; Lockwood, et al., 2020; Lockwood, 2022). Finally, the importance of educational and technological tools should not be underestimated, from simulation software to educational apps, educational videos on YouTube, and versatile programming languages (Logofătu & Gruber, 2014) such as Python. There are interesting examples of connecting some aspects of mathematics to combinatorics using the GeoGebra software and Python.

DOI: 10.4018/979-8-3373-3089-1.ch014

An example of how it is possible to tackle non-trivial problems using a methodology that connects "playful" situations to theory and programming is as follows: consider a game in which you move along a structure by throwing a die starting from the central point.

Figure 1. Pyramid game

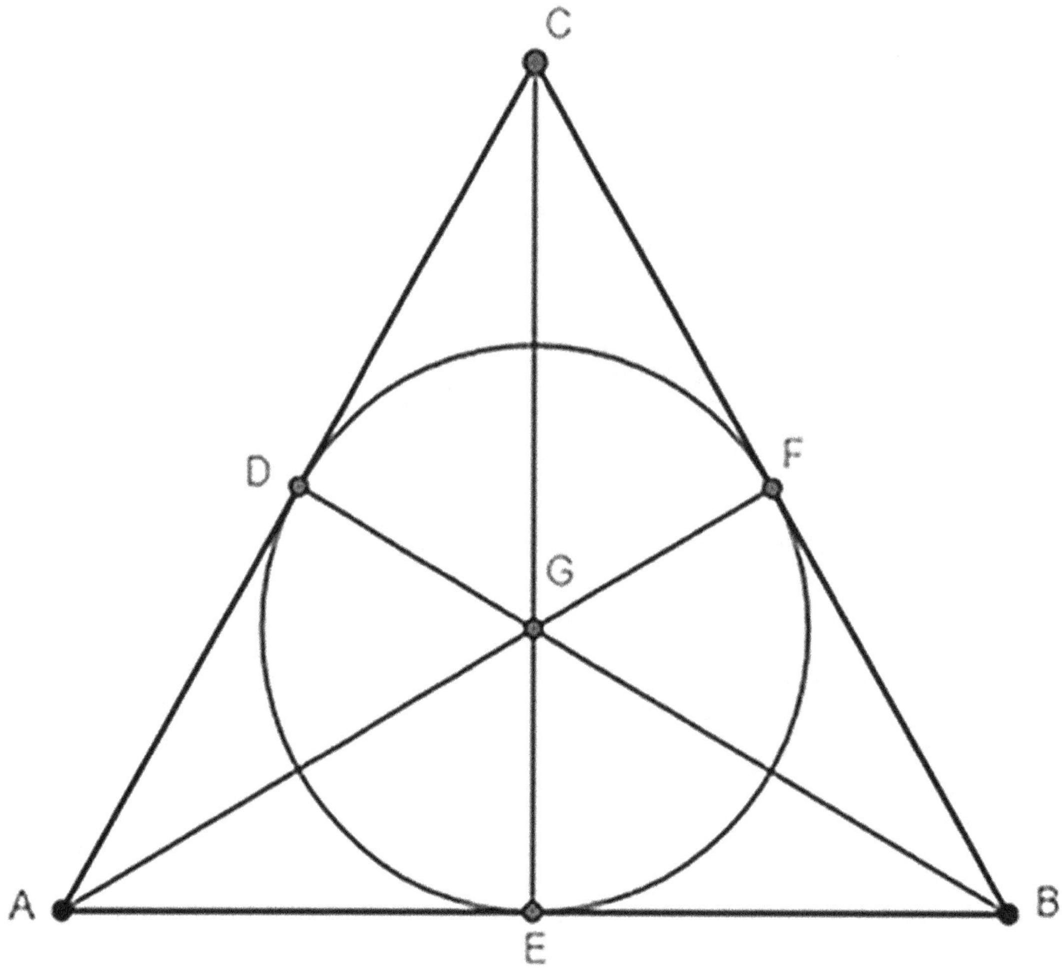

Depending on the result, we move to one of the external vertices. The game continues with a new roll of the dice. If the same number as the node comes up, the game ends. To view how the game works, one can use the following program:

Program 140 Pyramid Game

```
import matplotlib.pyplot as plt
import numpy as np
import random
# Function to draw the triangle, the inscribed circle, and the nodes numbered from 1 to 6
```

```
def draw_triangle_with_incircle_and_segments(ax, current_node, throws, colors):
ax.clear()
# Define the vertices of the equilateral triangle
P1 = np.array([0, 0]) # Node 1
P2 = np.array([1, 0]) # Node 2
P3 = np.array([0.5, np.sqrt(3)/2]) # Node 3
# Calculate the centroid (center of the inscribed circle)
O = (P1 + P2 + P3) / 3
# Calculate the radius of the inscribed circle
a = np.linalg.norm(P2 - P3) # Side opposite to P1
b = np.linalg.norm(P1 - P3) # Side opposite to P2
c = np.linalg.norm(P1 - P2) # Side opposite to P3
s = (a + b + c) / 2 # Semi-perimeter of the triangle
radius = np.sqrt(s * (s - a) * (s - b) * (s - c)) / s
# Calculate the midpoints of the sides (nodes 4, 5, 6)
P4 = (P2 + P3) / 2 # Node 4, midpoint of side P2P3
P5 = (P3 + P1) / 2 # Node 5, midpoint of side P3P1
P6 = (P1 + P2) / 2 # Node 6, midpoint of side P1P2
# Positions of the nodes
nodes = {
1: P1,
2: P2,
3: P3,
4: P4,
5: P5,
6: P6,
0: O
}
# Draw the triangle
triangle = plt.Polygon([P1, P2, P3], closed=True, fill=None, edgecolor='black', line-
width=2)
ax.add_patch(triangle)
# Draw the inscribed circle
circle = plt.Circle(O, radius, color='gray', fill=False, linestyle='-', linewidth=2)
ax.add_patch(circle)
# Draw the segments from the nodes to the opposite nodes
ax.plot([P1[0], P4[0]], [P1[1], P4[1]], 'k-') # Segment 1-4
ax.plot([P2[0], P5[0]], [P2[1], P5[1]], 'k-') # Segment 2-5
ax.plot([P3[0], P6[0]], [P3[1], P6[1]], 'k-') # Segment 3-6
# Draw the nodes with the specified colors
for node, pos in nodes.items():
size = 50
ax.scatter(*pos, s=size, c=colors[node])
# Annotate the nodes
offset = 0.05 # Distance between the letter and the point
ax.text(P1[0] - offset, P1[1] - offset, '1', fontsize=12, ha='right', va='top')
ax.text(P2[0] + offset, P2[1] - offset, '2', fontsize=12, ha='left', va='top')
ax.text(P3[0], P3[1] + offset, '3', fontsize=12, ha='center', va='bottom')
ax.text(P4[0] + offset, P4[1], '4', fontsize=12, ha='left', va='center')
```

```
ax.text(P5[0] - offset, P5[1], '5', fontsize=12, ha='right', va='center')
ax.text(P6[0], P6[1] - offset, '6', fontsize=12, ha='center', va='top')
ax.text(O[0] - offset, O[1] - offset, 'O', fontsize=12, ha='right', va='top', col-
or='blue')
# Display the number of rolls if the game is over
if current_node != 0:
ax.text(0.5, -0.4, f"Number of rolls: {throws}", fontsize=14, ha='center', va='center',
color='red')
# Graphic settings
ax.set_aspect('equal')
ax.axis('off') # Removes the axes
# Function to simulate the game and trace the sequence
def simulate_game_with_visualization():
current_node = 0 # Central node O
throws = 0
colors = {node: 'black' for node in range(7)} # Initial color black for all nodes
colors[0] = 'blue' # Central node colored blue
# Set up the graphical window
fig, ax = plt.subplots()
while True:
# Draw the current situation
draw_triangle_with_incircle_and_segments(ax, current_node, throws, colors)
plt.pause(1) # 1-second pause to show the current position
# Roll the dice
throws += 1
dice_roll = random.randint(1, 6)
# If the current node matches the dice roll, end the game
if current_node == dice_roll:
colors[dice_roll] = 'red' # Final node in red
draw_triangle_with_incircle_and_segments(ax, current_node, throws, colors)
plt.pause(2) # Show the final result for 2 seconds
break
# Update the colors
if current_node != 0: # Make the previous node black if it's not the central node
colors[current_node] = 'black'
colors[dice_roll] = 'red' # The reached node becomes red
current_node = dice_roll
plt.close(fig) # Close the graphical window at the end of the game
print(f"The game ended after {throws} rolls of the dice.") # Print the number of rolls
# Run the game with visualization
simulate_game_with_visualization()
```

At this point we can ask ourselves after how many shots, on average, the game ends.

Program 141 Pyramid Game Simulations

```
import random
# Function to simulate the game
```

```
def simulate_game():
current_node = 0 # Central node O
throws = 0
while True:
# Roll the dice
throws += 1
dice_roll = random.randint(1, 6)
# If the current node matches the dice roll, end the game
if current_node == dice_roll:
return throws # Return the number of throws when the game ends
# Update the current node
current_node = dice_roll
# Function to run 1000 simulations and calculate the average
def simulate_multiple_games(num_simulations=100000):
total_throws = 0
for _ in range(num_simulations):
total_throws += simulate_game()
average_throws = total_throws / num_simulations
return average_throws
# Run the simulation and calculate the average
average_throws = simulate_multiple_games()
print(f"Average number of throws after 100000 simulations: {average_throws:.2f}")
```

The drawing reproduces the Fano plan, which represents a finite projective plane with seven points, seven lines, and seven intersections.

The name comes from the mathematician Gino Fano (1871-1952), who first described it in 1892.

Figure 2. Fano plan

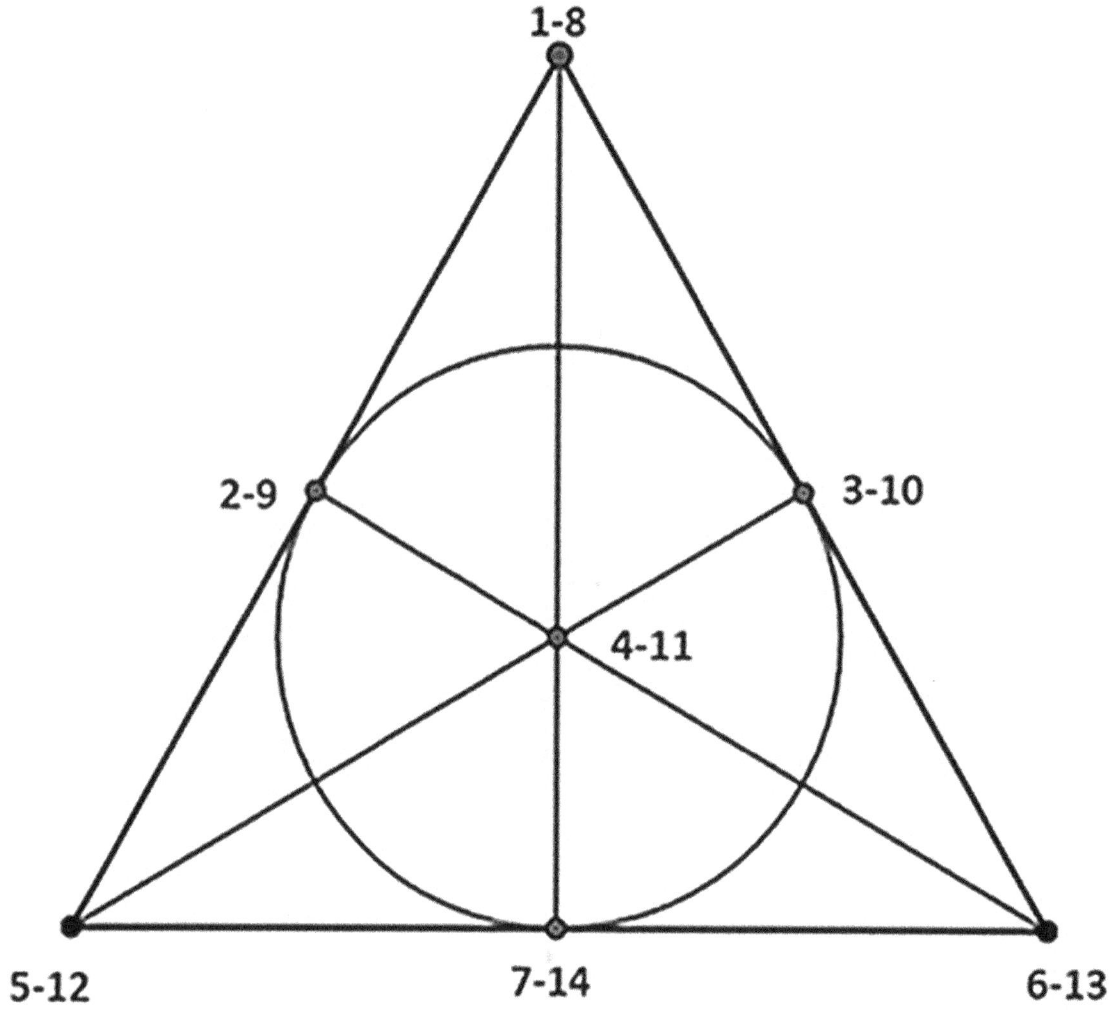

It is an example of a finite projective manifold with applications in various areas of mathematics.

The Fano plane has a configuration of 7 points and 7 lines (the curve is also considered a line), with the following properties: each line passes through 3 distinct points, and each point belongs to 3 lines.

This satisfies one of the formal definitions of a combinatorial configuration, where relationships are given between objects (in this case, points and lines) that cross each other according to specific rules.

If we want to give a formal description of the Fano plane, there are seven points and seven lines; each line contains precisely 3 points, and each point is contained in exactly 3 lines. The lines and points are arranged so that each pair of lines intersects at a point, and each triple point is contained in a line. In summary, the Fano plane is an example of a finite combinatorial configuration that satisfies particular geometric and combinatorial properties, including the incidence of points and lines.

Therefore, we can transition from play to combinatorial structure and, subsequently, arrive at practical applications. Let us provide a couple of examples..

So, we can move from play to combinatorial structure and, later, get to practical applications. Let us give a couple of examples.

In the first case, seven nurses must be assigned to night shifts that involve the presence of three people. How should these shifts be organized?

The Fano plan solves the problem. If we consider the points, such as nurses, and the lines, such as night shifts, we can organize the desired triples by choosing the points (nurses) that are on each line (night shifts).

Table 1. Shift

Monday	**AEB**
Tuesday	ADC
Wednesday	AGF
Thursday	BFC
Friday	BGD
Saturday	CGE
Sunday	DEF

In this way, three nurses will always be present on night shifts every day of the week. This scheme ensures that no nurse is overburdened compared to the others and that there is an equal rotation of shifts.

In the second case, consider a lottery where the player selects three values between 1 and 14, and three are drawn randomly. The player wins if two of their choices match the drawn results. The question is: How many tickets must be purchased to guarantee a win?

This problem can be solved using the Fano plane, where each point is assigned a value between 1 and 7 and another between 8 and 14.

Figure 3. Lottery

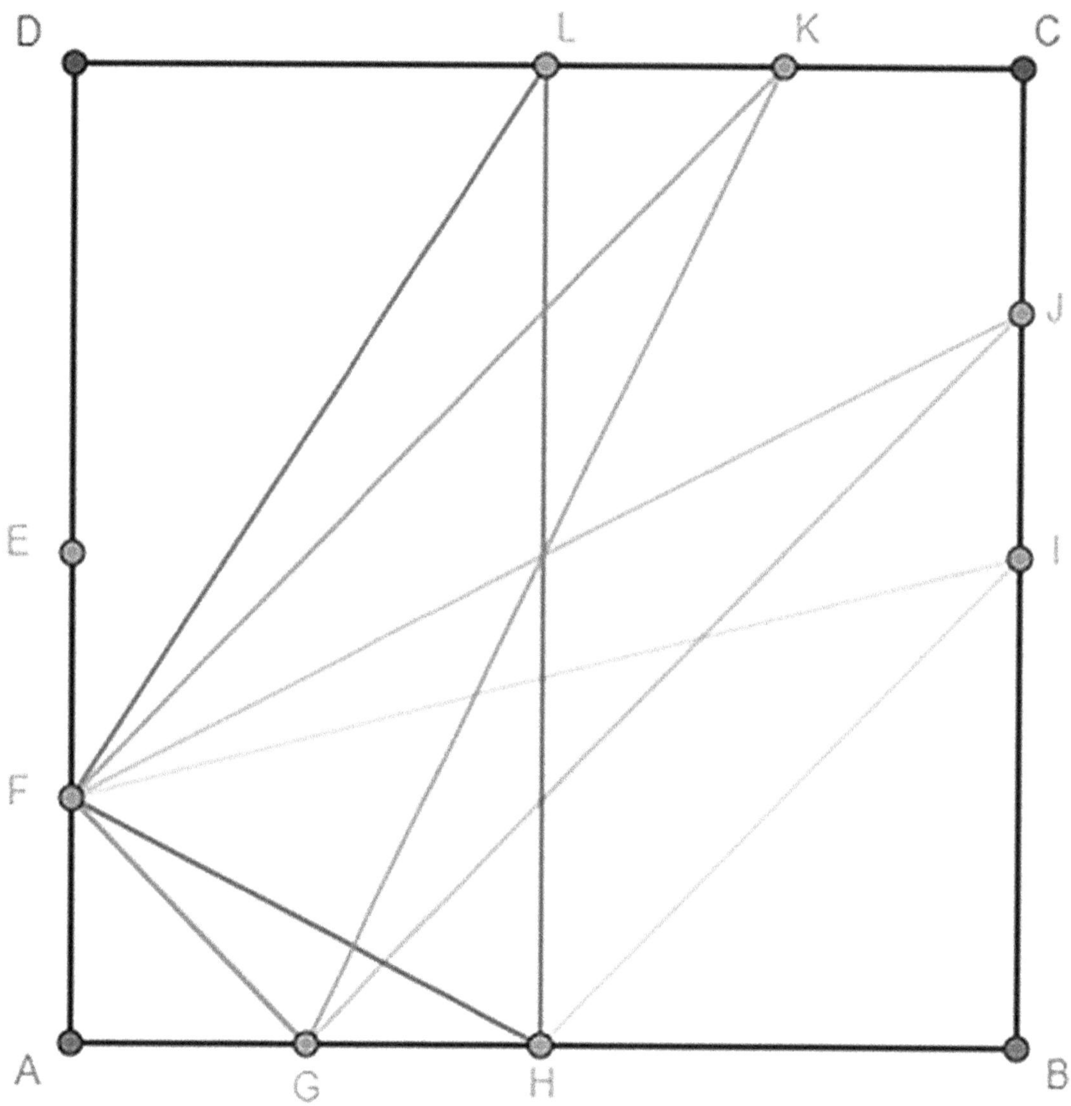

Thus, we have two sets:

Table 2. The set

Low	1-2-5	1-3-6	1-4-7	2-3-7	2-4-6	3-4-5	5-6-7
High	8-9-12	8-10-13	8-11-14	9-10-14	9-11-13	10-11-12	12-13-14

The solution is to buy 14 tickets with the above triples. If (a, b, c) is the winning ticket, at least two of these numbers are in the range [1, 7] or [8, 14].

Another example that can be presented as a game is a neuropsychological test. A crucial aspect in this context is that one can also analyse from an educational perspective. The *Tower of London Test* (TOL) introduces concepts of mathematical logic, algorithms and combinatorics (Boccia et al., 2017). The task consists of analysing a situation and finding an optimal or possible solution. It is an example of problem-solving, which involves developing the ability to break down a problem into smaller, more manageable subproblems. It is one of the most used tests in neuropsychology to evaluate planning and monitoring skills during the execution of a task (Shallice, 1982).

The test is based on a device consisting of three pegs of different lengths, designed to contain one, two or three spheres, respectively. The three spheres are distinguishable by colour: red, green and blue. The examinee's task is to rearrange an initial configuration of the spheres, indicated by the examiner, to reach a predefined final configuration, respecting a maximum number of moves and following specific rules:

1. One can move only one sphere at a time.
2. One cannot place spheres outside the pegs.
3. One must respect the capacities of the pegs (that is, a peg can contain a maximum of one, two or three spheres, depending on its length).

Considering an initial configuration with only one sphere already positioned, it is possible to calculate the total number of configurations that the examiner could propose. The calculation provides three main steps:

- Different capacities of the pegs (one can contain three spheres, one two and one only 1);
- Permutations of the coloured spheres (red, green, blue);
- Possible distributions of the spheres between the pegs, respecting their maximum capacities.

For each configuration, we consider that:

- There are three choices for the sphere to be placed on the peg with capacity 1 (one of R, G, B);
- The two remaining spheres must be distributed on the pegs with capacities 2 and 3;
- The vertical order of the spheres on the pegs matters (e.g. red above green $\neq$ green above red).

By examining all the admissible combinations, we identify six distributions, which represent all the possible ways of occupying the pegs according to their capacities: (3,0,0), (2,1,0), (2,0,1), (1,2,0), (1,1,1), and (0,2,1). For each distribution, we calculate the internal permutations of the spheres on the occupied pegs. Summing all the possibilities, we obtain a total number of 36 distinct configurations. Visualising these configurations is essential for a thorough understanding of this combinatorial concept.

Program 142 TOL

```
import tkinter as tk
from itertools import permutations
def generate_tower_of_london_configurations():
balls = ['R', 'G', 'B'] # The balls
peg_limits = [1, 2, 3] # Capacity limits of the pegs
configurations = set()
# Generate all possible assignments of the balls to the pegs
for assignment in permutations(balls, len(balls)):
# Try all possible partitions of the balls among the three pegs
for i in range(len(balls)+1):
for j in range(i, len(balls)+1):
```

```
peg1 = assignment[:i]
peg2 = assignment[i:j]
peg3 = assignment[j:]
# Check that the assignments respect the peg capacity limits
if len(peg1) <= peg_limits[0] and len(peg2) <= peg_limits[1] and len(peg3) <= peg_lim-
its[2]:
configurations.add((tuple(peg1), tuple(peg2), tuple(peg3)))
return list(configurations)
def visualize_configurations(configurations):
# Index of the current configuration
current_index = [0] # Used as a list so it can be modified within the functions
def show_configuration(index):
# Clear the previous window
for widget in window.winfo_children():
widget.destroy()
# Retrieve the current configuration
config = configurations[index]
# Create a canvas to visualize the configuration
canvas = tk.Canvas(window, width=400, height=400, bg="white")
canvas.pack()
# Coordinates and lengths of the pegs
peg_x = [100, 200, 300]
peg_y_top = [250, 200, 150] # Height of the pegs (corresponding to their capacity)
peg_y_bottom = 300 # Base height of all pegs
# Draw the pegs as line segments
for x, y_top in zip(peg_x, peg_y_top):
canvas.create_line(x, y_top, x, peg_y_bottom, width=5, fill="brown")
# Draw the balls
colors = {"R": "red", "G": "green", "B": "blue"}
for i, peg in enumerate(config):
for j, ball in enumerate(peg):
ball_x = peg_x[i]
ball_y = peg_y_bottom - j * 30 # Position of the ball
canvas.create_oval(ball_x - 15, ball_y - 15, ball_x + 15, ball_y + 15, fill=colors[ball])
# Button to navigate to the next configuration
button_frame = tk.Frame(window)
button_frame.pack()
if index > 0:
prev_button = tk.Button(button_frame, text="Back", command=lambda: show_configuration(in-
dex - 1))
prev_button.pack(side=tk.LEFT)
if index < len(configurations) - 1:
next_button = tk.Button(button_frame, text="Next", command=lambda: show_configuration(in-
dex + 1))
next_button.pack(side=tk.RIGHT)
# Create the main window
window = tk.Tk()
window.title("Tower of London Configurations")
# Show the first configuration
```

```
show_configuration(current_index[0])
# Start the Tkinter window
window.mainloop()
# Generate the configurations
configurations = generate_tower_of_london_configurations()
# Visualize the configurations with manual navigation
visualize_configurations(configurations)
configurations = generate_tower_of_london_configurations()
print(f"Total number of configurations: {len(configurations)}")
```

An example from an AIP conference: “There is a square ABCD in the plane. On each side of this square, *n*, inner points are given. Determine the number of all triangles whose vertices coincide with the given points and which vertex lies on a different side of the given square ABCD.”

The first four triangles can be easily drawn with GeoGebra (Břehovský et al., 2022):

Figure 4. Some triangles on the square

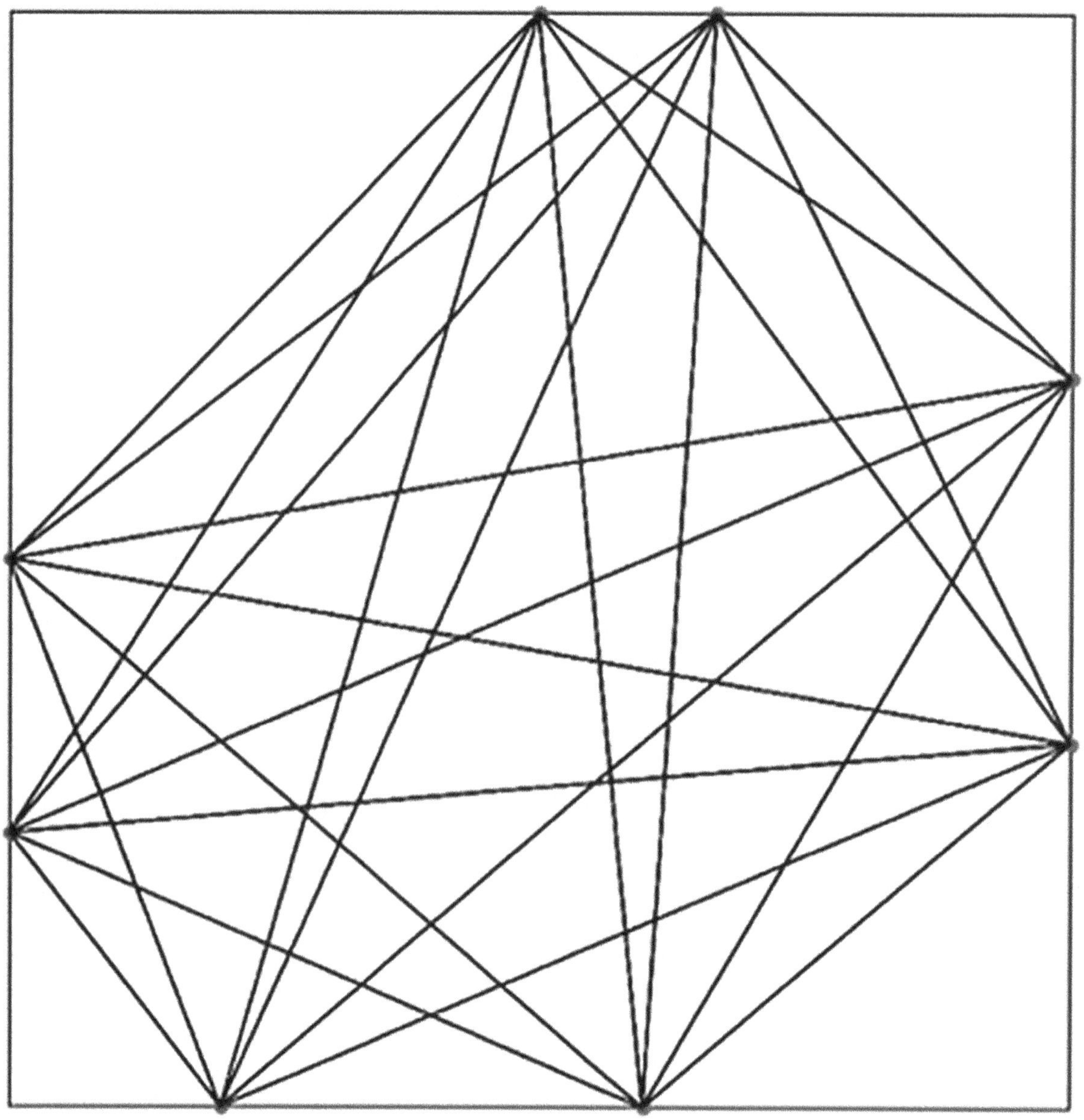

The figure gradually becomes more complicated as 32 triangles can be generated. Each triangle must have a point on three distinct sides of the square. We choose three distinct sides on which to place the stitches, and for each combination of sides $C_{4,3}$, we select one stitch from each side. For each side, there are two points available. So, for every combination of 3 sides, there are 8 distinct triangles. Consequently, the total number of triangles is 4 • 8 = 32. A Python program simplifies the construction of all solutions:

Program 143 triangles on square

```
import matplotlib.pyplot as plt
import itertools
def draw_square_and_triangles():
# Definition of the square's points
square = [(0, 0), (1, 0), (1, 1), (0, 1)]
# Definition of the points on each side
points = [
[(0, 1/4), (0, 1/2)], # Left side
[(1/4, 0), (1/2, 0)], # Bottom side
[(1, 1/2), (1, 3/5)], # Right side
[(1/2, 1), (3/4, 1)] # Top side
]
# Generation of all possible triangles
triangles = []
for i, j, k in itertools.combinations(range(4), 3):
for p1, p2, p3 in itertools.product(points[i], points[j], points[k]):
triangles.append([p1, p2, p3])
def show_triangle(index, axes):
axes.clear()
# Draw the square
axes.plot([p[0] for p in square + [square[0]]],
[p[1] for p in square + [square[0]]], 'k-')
# Draw the points on the sides
for side in points:
axes.plot([p[0] for p in side], [p[1] for p in side], 'ro')
# Draw the current triangle
triangle = triangles[index]
axes.plot([triangle[0][0], triangle[1][0], triangle[2][0], triangle[0][0]],
[triangle[0][1], triangle[1][1], triangle[2][1], triangle[0][1]],
color='black') # Draw the triangle in black
axes.set_title(f'Triangle {index + 1} of {len(triangles)}')
axes.set_aspect('equal', 'box')
axes.axis('off')
plt.draw()
plt.pause(1) # Pause for 1 second to show the triangle
# Create the plot
fig, axes = plt.subplots(figsize=(10, 10))
# Display one triangle at a time
for i in range(len(triangles)):
show_triangle(i, axes)
# Display all triangles at the end
axes.clear()
axes.plot([p[0] for p in square + [square[0]]],
[p[1] for p in square + [square[0]]], 'k-')
for side in points:
axes.plot([p[0] for p in side], [p[1] for p in side], 'ro')
for triangle in triangles:
axes.plot([triangle[0][0], triangle[1][0], triangle[2][0], triangle[0][0]],
[triangle[0][1], triangle[1][1], triangle[2][1], triangle[0][1]],
```

```
color='black') # Draw all triangles in black
axes.set_title(f'All {len(triangles)} possible triangles')
axes.set_aspect('equal', 'box')
axes.axis('off')
plt.show()
# Run the function
draw_square_and_triangles()
```

Figure 5. All triangles on the square

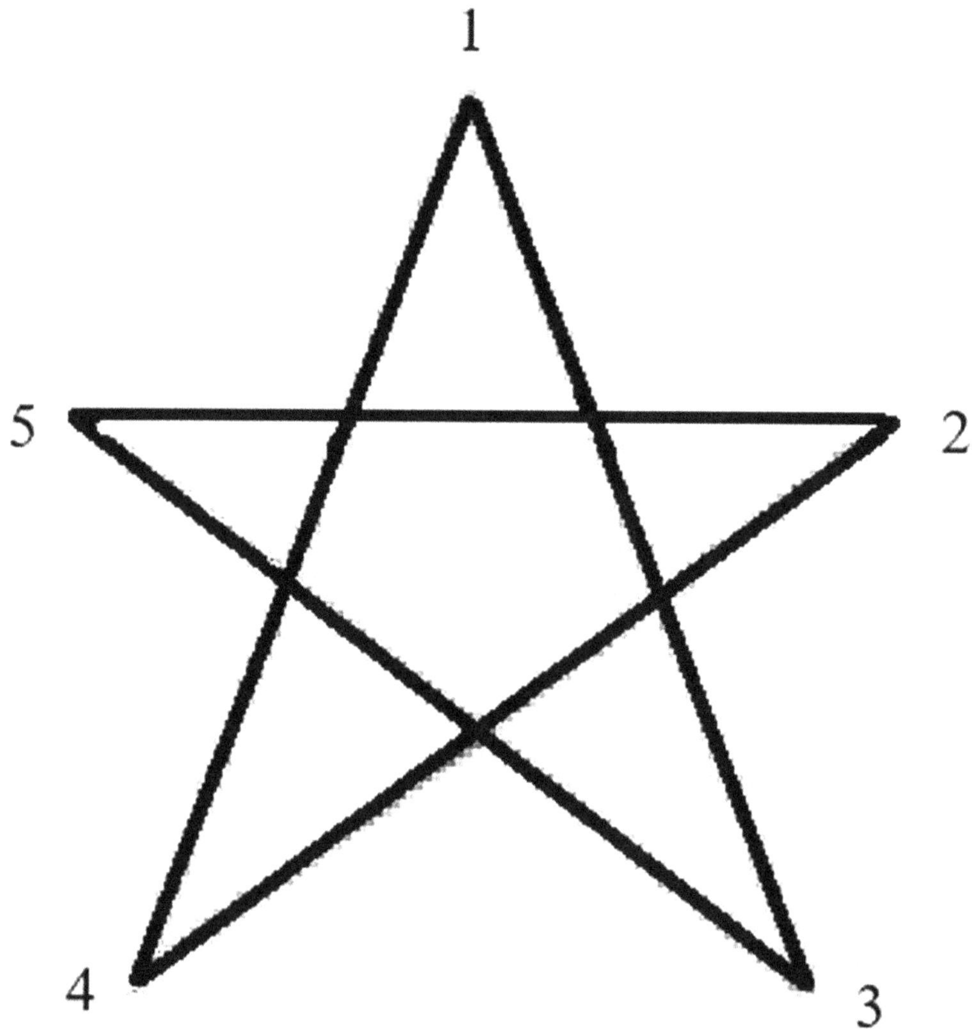

A graphical approach is stimulating to approach combinatorial problems and, more generally, mathematics. This example shows how to transform a pastime into a tool for analysing the intersections between different parts of mathematics. Imagine drawing a star without ever lifting the pencil from the paper and without going over the lines already drawn. This type of drawing is an example of an Eulerian path, which in mathematics indicates a path that passes through each segment exactly once. To better understand how to draw a star, we can number its vertices clockwise. Let us take the classic five-pointed star as an example. We number its vertices from 1 to 5 and identify the sequences to connect them.

Figure 6. Five-pointed star

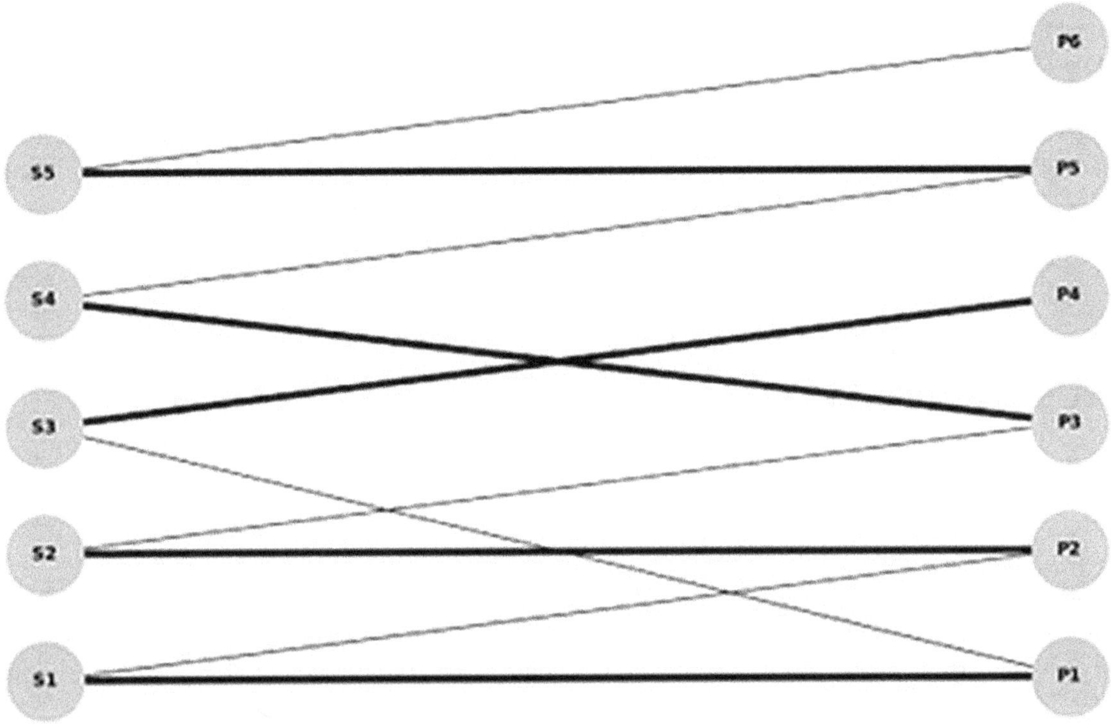

Follow a consistent order and analyse the differences between consecutive numbers. We can find interesting patterns:

1. 1, 3, 5, 2, 4, 1; Differences: +2, +2, −3, 2, -3
2. 1, 4, 2, 5, 3, 1; Differences: +3, −2, +3, −2, −2

At first glance, there do not appear to be any patterns, but things change if we consider the differences in modulo 5. In the first case, the differences correspond to +2 ($-3 \equiv 2 \pmod 5$ because $-3 + 5 = 2$), and in the second case, to +3. We can build a general rule for drawing stars based on the numerical differences between vertices. At the combinatorial level, we can consider how we can connect vertices without retracing the lines already drawn. Since we can connect each vertex to two possible subsequent vertices, we can analyse the total number of valid paths through the principle of multiplication and combinatorial calculus of permutations with constraints. Each time we start from a vertex, we can choose two directions (skip one or two vertices). Since there are five vertices, the possible paths will be 10, calculated as a combination of the possible movements without repetitions. In general, for a star with n vertices, the number of possible valid paths can be determined by considering the number of closed cycles that respect the constraint of not retracing the same segment. We can apply the same reasoning to stars with more points. With seven vertices, we could check which differences allow us to trace the star without interruptions and without passing over the drawn lines.

The method of modular differences helps to find the possible valid sequences and can be studied as a more complex combinatorial problem. This study is not limited to geometry. Eulerian paths and combinatorics concepts are in computer science, cryptography, network theory, and optimized problem-solving. Drawing stars may seem like a simple exercise, but this activity has a mathematical content. By studying the paths and connection rules, connections with graph theory, combinatorics, and other areas of applied mathematics emerge.

In the previous chapters, it is possible to trace these elements, even if, in several cases, we have presented, for completeness, concepts and resources that generally draw on a higher level.

Putting students in front of non-standard situations, however, is an essential step, the important thing is that they are faced with a concrete problem.

Consider, for example, a perfect matching in a bipartite graph. Let us imagine a school with five students and five projects. Every student has expressed interest in one or more of the projects. One wants to find a way to assign each student to a project so that all projects are assigned and each student receives a project they want. We can represent the problem using a bipartite graph. This chart has two groups of nodes: one for students and one for projects. There is a link (edge) between a student and a project if that student is interested. A matching is a set of links that assigns each student to a project, ensuring that no two students share the same project. One can model student-project assignments by creating a network in which students are linked to projects according to their interests. To ensure that every student can obtain a project, each group of students must have enough projects to access. If each group of k students can access at least k projects, we can find a perfect match (Hall's Theorem). Of course, someone will be excluded if the projects are too numerous.

All of this can be made evident by a program like:

Program 144 projects

```
import networkx as nx
import matplotlib.pyplot as plt
def find_matching(students, projects, interests):
matching = {}
used_projects = set()
def assign(student_index):
if student_index == len(students):
return True
student = students[student_index]
for project in interests[student]:
if project not in used_projects:
matching[project] = student
used_projects.add(project)
if assign(student_index + 1):
return True
del matching[project]
used_projects.remove(project)
return False
if assign(0):
return matching
else:
return None
# Data
```

```
students = ['S1', 'S2', 'S3', 'S4', 'S5']
projects = ['P1', 'P2', 'P3', 'P4', 'P5', 'P6']
interests = {
'S1': ['P1', 'P2'],
'S2': ['P2', 'P3'],
'S3': ['P1', 'P4'],
'S4': ['P3', 'P5'],
'S5': ['P5', 'P6']
}
# Find the perfect matching
matching = find_matching(students, projects, interests)
# Create the bipartite graph
G = nx.Graph()
G.add_nodes_from(students, bipartite=0)
G.add_nodes_from(projects, bipartite=1)
# Add edges based on interests
for student, project_interests in interests.items():
for project in project_interests:
G.add_edge(student, project)
# Draw the graph
pos = {}
pos.update((n, (1, i)) for i, n in enumerate(students)) # positions for student nodes
pos.update((n, (2, i)) for i, n in enumerate(projects)) # positions for project nodes
plt.figure(figsize=(10, 6))
nx.draw(G, pos, with_labels=True, node_color='lightblue', node_size=2000, font_size=10,
font_weight='bold')
# Highlight the perfect matching in black
if matching:
matching_edges = [(student, project) for project, student in matching.items()]
nx.draw_networkx_edges(G, pos, edgelist=matching_edges, edge_color='black', width=4)
plt.title("Perfect Matching between Students and Projects")
plt.show()
```

In this case, the result will be given by the following graph:

Figure 7. Matching

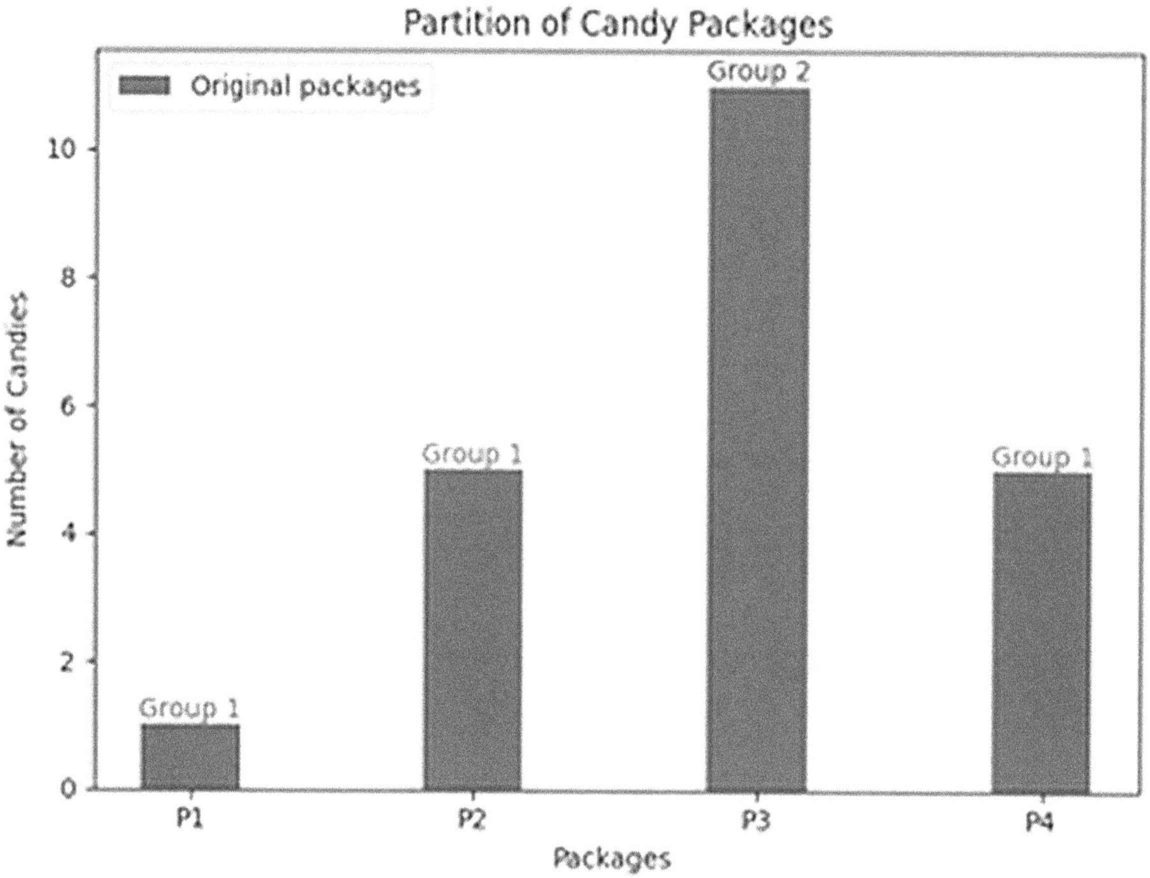

Another example is the problem of partitioning a set: given a set S of n integers, we want to determine if there is a subset whose sum equals half the sum of all the elements of S. In other words, given a set of integers, we want to determine whether we can divide the set into two subsets such that the sum of the elements in each subset is the same. To provide a concrete example of the problem, consider a group of friends who have a few packets of candies with varying amounts in each. Friends want to split the candy packets into two groups with the same number of candies. Despite the apparent simplicity of the definition, it is a complete NP problem involving combinatorial dynamics techniques and dynamic programming.

Program 145 Partition Sets

```
def can_divide_equally(packages, max_diff):
def divide_recursively(packages, groups, current_group, remaining_packages):
if not remaining_packages:
diff = abs(sum(groups[0]) - sum(groups[1]))
if diff <= max_diff:
return groups
```

```
else:
return None
current_package = remaining_packages.pop()
groups[current_group].append(current_package)
for next_group in range(2):
result = divide_recursively(packages, groups.copy(), next_group, remaining_packages.
copy())
if result:
return result
groups[current_group].remove(current_package)
remaining_packages.append(current_package)
return None
groups = [[], []]
return divide_recursively(packages, groups, 0, packages.copy())
# Input packages from user
num_packages = int(input("Enter the number of candy packages: "))
packages = []
for i in range(num_packages):
packages.append(int(input(f"Enter the number of candies in package {i+1}: ")))
# Input maximum allowed difference
max_diff = int(input("Enter the maximum allowed difference: "))
# Call the function
solution = can_divide_equally(packages, max_diff)
# Print the result
if solution:
print("Solution:")
for group in solution:
print(group)
else:
print("It is not possible to divide the packages with the specified absolute difference.")
```

The program accepts the data as input and then verifies the optimal subdivision. Since this is not always possible, a maximum difference between the two groups is expected to be allowed to obtain a suboptimal solution.

It may be interesting to visualize the possible partition:

Program 146 Partition Sets Chart

```
import matplotlib.pyplot as plt
def can_divide_candies(packages):
total_sum = sum(packages)
# Check if the total sum is even
if total_sum % 2 != 0:
return False
target = total_sum // 2
# Helper function for backtracking
def backtrack(index, sum1, sum2):
# If we found a valid partition
if sum1 == target and sum2 == target:
```

```
return True
# If we exceed the target sum or run out of packages
if sum1 > target or sum2 > target or index >= len(packages):
return False
# Try adding the current package to the first group
if backtrack(index + 1, sum1 + packages[index], sum2):
return True
# Try adding the current package to the second group
if backtrack(index + 1, sum1, sum2 + packages[index]):
return True
# Try skipping the current package
return backtrack(index + 1, sum1, sum2)
return backtrack(0, 0, 0)
def find_partition(packages):
total_sum = sum(packages)
# Check if the total sum is even
if total_sum % 2 != 0:
return None
target = total_sum // 2
n = len(packages)
# List to store the solution found
solution = []
# Helper function for backtracking
def backtrack(index, sum1, sum2, subset1, subset2):
nonlocal solution
# If we found a valid partition
if sum1 == target and sum2 == target:
solution.append((subset1[:], subset2[:]))
return True
# If we exceed the target sum or run out of packages
if sum1 > target or sum2 > target or index >= n:
return False
# Try adding the current package to the first group
subset1.append(packages[index])
if backtrack(index + 1, sum1 + packages[index], sum2, subset1, subset2):
return True
subset1.pop()
# Try adding the current package to the second group
subset2.append(packages[index])
if backtrack(index + 1, sum1, sum2 + packages[index], subset1, subset2):
return True
subset2.pop()
# Try skipping the current package
return backtrack(index + 1, sum1, sum2, subset1, subset2)
# Start backtracking from index 0 with both initial sums set to 0 and empty subsets
backtrack(0, 0, 0, [], [])
if not solution:
return None
# Return the first solution found
```

```
return solution[0]
def visualize_partition(packages):
partition = find_partition(packages)
if partition is None:
print("It is not possible to divide the candy packages into two groups with the same sum.")
return
subset1, subset2 = partition
# Create the graph
fig, ax = plt.subplots()
bar_width = 0.35
index = range(len(packages))
# Display the original packages
plt.bar(index, packages, bar_width, label='Original packages')
# Annotate the groups
for i, p in enumerate(packages):
if p in subset1:
plt.text(i, p + 0.1, 'Group 1', ha='center', color='red')
elif p in subset2:
plt.text(i, p + 0.1, 'Group 2', ha='center', color='blue')
plt.xlabel('Packages')
plt.ylabel('Number of Candies')
plt.title('Partition of Candy Packages')
plt.xticks(index, ['P{}'.format(i+1) for i in index])
plt.legend()
plt.tight_layout()
plt.show()
# Example usage
packages = [1, 5, 11, 5]
visualize_partition(packages)
```

Figure 8. Partition

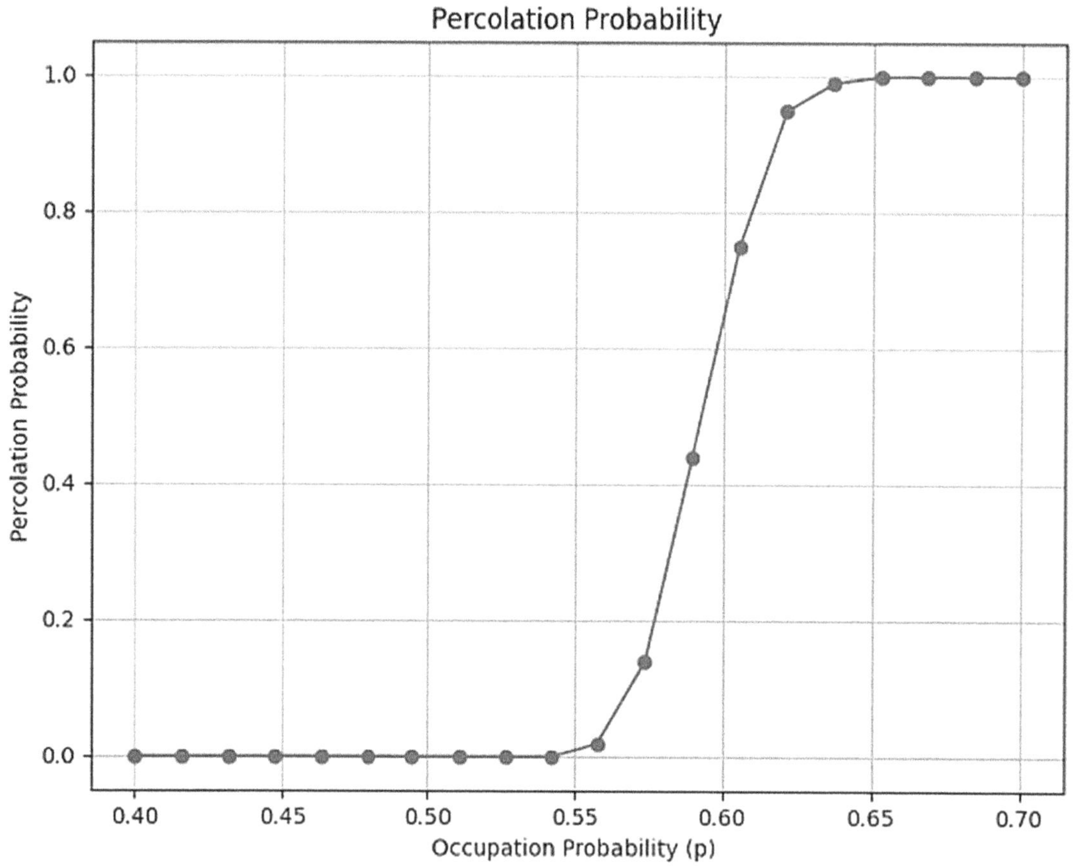

Another example related to the partition problem might concern natural numbers. Explaining the partition of integers to students can be an enjoyable and valuable activity to develop their understanding of number theory. A partition of a positive integer n is a way of writing *n* as the sum of positive integers, where the order of the numbers does not matter (Andrews, 1998). A concrete example: imagine that we need to organise the arrangement of 8 scientists across 3 research centres. Each centre must accommodate at least one scientist, but there are no limits on the maximum number of people a research centre can receive. How many ways can we distribute n = 8 in k = 3?

Program 147 Partition

```
from itertools import combinations_with_replacement
def partitions(n, k):
def get_partitions(n, k, start=1):
if k == 1:
if n >= start:
yield (n,)
return
for i in range(start, n // k + 1):
```

```
for tail in get_partitions(n - i, k - 1, i):
yield (i,) + tail
return list(get_partitions(n, k))
# Request input from the user
n = int(input("Enter the number to partition (n): "))
k = int(input("Enter the number of parts (k): "))
# Find all partitions of n into k parts
result = partitions(n, k)
# Print the result
print(f"All partitions of {n} into {k} parts are:")
for part in result:
print(part)
```

A classic example is the carpooling problem (Hartman et al., 2014): Suppose six teachers decide to share a car ride to work to save money. Every day, they have to split into two cars containing three people. How many different ways can they arrange travel for a working week (5 days), assuming all six can drive? It is a seemingly effortless problem for a computer program; in reality, several aspects must be considered. The main difficulty lies in the approach, and the assumptions made. We could calculate the number of ways to divide six people into two groups of 3 for one day, assuming that each day is independent of the others, and then raise this number to the fifth power to get the result for 5 days.

A second approach can generate all possible combinations of two groups of three people for all 5 days, check that there are different divisions and count only the cases that meet this condition. The key difference is that an additional constraint is imposed in the second hypothesis: it requires that the division into groups be different for each day of the week. To obtain the correct result according to the problem's indications, we need to decide whether it is acceptable for the same division to be repeated across different days or if we want each day to have a unique division. Different programs, different solutions.

Program 148 carpooling1

```
from itertools import combinations
from math import factorial
def car_pooling(teachers=6, days=5):
# Calculate the number of ways to divide 6 people into two groups of 3
daily_arrangements = len(list(combinations(range(teachers), 3))) // 2
# Calculate the total number of arrangements for the week
total = daily_arrangements ** days
return total
# Calculate and print the result
result = car_pooling()
print(f"The number of different ways to organize carpooling trips for a week is: {re-
sult}")
```

Program 149 carpooling2

```
from itertools import combinations, permutations
def carpool():
# List of the six teachers
```

```
teachers = ['A', 'B', 'C', 'D', 'E', 'F']
# Calculate all possible combinations of two groups of three people
combinations_list = list(combinations(teachers, 3))
# Counter for the different ways
count = 0
# Generate all possible permutations of these combinations for five days
for perm in permutations(combinations_list, 5):
# Check that each day has two different groups of three people
if len(set(perm)) == 5:
count += 1
return count
# Calculate the total number of different ways
total = carpool()
print(f"Number of different ways to organize the carpools: {total}")
```

REFERENCES

Andrews, G. E. (1998). *The theory of partitions* (No. 2). Cambridge university press.

Andriano, V., Capone, R., Manolino, C., Minisola, R., & Robutti, O. (2023). Teachers' awareness on combinatorics during Lesson Study discussion. In *Proceedings of the Thirteenth Congress of European Research in Mathematics Education* (pp. 4864-4871). Alfréd Rényi Institute of Mathematics and ERME.

Batanero, C., Navarro-Pelayo, V., & Godino, J. D. (1997). Effect of the implicit combinatorial model on combinatorial reasoning in secondary school pupils. *Educational Studies in Mathematics*, *32*(2), 181–199. DOI: 10.1023/A:1002954428327

Boccia, M., Marin, D., D'Antuono, G., Ciurli, P., Incoccia, C., Antonucci, G., Guariglia, C., & Piccardi, L. (2017). The Tower of London (ToL) in Italy: Standardization of the ToL test in an Italian population. *Neurological Sciences*, *38*(7), 1263–1270. DOI: 10.1007/s10072-017-2957-y PMID: 28432516

Braithwaite, D. W., & Goldstone, R. L. (2013). Integrating formal and grounded representations in combinatorics learning. *Journal of Educational Psychology*, *105*(3), 666–682. DOI: 10.1037/a0032095

Břehovský, J., Bímová, D., & Pirklová, P. (2022, September). Using GeoGebra for solving geometric-combinatorial problems. In *AIP Conference Proceedings* (Vol. 2505, No. 1). AIP Publishing.

Hart, E. W., & Martin, W. G. (2018). Discrete mathematics is essential mathematics in a 21st century school curriculum. *Teaching and learning discrete mathematics worldwide: Curriculum and research*, 3-19.

Hartman, I. B. A., Keren, D., Dbai, A. A., Cohen, E., Knapen, L., & Janssens, D. (2014). Theory and practice in large carpooling problems. *Procedia Computer Science*, *32*, 339–347. DOI: 10.1016/j.procs.2014.05.433

Kenney, M. J., & Hirsch, C. R. (1991). *Discrete Mathematics across the Curriculum, K-12. 1991 Yearbook*. National Council of Teachers of Mathematics, Inc., 1906 Association Drive, Reston, VA 22091.

Lockwood, E. (2013). A model of students' combinatorial thinking. *The Journal of Mathematical Behavior*, *32*(2), 251–265. DOI: 10.1016/j.jmathb.2013.02.008

Lockwood, E. (2022). Leveraging prediction and reflection in a computational setting to enrich undergraduate students' combinatorial thinking. *Cognition and Instruction*, *40*(3), 413–455. DOI: 10.1080/07370008.2021.2020793

Lockwood, E., Wasserman, N. H., & Tillema, E. S. (2020). A case for combinatorics: A research commentary. *The Journal of Mathematical Behavior*, *59*, 100783. DOI: 10.1016/j.jmathb.2020.100783

Logofătu, D., & Gruber, M. (2014). Learning by Doing: An Educational Framework for Teaching Combinatorics for Engineers Using Programming Tasks.

Mashiach Eizenberg, M., & Zaslavsky, O. (2004). Students' verification strategies for combinatorial problems. *Mathematical Thinking and Learning*, *6*(1), 15–36. DOI: 10.1207/s15327833mtl0601_2

Medová, J., Bulková, K. O., & Čeretková, S. (2020). Relations between generalization, reasoning and combinatorial thinking in solving mathematical open-ended problems within mathematical contest. *Mathematics*, *8*(12), 2257. DOI: 10.3390/math8122257

Nopriana, T., Hermanand, T., Suryadi, D., & Martadiputra, B. A. P. (2023, August). Combinatorics problems for undergraduate students: Identification of epistemological obstacles and model of combinatorial thinking. In *AIP Conference Proceedings* (Vol. 2811, No. 1). AIP Publishing.

Semanišinová, I. (2021). Multiple-Solution Tasks in Pre-Service Teachers Course on Combinatorics. *Mathematics*, *9*(18), 2286. DOI: 10.3390/math9182286

Shallice, T. (1982). Specific impairments of planning. *Philosophical Transactions of the Royal Society of London. Series B, Biological Sciences*, *298*(1089), 199–209. PMID: 6125971

Simamora, R. E., & Zunaiedy, D. A. (2021). Improving Students' Combinatorial Thinking Ability through Numbereds Head Together. *learning, 14*(1).

Tillema, E. S., & Burch, L. J. (2022). Using combinatorics problems to support secondary teachers understanding of algebraic structure. *ZDM Mathematics Education*, *54*(4), 777–793. DOI: 10.1007/s11858-022-01359-1

Vásquez, S., Barquero, B., & Bosch, M. (2021). Teaching and learning combinatorics in secondary school: A modelling approach based on the Anthropological Theory of the Didactic. *Quadrante*, *30*(2), 200–219.

Chapter 15
Insights and Conclusions

ABSTRACT

This chapter concludes the combinatorics analysis, highlighting its evolution, modern applications, and prospects. The discipline finds applications in many fields, demonstrating its interdisciplinary value. The importance of multiple representations in teaching combinatorics is emphasized to promote deeper understanding and critical thinking. New research areas such as asymptotic combinatorics, combinatorial bioinformatics, and complex network theory are also explored, showcasing their expanding relevance. Advanced examples are presented, including Stirling's approximation for factorials, percolation theory for studying diffusion phenomena, and using Feynman diagrams in quantum physics and statistical mechanics. Furthermore, the chapter examines the application of combinatorics in social network models, workforce shift optimization, and combinatorial game theory. The chapter raises questions about the discipline's future, suggesting that combinatorics represents a proper "way of thinking" capable of tackling complex problems across various fields.

INSIGHTS AND CONCLUSIONS

Those who dare to teach must never stop learning
John Cotton Dana

After providing a general overview of the discipline and its objectives, we will analyse several techniques and strategies for solving combinatorial problems. The previous examples have allowed us to see the practical application of the theories and methods studied. The sub-disciplines and their examples gave us a detailed look at the various specialized areas within combinatorics, demonstrating the breadth and diversity of this field. We explored how combinatorial methods are crucial in various fields, including algorithms (Mehlhorn & Sanders, 2007) and graph theory, as well as recursive counting and ordinary generating functions. Combinatorial games and random combinations have demonstrated the importance of combinatorics in playful and casual contexts.

The didactics of combinatorics offered insights on effectively teaching this discipline, emphasising the importance of a clear and structured approach to facilitate learning.

Additional guidance is provided by Maher and Martino (1996a). They believe it is essential that students are encouraged to construct and use more representations while working on problems. Using different tools to build and express ideas enables students to create connections between various representations, thereby enhancing their understanding of the math they are learning. This allows observers (teachers and classmates) to understand better the ideas formulated by students.

At this point, a question arises: What are the prospects? We know that combinatorics continues to evolve, with new applications emerging in various fields. Numerous open problems in combinatorics pose fascinating challenges for researchers. These include fundamental issues in more traditional topics, such as combinatorial counting and graph theory (Van Steen, 2010; Bollobás, 2011), but also in fields yet to be explored, such as asymptotic combinatorics, a branch of combinatorics that deals with the study of the behaviour of sequences of numbers or functions, particularly combinatorial counts, when the parameters in question tend towards infinity or algorithmic combinatorics that relies on artificial intelligence (Russell & Norvig, 2016) and Machine Learning. In addition, new research directions are emerging

DOI: 10.4018/979-8-3373-3089-1.ch015

in areas such as topological combinatorics, complex network theory, and combinatorial bioinformatics. The most classic example of asymptotic combinatorics refers to the Stirling formula:

$$n! \approx \sqrt{2\pi n} \cdot (\tfrac{n}{e})^n$$

It provides an approximation for the factorial of a large number. The following program displays the comparison of exact factorials and Stirling approximations on a logarithmic scale for n up to 50 and prints the error.

Program 150 Stirling

```
import math
import matplotlib.pyplot as plt
def stirling_approximation(n):
return math.sqrt(2 * math.pi * n) * (n / math.e) ** n
def factorial(n):
return math.factorial(n)
def compare_factorials_graph(max_n):
n_values = list(range(1, max_n + 1))
exact_factorials = [factorial(n) for n in n_values]
stirling_approximations = [stirling_approximation(n) for n in n_values]
relative_errors = [abs(exact_factorials[i] - stirling_approximations[i]) / exact_factori-
als[i] * 100 for i in range(max_n)]
# Plot comparing exact factorials and Stirling approximation
plt.figure(figsize=(14, 8))
plt.plot(n_values, exact_factorials, label="Exact Factorial", marker='o', linestyle='-',
color='b')
plt.plot(n_values, stirling_approximations, label="Stirling Approximation", marker='x',
linestyle='--', color='r')
plt.yscale('log')
plt.xlabel('n')
plt.ylabel('Value')
plt.title('Comparison between Exact Factorial and Stirling Approximation')
plt.legend()
plt.grid(True)
plt.show()
# Plot of the relative error
plt.figure(figsize=(14, 8))
plt.plot(n_values, relative_errors, label="Relative Error (%)", marker='s', linestyle='-',
color='g')
plt.yscale('log')
plt.xlabel('n')
plt.ylabel('Relative Error (%)')
plt.title('Relative Error of Stirling Approximation compared to Exact Factorials')
plt.legend()
plt.grid(True)
plt.show()
# Modify the value of max_n to see the comparison up to a specific value of n
compare_factorials_graph(50)
```

As we have mentioned, combinatorics is experiencing strong growth due to its ability to solve problems through an interdisciplinary approach (Zeitz, 2017). In the future, it can be expected to remain a key element for innovation in cutting-edge areas such as data science, computational biology, cryptography, mathematical physics, and quantum computing.

A couple of examples give an idea.

The first refers to the percolation theory (Li et al., 2021). It studies how processes propagate over discrete networks, such as the diffusion of a liquid through a porous material or the transmission of diseases. In this field, combinatorics plays an important role as it helps us to calculate the probability of cluster formation for discrete configurations. The following program implements the site percolation model, in which each site (node) of the lattice is occupied with a probability p and tries to determine if a percolating cluster connects the top side with the bottom side of the lattice (Malthe-Sørenssen, 2024).

The graph obtained represents the probability of percolation, i.e., the frequency with which a percolating cluster is observed that connects the upper and lower sides of the lattice during simulations as a function of the probability of occupancy. This parameter indicates how likely each site in the grid is occupied. For low p-values, the probability of percolation is close to zero, which means that it is unlikely to find a connected cluster crossing the lattice. By increasing p, the probability of percolation grows rapidly around a critical value, characterizing a phase transition: the system goes from a state in which it does not percolate to one in which percolation is highly probable. In conclusion, the graph shows how percolation emerges in a system when a certain threshold of pc density is exceeded. There are possible 2^{n^2}2 $^{n^2}$ configurations for a n × n size reticle, but the model considers only a subset of the relevant configurations.

Program 151 percolation

```
import numpy as np
import matplotlib.pyplot as plt
from scipy.ndimage import label
def generate_lattice(size, probability):
"""

Generates a 2D lattice of size `size` x `size`.
Each site is occupied with probability `probability`.

"""
lattice = np.random.rand(size, size) < probability
return lattice
def find_clusters(lattice):
"""

Finds clusters in the lattice using site percolation.
Uses the `label` function from scipy.ndimage to identify connected clusters.

"""
clusters, num_clusters = label(lattice)
return clusters, num_clusters
def check_percolation(clusters):
"""

Checks if there exists a cluster that connects the top and bottom sides.

"""
```

```
top_clusters = np.unique(clusters[0,:]) # Clusters on the top side
bottom_clusters = np.unique(clusters[-1,:]) # Clusters on the bottom side
# Checks if there's a common cluster between top and bottom
percolating_clusters = np.intersect1d(top_clusters, bottom_clusters)
return percolating_clusters[percolating_clusters > 0] # Excludes cluster 0 (not occupied)
def simulate_percolation(size, probability, visualize=False):
"""
```

Simulates percolation on a lattice of size `size` with occupation probability `probability`.
Returns True if the system percolates, False otherwise.

```
"""
lattice = generate_lattice(size, probability)
clusters, num_clusters = find_clusters(lattice)
if visualize:
# Visualizing the lattice
plt.figure(figsize=(6, 6))
plt.title(f"Lattice with p = {probability}")
plt.imshow(lattice, cmap="binary", interpolation="nearest")
plt.colorbar(label="Occupation")
plt.show()
# Visualizing the clusters in grayscale
plt.figure(figsize=(6, 6))
plt.title(f"Clusters in the lattice with p = {probability}")
plt.imshow(clusters, cmap="gray", interpolation="nearest")
plt.colorbar(label="Cluster ID")
plt.show()
percolating_clusters = check_percolation(clusters)
return len(percolating_clusters) > 0
def percolation_probability(size, probability_values, trials=100):
"""
```

Calculates the percolation probability for a series of `probability` values.

```
"""
probabilities = []
for site_probability in probability_values:
successes = 0
for _ in range(trials):
if simulate_percolation(size, site_probability):
successes += 1
probabilities.append(successes / trials)
return probabilities
# Model parameters
size = 100 # Increase the lattice size for better visualization
probability_values = np.linspace(0.4, 0.7, 20) # Increase the occupation probability
trials = 100 # Number of simulations for each probability value
# Calculating the percolation probability
probabilities = percolation_probability(size, probability_values, trials)
```

```
# Visualizing the results
plt.figure(figsize=(8, 6))
plt.plot(probability_values, probabilities, marker='o')
plt.title("Percolation Probability")
plt.xlabel("Occupation Probability (p)")
plt.ylabel("Percolation Probability")
plt.grid(True)
plt.show()
# Running a simulation to visualize an example
simulate_percolation(size, 0.7, visualize=True)
```

The program also produces two additional visualizations. The grid graph illustrates the distribution of occupied and empty sites, with the density of occupied sites varying according to the probability of occupation. The cluster chart, on the other hand, reveals groups of connected sites; the appearance of a spanning cluster—from one edge of the grid to the opposite—indicates the onset of percolation.

Together, these visuals offer complementary insights: the grid chart illustrates spatial distribution, while the cluster chart helps determine whether the system has reached the percolation threshold.

Percolation theory is deeply connected to combinatorics through configuration enumeration, the calculation of percolation probability, and the analysis of graph properties.

A second example refers to *quantum field theory* (QFT), a mathematical framework that describes the fundamental interactions between the elementary particles. A crucial aspect of this theory is the description of interactions using Feynman diagrams, which are graphical representations of contributions to perturbative developments of physical quantities. A Feynman diagram comprises vertices, which represent the points of interaction between particles, and arcs, which connect the vertices and represent the virtual particles that mediate the interactions. In general, the construction of these diagrams involves a series of discrete choices: how many vertices to consider, which vertices to connect, and which virtual particles to pass through the connections. The combinatorial problem then becomes counting the number of ways in which we can construct graphs by satisfying these physical restrictions.

Program 152 Feynman diagrams

```
import numpy as np
import matplotlib.pyplot as plt
import itertools
import networkx as nx # Import networkx to draw the graphs
# Function to generate possible Feynman diagram configurations with a variable number of
interactions
def generate_diagrams(n_vertices, max_interactions):
"""
```

Generates all possible configurations of Feynman diagrams with a variable number of interactions between vertices (up to max_interactions).

Each configuration is a graph with n_vertices nodes and a variable number of edges.

```
"""
vertices = list(range(n_vertices)) # Indices of the vertices
diagrams = []
# Random number of interactions for each diagram
for n_interactions in range(1, max_interactions + 1):
```

```
# Create all possible combinations of edges (interactions) between the vertices
possible_edges = list(itertools.combinations(vertices, 2))
# Randomly select n_interactions edges
diagram = list(np.random.choice(len(possible_edges), n_interactions, replace=False))
diagrams.append([possible_edges[i] for i in diagram])
return diagrams
# Function to calculate the total energy of a diagram
def calculate_energy(diagram):
"""
```

Calculates the total energy associated with a given diagram.
For simplicity, assume each interaction has a unitary energy.

```
"""
return len(diagram) # Each interaction has unitary energy
# Function to visualize a diagram
def visualize_diagram(diagram, n_vertices):
"""
```

Visualizes a diagram as a graph.

```
"""
G = nx.Graph()
G.add_nodes_from(range(n_vertices))
G.add_edges_from(diagram)
plt.figure(figsize=(6, 6))
nx.draw(G, with_labels=True, node_color="gray", edge_color="black", node_size=700, font_
size=12)
plt.title(f"Feynman Diagram with {len(diagram)} interactions")
plt.show()
# Input parameters
n_vertices = 4 # Number of vertices (particles)
max_interactions = 4 # Maximum number of interactions (edges)
num_diagrams = 5 # Number of diagrams to generate
# Generate Feynman diagrams with a variable number of interactions
possible_diagrams = generate_diagrams(n_vertices, max_interactions)
# Select a maximum number of diagrams, but not more than how many were generated
num_selected_diagrams = min(num_diagrams, len(possible_diagrams))
# Select a subset of diagrams to visualize
selected_diagrams = np.random.choice(len(possible_diagrams), num_selected_diagrams, re-
place=False)
# Visualization and energy calculation for each selected diagram
for i in selected_diagrams:
diagram = possible_diagrams[i]
energy = calculate_energy(diagram)
# Visualize the diagram
visualize_diagram(diagram, n_vertices)
# Show the energy
print(f"Diagram {i + 1}: Energy = {energy} units")
```

The examples could be multiplied as combinatorics represents a fundamental pillar of modern mathematics, with applications ranging from computer science to physics, biology to finance.

Once again, it must be emphasized that programs that use relatively simple language are important for studying and deepening some combinatorial topics. A few more examples will reinforce this belief.

The following program utilises the Erdős-Rényi model (Lyzinski et al., 2014; Yuan, 2024) to handle graphs and calculates essential parameters, including the degree distribution and the diameter of the graph.

Program 153 Random graph

```
import networkx as nx
import matplotlib.pyplot as plt
# Function to get a valid input for an integer
def input_valid_integer(prompt):
while True:
try:
value = int(input(prompt))
if value > 0:
return value
else:
print("Please, enter a positive integer.")
except ValueError:
print("Please, enter a valid integer.")
# Function to get a valid input for a probability
def input_valid_probability(prompt):
while True:
try:
value = float(input(prompt))
if 0 <= value <= 1:
return value
else:
print("Please, enter a probability between 0 and 1.")
except ValueError:
print("Please, enter a valid number for the probability.")
# Get user input
n = input_valid_integer("Number of nodes: ") # Number of nodes
p = input_valid_probability("Probability that there is an edge between any two nodes (0-
1): ")
# Create a random graph using the Erdős-Rényi model
G = nx.erdos_renyi_graph(n, p)
# Visualize the graph
plt.figure(figsize=(10, 10))
nx.draw(G, with_labels=True, node_size=500, node_color='lightblue', edge_color='gray',
font_weight='bold', font_size=10)
plt.title("Erdős-Rényi Random Graph")
plt.show()
# Calculate and print some graph properties
num_nodes = G.number_of_nodes()
num_edges = G.number_of_edges()
```

```
avg_degree = sum(dict(G.degree()).values()) / num_nodes
print("Number of nodes:", num_nodes)
print("Number of edges:", num_edges)
print("Average degree:", avg_degree)
# Calculate the degree distribution
degree_sequence = sorted([d for n, d in G.degree()], reverse=True)
plt.figure(figsize=(10, 6))
plt.hist(degree_sequence, bins=range(min(degree_sequence), max(degree_sequence) + 1), col-
or='blue', alpha=0.7, edgecolor='black')
plt.title("Degree Distribution")
plt.xlabel("Degree")
plt.ylabel("Number of Nodes")
plt.xticks(range(min(degree_sequence), max(degree_sequence) + 1))
plt.grid(True)
plt.show()
# Calculate connected components
components = list(nx.connected_components(G))
num_components = len(components)
largest_component = max(components, key=len)
size_largest_component = len(largest_component)
print("Number of connected components:", num_components)
print("Size of the largest connected component:", size_largest_component)
# Calculate the diameter of the largest connected component
if num_components > 1:
subgraph = G.subgraph(largest_component)
try:
diameter = nx.diameter(subgraph)
print("Diameter of the largest connected component:", diameter)
except nx.NetworkXError:
print("Unable to calculate the diameter: the largest connected component is not fully con-
nected.")
else:
# For a fully connected graph, calculate the diameter
diameter = nx.diameter(G)
print("Diameter of the graph:", diameter)
```

To illustrate the relevance of the combinatorics presented above, we aim to analyse the robustness of a social network represented as a graph, where nodes represent individuals and edges represent the connections between them. We want to understand how likely it is that the network will remain connected if we randomly remove connections. We can model the network as a graph of permutations and analyse the distribution of cycles to predict the probability that the network will remain connected. The resulting plot visually illustrates the distribution of connected components in the generated random networks (Van Der Hofstad, 2024). If most values cluster around x = 1, indicating a single connected structure, the network can be considered robust. Conversely, if the bars extend toward higher x-values—representing multiple disconnected clusters—it suggests that the generated networks tend to be more fragmented. This may indicate a more fragile network, where the random removal of some arcs could lead to the formation of many small, isolated components.

Program 154 Erdős-Rényi model

```
import numpy as np
import matplotlib.pyplot as plt
import networkx as nx
def generate_random_graph(n, p):
"""Generates a random graph using the Erdős-Rényi model."""
return nx.erdos_renyi_graph(n, p)
def count_connected_components(G):
"""Counts the number of connected components in the graph."""
return len(list(nx.connected_components(G)))
def connected_components_distribution(n, p, num_graphs):
"""Calculates the distribution of the number of connected components for num_graphs random
graphs."""
components_count = np.zeros(n)
for _ in range(num_graphs):
G = generate_random_graph(n, p)
num_components = count_connected_components(G)
if num_components <= n:
components_count[num_components - 1] += 1
return components_count / num_graphs
# Function to get a valid integer input
def input_valid_integer(prompt):
while True:
try:
value = int(input(prompt))
if value > 0:
return value
else:
print("Please, enter a positive integer.")
except ValueError:
print("Please, enter a valid integer.")
# Function to get a valid probability input
def input_valid_probability(prompt):
while True:
try:
value = float(input(prompt))
if 0 <= value <= 1:
return value
else:
print("Please, enter a probability between 0 and 1.")
except ValueError:
print("Please, enter a valid number for the probability.")
# Get input from the user
n = input_valid_integer("Number of nodes: ") # Number of nodes
p = input_valid_probability("Probability that there is an edge between any two nodes (0-
1): ")
num_graphs = 1000 # Number of graphs to generate
# Calculate the distribution of connected components
distribution = connected_components_distribution(n, p, num_graphs)
# Visualize the distribution
```

```
x = np.arange(1, n + 1)
plt.figure(figsize=(10, 6))
plt.bar(x, distribution, color='blue', alpha=0.7)
plt.xlabel("Number of Connected Components")
plt.ylabel("Probability")
plt.title(f"Distribution of the Number of Connected Components in Graphs with {n} Nodes")
plt.xticks(x) # Add x-tick values
plt.grid(axis='y', linestyle='--', alpha=0.7)
plt.show()
```

An interesting and relevant example in asymptotic combinatorics is the analysis of permutations and cycles. A typical case is to count the permutations with a particular asymptotic property, such as the average number of permutation cycles. The generated distribution shows the probability that a random permutation of *n* elements has a certain number of cycles. For large values of n, the distribution of cycles tends to a specific distribution predicted by the theory of asymptotic combinatorics.

Program 155 random permutations

```
import numpy as np
import matplotlib.pyplot as plt
import sympy as sp
def generate_random_permutation(n):
# Generates a random permutation of n elements
perm = np.random.permutation(n)
return perm
def count_cycles(perm):
# Counts the number of cycles in the permutation
n = len(perm)
visited = [False] * n
cycle_count = 0
for i in range(n):
if not visited[i]:
cycle_count += 1
j = i
while not visited[j]:
visited[j] = True
j = perm[j]
return cycle_count
def cycle_distribution(n, num_permutations):
# Calculates the distribution of the number of cycles for num_permutations permutations
cycle_count = np.zeros(n)
for _ in range(num_permutations):
perm = generate_random_permutation(n)
num_cycles = count_cycles(perm)
cycle_count[num_cycles - 1] += 1
return cycle_count / num_permutations
# Parameters
n = int(input("Number of elements in the permutation: "))
```

```
num_permutations = 10000 # Number of permutations to generate
# Calculate the cycle distribution
distribution = cycle_distribution(n, num_permutations)
# Display the distribution
x = np.arange(1, n + 1)
plt.bar(x, distribution, color='blue', alpha=0.7)
plt.xlabel("Number of cycles")
plt.ylabel("Probability")
plt.title(f"Cycle Distribution in Permutations of {n} Elements")
plt.show()
```

As a further in-depth study, we can examine shift planning (Caprara et al., 2003), a more complex combinatorial optimization. In particular, it is necessary to solve a problem of integer linear programming (Al-Rawi & Mukherjee, 2019), which aims to minimize costs by satisfying a series of constraints. Suppose we have the following data: the costs of the seven shifts are 600, 500, 450, 500, 500, 550, 800.

The staff requirement is as follows: Monday 23, Tuesday 17, Wednesday 13, Thursday 14, Friday 15, Saturday 18, Sunday 24.

Program 156 shift planning

```
from pulp import LpMinimize, LpProblem, LpVariable, lpSum, value
def solve_scheduling_problem():
# Create the problem
prob = LpProblem("Shift_Scheduling", LpMinimize)
# Define the variables
x = [LpVariable(f"x{i}", lowBound=0, cat='Integer') for i in range(1, 8)]
# Shift costs
costs = [400, 460, 500, 500, 500, 500, 400]
# Objective function
prob += lpSum(costs[i] * x[i] for i in range(7))
# Constraints for each day of the week
demand = [23, 17, 13, 14, 15, 18, 24]
shifts = [
[1, 1, 1, 1, 1, 0, 0], # Shift 1: rest on Sat and Sun
[1, 1, 1, 1, 0, 0, 1], # Shift 2: rest on Fri and Sat
[1, 1, 1, 0, 0, 1, 1], # Shift 3: rest on Thu and Fri
[1, 1, 0, 0, 1, 1, 1], # Shift 4: rest on Wed and Thu
[1, 0, 0, 1, 1, 1, 1], # Shift 5: rest on Tue and Wed
[0, 0, 1, 1, 1, 1, 1], # Shift 6: rest on Mon and Tue
[0, 1, 1, 1, 1, 1, 0], # Shift 7: rest on Sun and Mon
]
# Add the constraints for each day
for j in range(7):
prob += lpSum(x[i] * shifts[i][j] for i in range(7)) == demand[j], f"Demand_day_{j+1}"
# Solve the problem
prob.solve()
# Extract the solution
solution = [int(value(x[i])) for i in range(7)]
```

```
cost = value(prob.objective)
return solution, cost
# Run the optimization
solution, cost = solve_scheduling_problem()
# Print the total cost
print(f"Total cost: {cost}")
# Verify the solution
demand = [23, 17, 13, 14, 15, 18, 24]
shifts = [
[1, 1, 1, 1, 1, 0, 0], # Shift 1: rest on Sat and Sun
[1, 1, 1, 1, 0, 0, 1], # Shift 2: rest on Fri and Sat
[1, 1, 1, 0, 0, 1, 1], # Shift 3: rest on Thu and Fri
[1, 1, 0, 0, 1, 1, 1], # Shift 4: rest on Wed and Thu
[1, 0, 0, 1, 1, 1, 1], # Shift 5: rest on Tue and Wed
[0, 0, 1, 1, 1, 1, 1], # Shift 6: rest on Mon and Tue
[0, 1, 1, 1, 1, 1, 0], # Shift 7: rest on Sun and Mon
]
# Create and print the schedule table
days = ['Monday', 'Tuesday', 'Wednesday', 'Thursday', 'Friday', 'Saturday', 'Sunday']
for day, req in enumerate(demand):
staff = [solution[i] * shifts[i][day] for i in range(7)]
total = sum(staff)
print(f"\n{days[day]}")
for i in range(7):
if staff[i] > 0:
print(f"Shift {i+1}: {staff[i]} people")
print(f"Total: {total} people (minimum required: {req})")
```

The printout of the results confirms the correctness of the procedure:
Total cost: 12160.0

Monday
Shift 1: 1 people
Shift 2: 6 people
Shift 3: 4 people
Shift 4: 6 people
Shift 5: 6 people
Total: 23 people (minimum required: 23)
Tuesday
Shift 1: 1 people
Shift 2: 6 people
Shift 3: 4 people
Shift 4: 6 people
Total: 17 people (minimum required: 17)
Wednesday
Shift 1: 1 people
Shift 2: 6 people
Shift 3: 4 people
Shift 6: 2 people

Total: 13 people (minimum required: 13)
Thursday
Shift 1: 1 people
Shift 2: 6 people
Shift 5: 6 people
Shift 6: 2 people
Total: 15 people (minimum required: 14)
Friday
Shift 1: 1 people
Shift 4: 6 people
Shift 5: 6 people
Shift 6: 2 people
Total: 15 people (minimum required: 15)
Saturday
Shift 3: 4 people
Shift 4: 6 people
Shift 5: 6 people
Shift 6: 2 people
Total: 18 people (minimum required: 18)
Sunday
Shift 2: 6 people
Shift 3: 4 people
Shift 4: 6 people
Shift 5: 6 people
Shift 6: 2 people
Total: 24 people (minimum required: 24)

At this point, proposing some (partial) conclusions is essential. Some general aspects were examined in Chapters 1 and 2, exploring fundamental definitions and methods. From Chapter 3 to Chapter 5, the focus was on generating functions and combinatorial algorithms and addressing some practical problems. In Chapters 6-8, the exploration focused on the sub-disciplines and the role of graphs, highlighting how they form the basis of numerous combinatorial approaches. More specific applications were covered in Chapters 9-12, from combinatorial games to connections with probability and artificial intelligence. In chapters 13 and 14, we have attempted to examine the main methodological and didactic challenges.

Combinatorics, as we have seen, is a fundamental discipline that finds applications in various fields of mathematics and science. Combinatorics is not limited to pure mathematics, but it finds a wide range of applications. In conclusion, combinatorics is a rich and fascinating discipline that offers powerful tools for solving complex problems. Future research directions and practical applications continue to expand the boundaries of this discipline, making it increasingly relevant. Combinatorics is not only confirmed as a field of mathematical study but also as an authentic "style of thought" capable of crossing disciplinary boundaries, offering original interpretations to understand the complexity of the world.

All this leads us to formulate some fundamental questions:

1. What are the still unexplored frontiers of combinatorics?
2. How can combinatorics influence the development of artificial intelligence?
3. How can the teaching of combinatorics evolve?

These are complex questions, and it is not easy to find articulated and satisfactory answers, but some indications are possible:

1. One of the most promising prospects concerns the use of combinatorics to describe high-dimensional systems, such as those that emerge in the analysis of complex data (big data). At the same time, the interaction between combinatorics and theoretical physics, particularly in statistical mechanics and complex systems theory, could pave the way for entirely new models. Additionally, Probabilistic Combinatorics (Pearl, 2014): The branch known as probabilistic combinatorics combines various tools to study the properties of "random" combinatorial objects. For example, one can study the typical properties of a random graph or a random permutation.. For example, one can study the typical properties of a random graph or a random permutation.
2. Combinatorics could provide crucial tools for processing learning graphs, finding optimal strategies in combinatorial games, and designing optimized neural networks. In addition, many AI problems, such as planning, routing, and finding optimal solutions in complex search spaces, can be formulated as combinatorial optimization. Efficient combinatorial algorithms are, therefore, essential to solve these problems.
3. Digital tools, such as visualization software and interactive platforms, can improve the integration of combinatorics into school curricula. In addition, the link between combinatorics and play could be exploited to make learning more engaging, for example, through practical exercises based on puzzles, graphs, and mathematical games. Regarding the development of computational thinking (Branco & Colaço, 2023), it is necessary to integrate the teaching of combinatorics with activities that develop such thinking, such as the design of algorithms to generate combinatorial configurations or to solve it.

On the one hand, combinatorics provides solid theoretical foundations that continue to be explored; on the other, it addresses the challenges emerging from new technologies and global problems. The questions presented here should not only stimulate reflection on the role of combinatorics but also suggest that this discipline has a potential that is still partly unexplored.

REFERENCES

Al-Rawi, O. Y. M., & Mukherjee, T. (2019). Application of linear programming in optimizing labour scheduling. *Journal of Mathematical Finance*, *9*(3), 272–285. DOI: 10.4236/jmf.2019.93016

Bollobás, B. (2011). Random graphs. In *Modern graph theory* (pp. 215–252). Springer New York.

Branco, N., & Colaço, S. (2023, July). Computational thinking and mathematics teaching: An experience on initial teacher education. In *Thirteenth Congress of the European Society for Research in Mathematics Education (CERME13)* (No. 9). Alfréd Rényi Institute of Mathematics; ERME.

Caprara, A., Monaci, M., & Toth, P. (2003). Models and algorithms for a staff scheduling problem. *Mathematical Programming*, *98*(1-3), 445–476. DOI: 10.1007/s10107-003-0413-7

Li, M., Liu, R. R., Lü, L., Hu, M. B., Xu, S., & Zhang, Y. C. (2021). Percolation on complex networks: Theory and application. *Physics Reports*, *907*, 1–68. DOI: 10.1016/j.physrep.2020.12.003

Lyzinski, V., Fishkind, D. E., & Priebe, C. E. (2014). Seeded graph matching for correlated Erdös-Rényi graphs. *Journal of Machine Learning Research*, *15*(1), 3513–3540.

Malthe-Sørenssen, A. (2024). *Percolation theory using python*. Springer Nature. DOI: 10.1007/978-3-031-59900-2

Mehlhorn, K., & Sanders, P. (2007). Algorithms and data structures. *The Basic Toolbox, 295.*

Pearl, J. (2014). *Probabilistic reasoning in intelligent systems: networks of plausible inference*. Elsevier.

Russell, S. J., & Norvig, P. (2016). *Artificial intelligence: a modern approach*. Pearson.

Van Der Hofstad, R. (2024). *Random graphs and complex networks* (Vol. 54). Cambridge university press. DOI: 10.1017/9781316795552

Van Steen, M. (2010). Graph theory and complex networks. *An introduction, 144*(1).

Yuan, M. (2024). On the Rényi index of random graphs. *Statistische Hefte*, *65*(3), 1773–1803.

Zeitz, P. (2017). *The art and craft of problem solving*. John Wiley & Sons.

Glossary

Additive combinatorics: This branch of combinatorics deals with the study of the additive properties of sets of numbers.

Algebraic combinatorics: Connects combinatorics with algebra, using algebraic tools to solve combinatorial problems and studying the relationships between combinatorial functions and algebraic structures.

Algorithms: Algorithms designed to solve problems involving discrete structures such as graphs, sets, strings, or permutations. They are classified as exact (exploring the complete solution space), heuristic (finding good solutions based on combinatorial patterns), and analytical (using mathematical tools to model or simplify the combinatorial structure)

Arithmetic progression: Sequence of numbers in which the difference between two consecutive terms is constant.

Arrangements: Similar to permutations, but the order of elements is important, and not all elements of the set are necessarily used

Art gallery surveillance problem: The problem of determining the minimum number of cameras needed to monitor all the points in a polygon-shaped room.

Artificial intelligence (AI): The simulation of human intelligence processes by computer systems.

Bin packing problem: A combinatorial optimisation problem that involves arranging objects of different sizes into a limited number of containers or "bins", trying to minimise the bins used.

Binomial coefficient: Indicates the number of ways to choose k elements from a set of n elements without regard to order.

Boggle: A game in which players form words on a grid of letters.

Clustering: Grouping a set of data into subsets (clusters) so that elements within the same cluster are more similar to each other than to elements in other clusters.

Coding Theory: Applies combinatorics to designing and analysing codes for transmitting information.

Combinations: Ways to choose objects without considering their order.

Combinatorial Computation: The practical application of combinatorics used to solve specific problems, focusing on the development of algorithms and techniques for counting clusters.

Combinatorial Design Theory: This concerns the construction of combinatorial structures with particular properties, such as block designs, to plan experiments efficiently.

Combinatorial Game Theory: A class of abstract games involving strategies and decisions based on a finite set of options or configurations, focusing on the combination and arrangement of available objects or moves

Combinatorial fractals: Combinatorial structures that have infinite dimension and scale invariance properties.

Combinatorics: A broad discipline of mathematics that studies and analyses the various techniques of combining and organizing objects or elements in sets, focusing on discrete and finite structures. It is also defined as "the art of counting without counting".

Conjecture: A mathematical statement that is believed to be true but for which a proof has not yet been found.

Convex hull: Contour that encloses a set of points, useful for geometric optimisation problems.

Delaunay Triangulation: Technique for dividing a space into triangles, optimising the shape of the triangles.

Dirac theorem: A graph with n nodes has a Hamiltonian cycle if each node has a degree of at least n/2.

Discrete Mathematics: Deals with finite or countable objects, such as sequences and graphs, and uses counting and enumeration techniques.

Discrete vs. continuous reality: Philosophical debate about the fundamental nature of reality, whether it is composed of discrete units or varies continuously.

Enumerative Combinatorics: Deals with the counting of possible configurations of objects. This branch includes techniques such as the inclusion-exclusion principle and recurrence formulas

Erdős-Szekeres theorem: An example of a conjecture explored in combinatorics concerning monotone subsequences.

Experimental design: A link between combinatorics and the design of experiments, where it is necessary to combine several factors to obtain meaningful results.

Extremal Combinatorics: Investigates the properties of combinatorial structures of maximum or minimum dimensions.

Galton machine (Quincunx): Apparatus used to visualise the binomial distribution, demonstrating the connection between combinations and probabilities.

Game Theory: Study of strategic interactions between rational agents.

Generating functions: Powerful mathematical tools that allow manipulating numerical sequences by converting combinatorial problems into algebraic problems.

Genetic Algorithms: Optimisation and search algorithms inspired by the processes of biological evolution.

Geometric combinatorics: Study the combinatorial properties of geometric objects.

Graph: A mathematical structure representing relationships between objects, consisting of nodes (vertices) and arcs (connections).

Graph colouring: Assigning "colours" (labels) to the vertices of a graph so that adjacent vertices have different colours.

Green-Tao theorem: Theorem that arbitrarily long arithmetic progressions of prime numbers exist.

Hamiltonian cycle: A path in a graph that visits each node exactly once and return to the starting point.

Hasse diagrams: Graphical representations of a poset (partially ordered set), useful for visualising partial-order relationships.

Hidden Markov models (HMM): Statistical models in which the system is viewed as a Markov process with unobservable (hidden) states.

Inclusion-Exclusion Principle: A powerful tool used to calculate the cardinality of sets that can be obtained by joining smaller sets, considering their overlaps and alternating additions and subtractions.

LCS (Longest Common Subsequence): The longest common subsequence between two or more given sequences.

Local search: An optimisation algorithm that explores the solution space by iteratively moving toward better neighbouring solutions.

Machine Learning: A field of artificial intelligence that focuses on developing algorithms that can learn from data.

Markov chains: Stochastic models that describe a sequence of events in which the probability of each event depends only on the state of the previous event.

Maximum cut problem (Max-Cut): A combinatorial optimisation problem that consists of dividing the vertices of a graph into two sets so as to maximise the number of edges that pass through the cut.

Maximum flow problem: Given a directed graph with capacity on its edges, find the maximum flow from the source node to the destination node.

Monochromatic complete subgraph: A subgraph in which all edges have the same colour.

Oulipo (Ouvroir de Littérature Potentielle): French literary movement that explores new expressive possibilities through mathematical constraints and structures.

Partition of an integer: The division of an integer into sums of positive numbers, regardless of the order of the addends.

Pascal's Triangle: Graphical representation of binomial coefficients organized in a triangular shape, where each number is obtained by adding the two adjacent ones in the top row.

Permutations: Arrangements of objects in a given order.

Pigeonhole principle: If more elements are placed than there are bins, at least one bin must contain more than one element.

Planck length: A tiny unit of length, assumed to be the minimum space-time dimension.

Polya lattice: Structure on which the numbers of the shortest paths to go from one point to another are calculated, helpful in solving counting problems.

Poset (partially ordered set): A set of elements with a partial order relation.

Probabilistic Combinatorics: Analyzes the combinatorial properties of random objects, using probabilistic techniques to obtain results on combinatorial structures.

Ramsey theorem: Theorem that in every colouring of the edges of a complete graph with a sufficiently large number of vertices, it is possible to find a monochromatic complete subgraph.

Random Combinatorics: An unstructured approach that uses probabilistic methods to generate new selections of elements from a given set.

Recursive Counting: A fundamental technique that breaks down complex problems into smaller, more manageable subproblems, expressing the solution of a problem in terms of solutions to smaller problems of the same type

Regularity method: This approach is used in extreme graph theory, which states that a more straightforward structure can approximate every graph.

Reinforcement Learning: A type of machine learning in which an agent learns to make decisions in an environment to maximise a cumulative reward.

Simple combinatorial configurations: Classified by J.G. Dubois into three models: selections (sampling), distributions (mapping and partitioning objects), and partitions (dividing a set into disjoint subsets).

Simulated annealing: An optimisation algorithm inspired by the physical process of hardening metals.

Subsequence sum problem: Given a set of numbers, find the subsequence of maximum length whose sum is maximum.

Supervised Learning: A type of machine learning where the model is trained on labelled data to learn the mapping between inputs and outputs.

TF-IDF (Term Frequency-Inverse Document Frequency): Text analysis technique used to extract significant features from texts, evaluating the importance of a word within a document compared to a collection of documents.

Topological combinatorics: Build a bridge between combinatorics and the topological properties of the objects studied, exploring the combinatorial properties of topological structures.

Traveling Salesman Problem (TSP): A combinatorial optimisation problem in which one searches for the shortest path that visits a set of cities and returns to the origin.

Turán problem: The problem of determining the maximum number of edges in a graph that does not contain a complete Kr+1 subgraph.

Ulam spiral: A method for visualising prime, revealing geometric patterns.

Voronoi diagram: This diagram associates each point in a space with a region representing the area closest to that point with respect to the others.

Word Cloud: Visual representation of the words present in a text, in which the size of the words reflects their frequency or importance.

Zero-sum subset problem: Given a set of numbers, find a subset with zero sum.

About the Author

Alessio Drivet graduated from the University of Trento in 1975. He has taught mathematics, probability, and statistics at commercial institutes since 1976 and retired in 2011. He has published numerous books and articles on mathematics and computer science at various publishing houses and has been a speaker at several conferences. He has been a contract trainer for the University of Turin and several Public Institutions. He is currently on the scientific and organizational committee of DI.FI.MA (Didactics of Physics and Mathematics) and is a member of the GeoGebra Institute in Turin. Since 2013, he has been involved in disseminating mathematics among secondary school students.

Index

A

B

C

D

E

F

I

L

M

N

P

R

S

T

U

V

W

Z

www.ingramcontent.com/pod-product-compliance
Lightning Source LLC
LaVergne TN
LVHW060300011026
849161LV00011B/2747
* 9 7 9 8 3 3 7 3 3 0 8 9 1 *